T0310606

# Applied ethology 2019
## Animal Lives Worth Living

# ISAE 2019

## Proceedings of the
## 53rd Congress of the ISAE

5th–9th August, 2019
Bergen, Norway

# ANIMAL LIVES WORTH LIVING

edited by:

Ruth C. Newberry
Bjarne O. Braastad

OASES
Online Academic Submission and Evaluation System

EAN: 9789086863389
e-EAN: 9789086868896
ISBN: 978-90-8686-338-9
e-ISBN: 978-90-8686-889-6
DOI: 10.3920/978-90-8686-889-6

**First published, 2019**

# Welcome to ISAE 2019

We welcome you to Bergen, Norway, for the 53rd Congress of the International Society for Applied Ethology (ISAE). It is our sincere hope that you will have a rewarding and valuable experience participating in the congress. We also encourage you to enjoy Bergen and its surrounding mountains and fjords. Bergen was chosen as the congress venue partly because of its rich history, culture and nature, and partly for practical reasons, including accessibility, congress facilities and accommodation opportunities. This is the second time that an ISAE congress has been organised in Norway, the first one having been held in Lillehammer 20 years ago.

The main theme of this year's congress is 'Animal lives worth living'. This theme focuses on our responsibility for all animals kept or influenced by humans, to ensure that we can provide a life for them that takes into account all relevant aspects of animal welfare, aided by applied ethology as the key scientific discipline. This means considering not only means for avoiding and alleviating suffering but also means for promoting resilience and positive experiences. By monitoring and interpreting animal behaviour, we gain important insights into each of these aspects of quality of life.

Instead of organising the scientific programme according to species, we chose to focus the sessions on themes within applied ethology that contribute to optimal conditions for 'a life worth living'. Genetic selection of animals must be conducted in ways that accommodate behavioural function and reduce behavioural problems and adverse emotions. During the foetal period, the conditions for the pregnant mother must be suitable for avoiding long-term adverse effects on the offspring. After birth, the offspring must experience a safe but enriched environment that prepares them for a good life in the environment(s) to which they are destined. Throughout life, the physical and social environment must stimulate curiosity, facilitate social harmony and contribute to a sense of well-being. Boredom must be avoided, so providing positive cognitive stimulation must not be forgotten. The nutrition, and how food and water is offered, must positively affect behaviour and welfare. Because humans are a prominent feature of the environment, it is essential that interactions between humans and animals are experienced without fear and stress, but instead contribute to positive emotions and good health in both animals and humans. Applied ethology must also be forward thinking, predictive and proactive, providing directions for future ways of selecting and keeping animals instead of just attempting to fix unintended problems after they have become established.

A congress on applied ethology in Norway would not be complete without addressing the behaviour and welfare of fish used in aquaculture. A session on this topic is therefore included. This is also the reason for the choice of this year's David Wood-Gush Memorial Lecturer, Felicity Huntingford, who is a leading authority on fish behaviour. Additionally, we feature plenary talks by recipients of the ISAE Creativity Award, Per Jensen and Françoise Wemelsfelder, and the ISAE New Investigator Award, Irene Camerlink. Finally, we draw your attention to six exciting workshops organised by ISAE members, covering a broad spectrum of current interests within applied ethology.

We wish you all a memorable stay in Bergen!

*Bjarne O. Braastad and Ruth C. Newberry*

# Acknowledgements

## Local Congress Organising Committee

Bjarne O. Braastad (chair), Knut E. Bøe, Ruth C. Newberry, Inger Lise Andersen, Randi Oppermann Moe (all Norwegian University of Life Sciences)

## Local Scientific Committee

Ruth C. Newberry (chair), Inger Lise Andersen, Andrew M. Janczak, Janicke Nordgreen, Marco Vindas (all Norwegian University of Life Sciences), Cecilie Mejdell (Norwegian Veterinary Institute), Grete H.M. Jørgensen (Norwegian Institute of Bioeconomy Research), Tore Kristiansen (Institute of Marine Research), Guro Vasdal (Norwegian Meat and Poultry Research Centre)

## ISAE Ethics Committee

Alexandra Whittaker (chair), Franck Peron, Francesco De Giorgio, Cecilie Mejdell, Elize van Vollenhoven, Beth Ventura, Ellen Williams

## Web designer

Janne Karin Brodin

## Student Poster Competition chairs

Marko Ocepek, Inger Lise Andersen

## Student helpers

Natalie Solheim Bernales, Kim Iversen Bjørnson, Johanna Gjøen, Cecilie Blakstad Løkken, Jenny Kristine Runningen, Mari Smith, Kristina Svennekjær, Benedicte Marie Woldsnes, Fan Wu

## Professional congress organiser

First United AS, Norway (www.firstunited.no), headed by Harald Riisnæs

## Logos

A7 Print AS, Bergen

# Referees

Marta Alonso De La Varga
Inger Lise Andersen
Michael Appleby
Xavier Averós
Eddy Bokkers
Laura Boyle
Bjarne O. Braastad
Jennifer Brown
Oliver Burman
Knut E. Bøe
Irene Camerlink
Joao Cardoso Costa
Michael Cockram
Jonathan Cooper
Ingrid De Jong
Trevor Devries
Catherine Dwyer
Becca Franks
Brianna Gaskill
Conor Goold
Derek Haley
Alison Hanlon
Moira Harris
Lynette Hart
Marie Haskell
Paul H. Hemsworth
Mette Herskin
Regine Holt
Maria Hötzel
Anne Lene Hovland
Felicity Huntingford
Andrew Janczak
Margit Bak Jensen
Per Jensen
Grete H.M. Jørgensen
Linda Keeling
Ute Knierim
Tore Kristiansen
Jan Langbein
Alistair Lawrence
Don Lay Jr.
Lena Lidfors
Jens Malmkvist
Jeremy Marchant-Forde

Georgia Mason
Lindsay Matthews
Anne McBride
Cecilie Mejdell
Suzanne Millman
Daniel Mills
Lene Munksgaard
Christian Nawroth
Ruth C. Newberry
Lee Niel
Birte Nielsen
Janicke Nordgreen
Cheryl O'Connor
Marko Ocepek
Anna Olsson
Jean-Loup Rault
Bas Rodenburg
Vicky Sandilands
Yolande Seddon
James Serpell
Janice Siegford
Marek Špinka
Hans Spoolder
Mhairi Sutherland
Karen Thodberg
Stephanie Torrey
Michael Toscano
Cassandra Tucker
Simon Turner
Frank Tuyttens
Katsuji Uetake
Anna Valros
Judit Vas
Guro Vasdal
Marco Vindas
Eberhard Von Borell
Marina Von Keyserlingk
Susanne Waiblinger
Daniel Weary
Laura Webb
Françoise Wemelsfelder
Tina Widowski
Christoph Winckler

With funding from

**The Research
Council of Norway**

Norwegian Meat and Poultry Research Centre

 Nestlé PURINA

**Main sponsor**

**Silver sponsors**

**Bronze sponsors**

# Maps

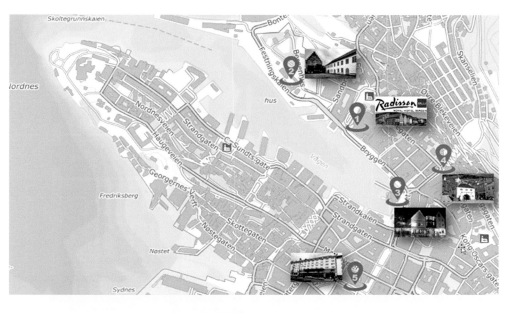

 Congress venue

 Welcome Reception – Håkonshallen

 ISAE 2019 Congress Pub at Piano Club Zachariasbryggen

 Funicular to dinner at Mount Fløien Restaurant

 Farewell Party

Central Bergen with indications of where the congress and social events will take place.

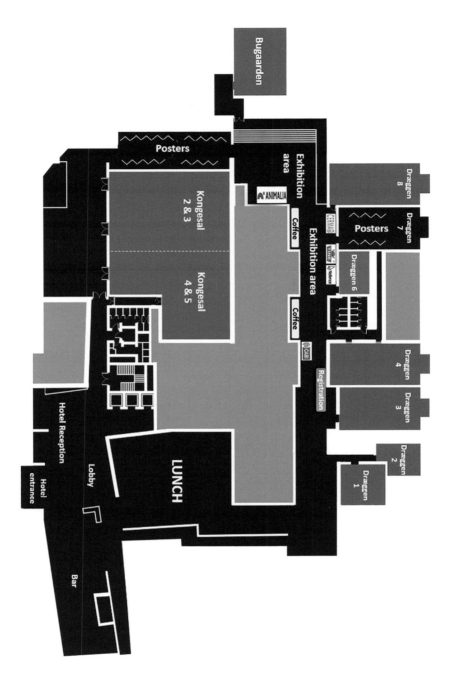

## Congress facilities at the Radisson Blu Royal Hotel

Plenary room: Kongesal 2-5
Parallel sessions: Kongesal 2-3 and Kongesal 4-5
Speaker's room and Congress Secretariat: Dræggen 2, behind the Registration Desk.
Press conference room: Dræggen 1

# General information

**Venue**
The Congress will be held at the Radisson BLU Royal Hotel, Dreggsallmenningen 1, NO-5003 Bergen, Norway. Phone: +47 55 54 30 00.

**Official language**
English is the official language of the ISAE 2019 Congress.

**Registration and Information/Secretariat desk**
The registration desk will be located at the entrance to the exhibition and break-out area on the Ground floor. There will be signage that will lead you from the hotel lobby to the registration desk.

Opening hours of the registration and information desk:
|  |  |
|---|---|
| Monday 5th August: | 15:00 – 21:00 |
| Tuesday 6th August: | 08:00 – 20:00 |
| Wednesday 7th August: | 08:30 – 17:00 |
| Thursday 8th August: | 08:00 – 18:00 |
| Friday 9th August: | 08:30 – 18:15 |

The desk will be staffed by the ISAE 2019 Professional Congress Organiser, First United. Tel: +47 55 23 00 70. E-mail: pco@firstunited.no

**Speaker's room/Submission of presentations**
Powerpoint files for oral presentations shall be on a USB memory stick submitted at the Speaker's room, Dræggen 2, close to the Registration desk, for morning sessions the day before, and for afternoon sessions before 10:00h. Student assistants will collect the presentations and you are advised to check that all slides and videos are viewable as intended. Take care to follow the guidelines for presentations at http://www.isae2019.com/guidelines-for-speakers/.

**Name badges & Tickets**
Name badges are required for admittance to the Congress sessions, coffee breaks, lunches, welcome reception, dinners and farewell party. Badges will be handed out at the Registration desk, along with tickets issued for all separately-priced social events and excursions.

## Posters

Posters will be shown in Dræggen 7, the Exhibition area, Kongesal 1, and the Foyer outside Kongesal 1-4. Poster placement and presentation by authors is scheduled as follows:

### Group A

| | | |
|---|---|---|
| Monday 5th August: | 15:00 – 21:00 | Group A poster placement |
| Tuesday 6th August: | 08:00 – 09:00 | Group A poster placement |
| Tuesday 6th August: | 10:15 – 11:15 | Group A poster presentation (odd numbers) |
| Tuesday 6th August: | 14:45 – 15:45 | Group A poster presentation (even numbers) |
| Wednesday 7th August: | 10:15 – 11:15 | Group A poster presentation (all) |
| Wednesday 7th August: | 13:30 – 14:00 | Group A poster removal |

### Group B

| | | |
|---|---|---|
| Thursday 8th August: | 08:00 – 09:00 | Group B poster placement |
| Thursday 8th August: | 10:45 – 11:45 | Group B poster presentation (odd numbers) |
| Thursday 8th August: | 14:45 – 15:45 | Group B poster presentation (even numbers) |
| Friday 9th August: | 09:45 – 10:45 | Group B poster presentation (odd numbers) |
| Friday 9th August: | 14:15 – 15:15 | Group B poster presentation (even numbers) |
| Friday 9th August: | 17:45 – 18:15 | Group B poster removal |

## Exhibition area

Sponsor exhibitions are located in the Exhibition and break-out area.

## Internet access

Free high-speed, wireless Internet is available throughout the hotel; no access codes required.

## Coffee breaks

Tea, coffee, and refreshments will be served in the Exhibition and break-out area.

## Lunches

Lunches will be held in the hotel restaurant and additional rooms. Follow signs. Please wear your badge.

## Parking

Delegates who have vehicles may park in the parking garage underneath the hotel. The parking house is operated by third party, private operator APCOA https://www.apcoa.no/en/find-parking/bergen/radisson-blu-royal-hotell/).
- NOK 31 per hour
- NOK 225 per day

There are other public parking facilities available further away (e.g. KlosterGarasjen, NOK 36 per hour, NOK 200 per day; ByGarasjen, NOK 24 per hour, NOK 150 per day). See https://bergenparkering.no/

# Social program and excursions

**Welcome Reception, Monday August 5th, 19:00-20:00**

The Welcome Reception (free for all participants and registered accompanying persons) is hosted by the City of Bergen. It will take place on Monday 5th August from 19:00-20:00 in Håkonshallen, a few hundred meters from the conference hotel. Signs will be set up along the street between the conference hotel and Håkonshallen. See spot no. 2 on the city map.

**Excursions, Wednesday August 7th (for those with pre-booked tickets)**

*Tour 1:*
*SCENIC DRIVE TO THE FJORDS & VISIT TO THE MATRE RESEARCH STATION*
The excursion starts from the Radisson Blu Royal Hotel at 14:00 and ends at the Funicular station at 18:00.
The research station is located in Matredal, 80 km north of Bergen. The coach will depart from the hotel and drive north, over bridges and along the shores of the Osterfjord, then crossing the Romarheim Mountain before descending to the base of the Fensfjord and the fjord village of Matre.

*Tour 2:*
*3 HOUR NORWEGIAN FJORDS CRUISE*
Cross the street from the Radisson Blu Royal Hotel to board the boat at 15:00. The excursion ends here at 18.00. Those participating in the dinner at Mt. Fløien will then proceed to the Funicular station (walking distance 400 meters / 7 minutes).
Join this round-trip cruise from Bergen through spectacular fjord scenery. This tour takes you along the 27-kilometer Osterfjord.

*Tour 3:*
*THE CLASSICAL GUIDED TOUR OF BERGEN BY COACH*
The excursion starts from the Radisson Blu Royal Hotel at 15:00 and ends at the Funicular station at 18:00.
This tour will take you by private coach and local guide for an orientation tour of the city centre of Bergen, including visits to a stave church, the home of Edvard Grieg, and an open-air museum.

*Tour 4:*
*WALKING TOUR OF THE HANSA AREA OF BERGEN & VISIT TO MT. FLØIEN*
The excursion starts from the Radisson Blu Royal Hotel at 15:00 and ends at Mt. Fløien at 18:00. You will see and learn about the magnificent and unique architecture of this UNESCO World Heritage-listed Hansa City. The tour includes a visit to the Hanseatic Museum and Funicular ride.

### Dinner 'On the Roof Top of Bergen', Wednesday 7th August

For those with tickets to the dinner, take the 5-min walk from the Radisson Blu Royal Hotel to Fløien Funicular station (spot no. 4 on the map), from where you will take the funicular to the top of Mount Fløien. Meet at the Funicular station between 18:00-18:15. Tours 1, 3 and 4 will end by the Funicular Station at 18:00. Dinner starts at 18:30. You will be served a two-course dinner with a glass of wine/beer or alternative. During dinner, you will experience a taste of Norwegian classical music. After dinner, the funicular will take you back down the mountain to the city centre, or you can choose to take the scenic walking path instead.

### Banquet, Thursday 8th August

All those with pre-booked tickets are invited to the congress banquet, which will be held at Radisson Blu Royal Hotel. The banquet starts with an aperitif at 19:30 followed by a dinner at about 20:00. During and after the dinner, you will experience musical entertainment. This is followed by dancing for all; perhaps you will also learn a few Norwegian folkdances.

### Farewell Party, Friday 9th August

All those with pre-booked tickets are welcome to join us at the Farewell party, to be held at Ole Bull Scene in the middle of the city centre. See spot no. 5 on the map. The party will start at 19:00. You will be served various tapas-style Norwegian foods and two glasses of wine/beer etc. Afterwards a DJ will play dance music.

### ISAE 2019 Congress Pub at Piano Club Zachariasbryggen

We encourage ISAE participants and friends to mingle and enjoy the Bergen nightlife together by dropping in at the Piano Club Zachariasbryggen, situated at the fish market, spot no. 3 on the map. The club has a large open-air bar / restaurant facing the picturesque harbour front. The Piano Club bar is a section of the large Zachariasbryggen Restaurant House – with several types of restaurants and bars – all located in one building at the fish market. A great informal place to chill out at the start or end of any evening!

> Address: Zachariasbryggen, Torget 2
> Open every day 12:00 – 03:00.
> Tel: +47 55 55 31 55
> www.zabr.no/barer/pianobar/

# Getting to the congress

### Transportation to/from Bergen Airport

The Bergen Airport (BGO) is located some 16 km south of the city centre. The new terminal building is well connected with a range of international flights.

Transfer between the Bergen Airport (BGO) and Bergen city centre on the Bergen Light Rail Service costs NOK 38 each way. The train departs from the airport terminal and takes approx. 45 minutes to its terminal stop at Byparken in the city centre. From there to the conference venue, Radisson Blu Royal Hotel, walking time is approx. 12 minutes. The Light Rail runs every 5-10 minutes. Tickets can be bought from machines at the stations/stops. See www.skyss.no/en/timetable-and-maps/bergen-light-rail/ for more info and time schedule for the Light Rail.

There is also an airport express bus, 'Flybussen', with a line to the city centre (Bus station and Festplassen) taking 20-30 minutes, and a line to Bryggen (Dreggsallmenningen), just outside the Radisson Blu Royal Hotel, and taking 40-50 minutes. The single fare is NOK 115 if bought online, or NOK 145 upon boarding. Credit cards accepted. For more info and time schedule, see en.visitbergen.com/visitor-information/travel-information/airport-bus-p944603.

Taxi service is also available, taking about 25 minutes pending traffic and time of day, and costing approx. NOK 500-700 each way. Credit cards accepted.

### Visit Bergen

The city of Bergen has a lot to offer to visitors. It is a 'World Heritage City' and considered 'The Gateway to the Norwegian Fjords'. If you wish to extend your stay and explore further, have a look at www.visitbergen.com or contact First United PCO at the Registration desk for local recommendations!

# ISAE 2019 Programme

## Monday 05 August

| 15:00-21:00 | **Registration** | Radisson Blu Royal Hotel |

| 09:00-16:00 | **Council Meeting** | Bugaarden |

| 15:00-21:00 | Placement of Group A Posters | Poster areas |

| 19:00-20:00 | **Welcome reception** | Håkonshallen |

## Tuesday 06 August

| 08:00 | Registration & Placement of Group A Posters................................................ |

| 09:00 | **Opening Ceremony and Presidential Address** – Kongesal 2-5 |

| 09:30 | **Wood-Gush Memorial Lecture** – Kongesal 2-5 – Chair: Linda Keeling<br>Synergies between fundamental and applied behavioural science: lessons from a lifetime of fish watching. *Felicity Huntingford* |
| 10:15 | **Group A Posters – Coffee break** ........................................................<br>Fish Behaviour & Welfare, Pre- & Postnatal Effects, Behaviour & Nutrition, Environmental Enrichment, Human-Animal Interactions |

| | **Pre- & Postnatal Effects**<br>Kongesal 2-3 – Chair: Andrew Janczak | **Fish Behaviour & Welfare**<br>Kongesal 4-5 – Chair: Tore Kristiansen |
|---|---|---|
| 11:15 | Association between maternal social success during late pregnancy & offspring growth & behaviour in dairy cows. *Mayumi Fujiwara* | Identifying enriched housing conditions for zebrafish (*Danio rerio*) that vary along a scale of preference. *Michelle Lavery* |
| 11:30 | The effect of gilt rearing strategy on offspring behaviour. *Phoebe Hartnett* | Knowledge gaps in global aquaculture welfare: scope of the problem, current research & future directions. *Becca Franks* |
| 11:45 | Lameness in sows during pregnancy impacts welfare outcomes in their offspring. *Marisol Parada* | Behavioural predictors of stress in groups of fish. *Tanja Kleinhappel* |
| 12:00 | Oesophageal tube colostrum feeding suppresses oxytocin release & induces vocalizations in healthy newborn dairy calves. *Carlos E. Hernandez* | Increasing welfare through early stress exposure in aquaculture systems. *Marco Vindas* |
| 12:15 | Stress responses resulting from commercial hatching of laying hen chicks. *Louise Hedlund* | Does sudden exposure to warm water cause pain in Atlantic salmon? *Jonatan Nilsson* |

| 12:30 | Lunch ........................................................................................... | |

| | **Pre- & Postnatal Effects**<br>Kongesal 2-3 – Chair: Cecilie Mejdell | **Behaviour & Nutrition**<br>Kongesal 4-5 – Chair: Bas Rodenburg |
|---|---|---|
| 13:30 | Misdirected oral behaviors in orphaned neonatal kittens. *Mikel Delgado* | Dietary fibre content in the feed & its effect on feather pecking, performance & cecal microbiome composition. *Antonia Patt* |
| 13:45 | Identification of fear behaviours shown by kittens in response to novel stimuli. *Courtney Graham* | Early life microbiota transplantation affects behaviour & peripheral serotonin in feather pecking selection lines. *Jerine A.J. Van Der Eijk* |
| 14:00 | The role of age in the selection of police patrol dogs based on behaviour tests. *Kim I. Bjørnson* | Modulation of kynurenine pathway & behavior by probiotic bacteria in laying hens. *Claire Mindus* |
| 14:15 | Calf-cow contact during rearing improves health status in dairy calves – but is not a universal remedy. *Edna Hillmann* | Investigating the gut-brain axis: effects of prebiotics on learning & memory in pigs. *Else Verbeek* |
| 14:30 | Effects of a 3-month nursing period on fertility & milk yield of cows in their first lactation. *Kerstin Barth* | Can dietary magnesium improve pig welfare & performance?. *Emily V. Bushby* |

| 14:45 | **Group A Posters – Coffee break** .................................................................... | |

Fish Behaviour & Welfare, Pre- & Postnatal Effects, Behaviour & Nutrition, Environmental Enrichment, Human-Animal Interactions

| | **Cognition & Welfare**<br>Kongesal 2-3 – Chair: Oliver Burman | **Behaviour & Nutrition**<br>Kongesal 4-5 – Chair: Marina Von Keyserlingk |
|---|---|---|
| 15:45 | Domestic hens: affective impact of long-term preferred & non-preferred living conditions. *Elizabeth Paul* | Dietary modification of behaviour in pigs. *Jeremy N. Marchant-Forde* |
| 16:00 | Exploring attentional bias towards emotional faces in chimpanzees using the dot probe task. *Duncan Wilson* | Why drink milk replacer when you can suckle the sow? Foraging strategy in sow-reared large litters with milk replacer. *Cecilie Kobek-Kjeldager* |
| 16:15 | An automated & self-initiated judgement bias task based on natural investigative behaviour. *Michael Mendl* | Individual characteristics predict weaning age of dairy calves when weaned automatically based on solid feed intake. *Heather W. Neave* |
| 16:30 | Investigating animal affect & welfare using computational modelling. *Vikki Neville* | The effect of omega-3 enriched maternal diets on social isolation vocalisations in ISA Brown & Shaver White chicks. *Rosemary Whittle* |
| 16:45 | Effect of early life & current environmental enrichment & personality on attention bias in pigs. *Lu Luo* | Characterization of plant eating in cats. *Benjamin L. Hart* |

| 17:00 | **Break** ............................................................................................ | |

| | |
|---|---|
| 1. Visualizing and analysis of individual-level data within large group systems | Kongesal 2-3 |
| 2. Future trends in the prevention of damaging behaviour | Dræggen 3 |
| 3. Novel indicators of fish welfare | Dræggen 4 |
| 4. Engaging students in learning about production animal welfare assessment | Bugaarden |
| 5. Managing the dairy cow around the time of calving – can we do better? | Dræggen 8 |
| 6. Animal training – efficacy and welfare | Kongesal 4-5 |

# Wednesday 07 August

| | **Cognition & Welfare**<br>Kongesal 2-3 – Chair: Daniel Weary | **Environmental Enrichment**<br>Kongesal 4-5 – Chair: Inger Lise Andersen |
|---|---|---|
| 09:00 | Half full or half empty? Comparing affective states of dairy calves. *Katarína Bučková* | The development of stereotypic behaviours by two laboratory mouse strains housed in differentially enriched conditions. *Emily Finnegan* |
| 09:15 | Brain size & cognitive abilities in Red Junglefowl selected for divergent levels of fear of humans. *Rebecca Katajamaa* | Providing complexity: a way to enrich cages for group housed male mice without increasing aggression? *Elin M. Weber* |
| 09:30 | Thermal comfort in horses: their use of shelter & preference for wearing rugs. *Cecilie M. Mejdell* | Environmental controllability & predictability affect the behavioural adaptability of laying hen chicks in novel situations. *Lena Skånberg* |
| 09:45 | Can horses watch & learn? The good, the bad & the not so ugly evidence of social transmission in horses. *Maria Vilain Rørvang* | Environmental enrichment, group size & confinement duration affect play behaviour in goats. *Regine Victoria Holt* |
| 10:00 | Visual laterality in pigs & the emotional valence hypothesis. *Charlotte Goursot* | Assessing the behaviour of fast-growing broilers reared in pens with or without enrichment. *Zhenzhen Liu* |

| | |
|---|---|
| 10:15 | **Group A Posters – Coffee break** ....................................................<br>Fish Behaviour & Welfare, Pre- & Postnatal Effects, Behaviour & Nutrition, Environmental Enrichment, Human-Animal Interactions |

| | **Cognition & Welfare**<br>Kongesal 2-3 – Chair: Michael Mendl | **Environmental Enrichment**<br>Kongesal 4-5 – Chair: Heleen van de Weerd |
|---|---|---|
| 11:15 | Hemispheric specialisation for processing the communicative & emotional content of vocal signals in young pigs. *Lisette M.C. Leliveld* | Differences in laying behaviour associated with nest box design in commercial colony cage units. *Sarah Lambton* |
| 11:30 | Assessing the impact of lameness on the affective state of dairy cows using lateralisation testing. *Sarah Kappel* | Long-term effects of peat provision in broiler chicken flocks. *Judit Vas* |
| 11:45 | Development of a novel fear test in sheep – the startle response. *Hannah Salvin* | Which rooting materials make a weaner most happy? *Marko Ocepek* |

| 12:00 | Pecking in Go/No-Go task: motor impulsivity in feather pecking birds?. *Alexandra Harlander* | Nest-building material affects pre-partum sow behaviour & piglet survival in crate-free farrowing pens. *Ellen Marie Rosvold* |
| 12:15 | Pharmacological intervention of the reward system in the laying hen has an impact on anticipatory behaviour. *Peta Simone Taylor* | Brush use by dairy heifers. *Jennifer M.C. Van Os* |

**12:30**    **Lunch**............................................................................

13:30    Removal of Group A Posters

14:00    **Excursions**

18:00-21:00    **Dinner** at 18:30 - Fløien Folkerestaurant (meet 18:00-18:15 at Funicular station)

## Thursday 08 August

08:00    Placement of Group B Posters

**Plenary session: ISAE Creativity Award**
Kongesal 2-5 – Chairs: Birte Nielsen, Randi Oppermann Moe

09:00    Genetics & epigenetics of domesticated behaviour - what we can learn from chickens & dogs. *Per Jensen*

**09:45**    **Break** ...........................................................................

10:00    Science & sentience: assessing the quality of animal lives. *Françoise Wemelsfelder*

**10:45**    **Group B Posters – Coffee break** ...................................................
Cognition & Welfare, Behaviour & Genetics, Social Environment, Addressing Future Trends in Animal Production, Free Communications

| | **Behaviour & Genetics** Kongesal 2-3 – Chair: Tina Widowski | **Human-Animal Interactions** Kongesal 4-5 – Chair: Conor Goold |
|---|---|---|
| 11:45 | Parturition in two sow genotypes housed under free-range conditions. *Sarah-Lina A. Schild* | Awareness & use of canine quality of life assessment tools in UK veterinary practice. *Claire Roberts* |
| 12:00 | Growing slowly with more space: effects on 'positive behaviours' in broiler chickens *Annie Rayner* | Owner caregiving style & the behaviour of dogs when alone. *Luciana Santos De Assis* |
| 12:15 | Range use & plumage condition of two laying hen hybrids in organic egg production. *Fernanda M. Tahamtani* | What can short-term fostering do for the welfare of dogs in animal shelters?. *Lisa M. Gunter* |

| 12:30 | **Lunch – Eating with Ethologists** | |
|---|---|---|

| | **Social Environment**<br>Kongesal 2-3 – Chair: Jan Langbein | **Human-Animal Interactions**<br>Kongesal 4-5 – Chair: Hajime Tanida |
|---|---|---|
| 13:30 | Ranging patterns & social associations in laying hens. *Yamenah Gómez* | Farmer perceptions of pig aggression compared to animal-based measures of fight outcome. *Rachel Peden* |
| 13:45 | Socio-positive interactions in goats: prevalence & social network. *Claire Toinon* | Consistency of blue fox behaviour in three temperament tests. *Eeva Ojala* |
| 14:00 | Play fighting social network position does not predict injuries from later aggression between pigs. *Simon P. Turner* | Owner perspectives of cat handling techniques used in veterinary clinics. *Carly Moody* |
| 14:15 | Quantifying play contagion: low playing calves may depress play behaviour of pen-mates. *Verena Größbacher* | A survey of visitors' views on free-roaming cats living in the tourist town of Onomichi, Japan. *Aira Seo* |
| 14:30 | Salivary oxytocin is associated with ewe-lamb contact but not suckling in lactating ewes. *Cathy Dwyer* | Risk factors for aggression in adult cats that were fostered through a shelter program as kittens. *Kristina O'Hanley* |

| 14:45 | **Group B Posters – Coffee break** .........................................................<br>Cognition & Welfare, Behaviour & Genetics, Social Environment, Addressing Future Trends in Animal Production, Free Communications | |
|---|---|---|

| | **Social Environment**<br>Kongesal 2-3 – Chair: Margit Bak Jensen | **Human-Animal Interactions**<br>Kongesal 4-5 – Chair: Lynette Hart |
|---|---|---|
| 15:45 | Maternal protection behavior in Hereford cows. *Marcia Del Campo* | Bit vs. bitless bridle - what does the horse say?. *Inger Lise Andersen* |
| 16:00 | Do stocking density or a barrier affect calving location & labor length in group-housed Holstein dairy cows?. *Katherine Creutzinger* | Pet ownership among people with substance use disorder: implications for health & use of treatment services. *Andi Kerr-Little* |
| 16:15 | Sociability is associated with pre-weaning feeding behaviour & growth in Norwegian Red calves. *Laura Whalin* | Sleeping through anything: the effects of predictability of disruption on laboratory mouse sleep, affect & physiology. *Amy Robinson-Junker* |
| 16:30 | Effect of cow-calf contact on motivation of dairy cows to access their calf. *Margret L. Wenker* | Human beliefs & animal welfare: a cross-sectional survey on rat tickling in the laboratory. *Megan R. LaFollette* |

| 16:50 | **ISAE Annual General Meeting** – Kongesal 2-3 | |
|---|---|---|

| 17:50 | **Break** ........................................................................................... | |
|---|---|---|

| 19:30-24:00 | **ISAE Banquet** – Kongesal 2-5 | |
|---|---|---|

# Friday 09 August

| 09:00 | **Plenary session: ISAE New Investigator Award** – Kongesal 2-5 – Chair: Susanne Waiblinger<br>New perspectives for assessing the valence of social interactions. *Irene Camerlink* |
|---|---|

**09:45** Group B Posters – Coffee break ........................................................................
Cognition & Welfare, Behaviour & Genetics, Social Environment, Addressing Future Trends in
Animal Production, Free Communications

|  | **Addressing Future Trends in Animal Production**<br>Kongesal 2-3 – Chair: Janice Siegford | **Social Environment**<br>Kongesal 4-5 – Chair: Marko Ocepek |
|---|---|---|
| 10:45 | Role of animal behaviour in addressing future challenges for animal production. *Laura Boyle* | Vomeronasal organ alterations & their effects on social behaviour: results & perspectives after 10 years of research. *Pietro Asproni* |
| 11:00 | Association between locomotion behavior & *Campylobacter* load in broilers. *Sabine G. Gebhardt-Henrich* | Sweaty secrets: plantar gland secretions influence male mouse social behavior in the home cage. *Amanda Barabas* |
| 11:15 | Changes in activity & feeding behaviour as early-warning signs of respiratory disease in dairy calves. *Marie J. Haskell* | Can live with 'em, can live without 'em: implications of pair vs. single housing for the welfare of male C57BL/6J mice. *Luca Melotti* |
| 11:30 | Using calves' behavioural differences to design future control measures for important zoonotic pathogens. *Lena-Mari Tamminen* | The effect of LPS & ketoprofen on social behaviour & stress physiology in group-housed pigs. *Christina Veit* |
| 11:45 | Dairy cows with Johne's disease spend less time lying and increase feed intake around peak lactation. *Mark Rutter* | Housing gilts in crates prior to mating compromises welfare & exacerbates sickness behavior in response to LPS challenge. *Adroaldo José Zanella* |

| 12:00 | **Lunch - Discussion Topics Tables** |
|---|---|

|  | **Addressing Future Trends in Animal Production**<br>Kongesal 2-3 – Chair: Inma Estevez | **Free Communications**<br>Kongesal 4-5 – Chair: Lee Niel |
|---|---|---|
| 13:00 | Animal welfare: an integral part of the United Nations Sustainable Development Goals? *Linda Keeling* | Do human-dog interactions affect oxytocin concentrations in both species? *Lauren Powell* |
| 13:15 | Systems modelling of the UK pig industry: implications for pig health, behaviour & welfare *Conor Goold* | Prioritising cat welfare issues using a Delphi method. *Cathy Dwyer for Fiona C. Rioja-Lang* |
| 13:30 | Environmental impact of animal welfare improvement measures in dairy farming – model calculations for Austria *Anna Christina Herzog* | Dogs with canine atopic dermatitis exhibit differing behavioural characteristics compared to healthy controls. *Naomi D. Harvey* |

| | |
|---|---|
| 13:45 | Automated tracking of individual activity of broiler chickens. *Malou Van Der Sluis* | Hinged farrowing crates promote an increase in the behavioral repertoire of lactating sows. *Maria C. Ceballos* |
| 14:00 | Assessment of open-source programs for automated tracking of individual pigs within a group. *Kaitlin Wurtz* | Relative preference for wooden nests affects nesting behaviour in broiler breeders. *Anna C.M. Van Den Oever* |

**14:15** **Group B Posters – Coffee break** ........................................................
Cognition & Welfare, Behaviour & Genetics, Social Environment, Addressing Future Trends in Animal Production, Free Communications

| | **Addressing Future Trends in Animal Production** Kongesal 2-3 – Chair: Knut Bøe | **Free Communications** Kongesal 4-5 – Chair: Lena Lidfors |
|---|---|---|
| 15:15 | What do we know about the link between ill-health & tail biting in pigs? *Janicke Nordgreen* | Associations between qualitative behaviour assessments & measures of leg health, fear & mortality in Norwegian broilers. *Guro Vasdal* |
| 15:30 | Validation of accelerometers to automatically record postures & number of steps in growing lambs. *Niclas Högberg* | Individual laying hen mobility in aviary systems is linked with keel bone fracture severity. *Christina Rufener* |
| 15:45 | Have the cows hit the wall? Validation of contact mats to monitor dairy cow contact with stall partitions. *Elsa Vasseur* | Vocal changes as indicators of pain in harbour seal pups (*Phoca vitulina*). *Amelia Mari MacRae* |
| 16:00 | Welfare aspects related to virtual fences for cattle. *Silje Eftang* | Effect of hot-iron disbudding on rest & rumination in dairy calves. *Cassandra B. Tucker* |
| 16:15 | Evaluating unmanned aerial vehicles for observing cattle in extensive field environments. *John Scott Church* | Effects of disbudding on use of a shelter and activity in group-housed dairy calves. *Emily Miller-Cushon* |

**16:30** **Break** ........................................................

| | | |
|---|---|---|
| 16:45 | **Closing Ceremony** (with awards presentation) | Kongesal 2-5 |

17:45 Removal of Group B Posters

**18:15** **Break** ........................................................

| | | |
|---|---|---|
| 19:00-23:30 | **Farewell Party** | Ole Bull Scene |

# List of posters and workshops

ISAE 2019 Posters – Dræggen 7, Exhibition area, Kongesal 1, Kongesalen Foyer

| Group A Posters | Fish Behaviour & Welfare, Pre- & Postnatal Effects, Behaviour & Nutrition, Environmental Enrichment, Human-Animal Interactions (Tuesday 6th August - Wednesday 7th August) | |
| --- | --- | --- |
| **Fish Behaviour & Welfare** | | |
| Isabel Fife-Cook | Identifying potential play behaviors in fish: a study using online video analysis | A1 |
| Cécile Bienboire-Frosini | Impact of sea lice management methods on Atlantic salmon welfare: experimental findings & perspectives | A2 |
| Daniel Santiago Rucinque | Mechanical spiking as a killing method in Nile tilapia (*Oreochromis niloticus*) – A pilot study | A3 |
| **Pre- and Postnatal Effects** | | |
| Tina Widowski | Behavioural & physiological effects of prenatal stress in different genetic lines of laying hens | A4 |
| Elodie Merlot | Heat exposure of pregnant sows modulates behaviour & corticotrope axis responsiveness of their offspring after weaning | A5 |
| Andrew Janczak | Trends in early life conditions of pigs and laying hens in order to prevent damaging behaviour: a GroupHouseNet update | A6 |
| Friederike Katharina Warns | Identification of dominance in suckling piglets | A7 |
| Eddie Bokkers | Postnatal effects of colostrum quality & management, & hygiene practices, on immunity & mortality in Irish dairy calves | A8 |
| Franziska Hakansson | Prevalence & early life risk factors for tail damage in Danish long-tailed piglets prior to weaning - preliminary results | A9 |
| Julie Føske Johnsen | A cross-sectional study of associations between herd-level calf mortality, welfare & management on Norwegian dairy farms | A10 |
| Janja Sirovnik | The effect of maternal contact on behaviour patterns indicative of emotionality & social competence in young dairy calves | A11 |
| Anna Trevarthen | Associations between the early life husbandry & post-weaning behaviour of lambs | A12 |
| Daniel Asif | Maternal behaviour of Markhors (*Capra falconeri*) – a life worth living *ex situ* | A13 |
| Allison N. Pullin | Pullet rearing affects long-term perch use by laying hens in enriched colony cages | A14 |
| **Behaviour & Nutrition** | | |
| Hannah Phillips | The effect of outdoor stocking density & weather on the behavior of broiler chickens raised in mobile shelters on pasture | A15 |
| Terence Zimazile Sibanda | Outdoor feeder increased range usage but not bone quality in commercial free-range laying hens | A16 |

| Karen F. Mancera | Pilot study of grazing strategies of beef cattle in relation to woody elements percentage in a silvopastoral system | A17 |
|---|---|---|
| Andrew Fisher | The relationship between motion score & reproductive behaviour in dairy cows managed in pasture & feedpad environments | A18 |
| Robin E. Crossley | Do grazing management practices influence the behaviour of dairy cows at pasture? | A19 |
| Guilherme Amorim Franchi | Use of a feed frustration test to assess level of hunger in dairy cows subjected to various dry-off management routines | A20 |
| Margit Bak Jensen | The effect of step-down milk feeding & daily feeding frequency on behavior of male dairy calves | A21 |
| Blair C. Downey | Experience with hay influences the development of oral behaviors in pre-weaned dairy calves | A22 |
| Michal Uhrincat | Influence of water delivery method on dairy calf performance & inter-sucking after weaning | A23 |
| Laura Schneider | Feeding behavior of fattening bulls fed with an automatic feeding system | A24 |
| Laura Webb | The ins & outs of abomasal damage in veal calves | A25 |
| Marjorie Cellier | Individual differences in feeding behaviour of dairy goats | A26 |
| Anouschka Middelkoop | Foraging in the farrowing room to stimulate feeding behaviour | A27 |
| Jaime Figueroa | Effect of flavour variety on productive performance of fattening pigs | A28 |
| Mariia Tokareva | Influence of satiety on the motivation of stall-housed gestating sows to exit their stall | A29 |
| Leandro Costa | Feeding behavior of Iberian pigs in a multi-feeder system under high environmental temperatures | A30 |
| Hajime Tanida | Effect of long-term feeding on home range size & colony growth of free-roaming cats at a popular tourist site in Japan | A31 |

## Environmental Enrichment

| Rachel Park | Impact of environmental enrichment on circadian patterns of feedlot steer behavior | A32 |
|---|---|---|
| Xandra Christine Meneses | Brush use & displacement behaviors at a brush in Angus crossbred feedlot cattle | A33 |
| Pablo Pinedo | Behavior of recently weaned organic Holstein calves exposed to a mechanical calf brush | A34 |
| Jenny Stracke | Behavioural differences in a Novel Object Test between male & female turkeys – a pilot study | A35 |
| Midian Nascimento Dos Santos | The influence of enrichment on leg parameters in a conventional strain of broiler chicken | A36 |
| Daniel Rothschild | The effect of enrichment on organ growth, cardiac myopathies & bursal atrophy in a conventional strain of broiler chicken | A37 |
| Jennifer Brown | Effects of enrichment objects on piglet growth & behaviour | A38 |
| Benedicte M. Woldsnes | A combination of rooting stimuli reduces fear of novelty & enhances collaboration in groups of weaned pigs | A39 |
| Jen-Yun Chou | Individual variation in enrichment use in finishing pigs & its relationship with damaging behaviour | A40 |
| Emma Fàbrega | Economic impact of an on-farm innovation to enhance novelty of enrichment materials for fattening pigs | A41 |

| Martin Fuchs | Nest-building behaviour in crated sows provided with a jute-bag – an exploratory case study | A42 |
|---|---|---|
| Hyun-jung Jung | Effects of environmental enrichment on reproductive performance & behavior patterns in gestating sows | A43 |
| Maggie Creamer | Preference for & behavioral response to environmental enrichment in a small population of sexually-mature, commercial boars | A44 |
| Thiago Bernardino | Endotoxin challenge & environmental enrichment changes the behaviour of crated boars | A45 |
| Kate Norman | Experience of ramps at rear is beneficial for commercial laying hens | A46 |
| Anette Wichman | Do laying hens have a preference for "jungle" light? | A47 |
| Oluwaseun Iyasere | Welfare of FUNAAB Alpha chickens as affected by environmental enrichment | A48 |
| Monique Bestman | Role of range use in infections with parasites in laying hens | A49 |
| Arantxa Villagra | Are breeding rabbits motivated for bigger cages? | A50 |
| Miriam Gordon | Effect of environmental enrichment rotation on juvenile farmed mink behaviours | A51 |
| Megumi Fukuzawa | Physiological response of pet dogs to different methods of provision of classical music | A52 |
| Christine Arhant | Effect of different forms of presentation of an interactive dog toy on behaviour of shelter dogs | A53 |
| Marta Elena Alonso | Evaluation of environmental enrichment influence on animal welfare in captive *Panthera onca* through non-invasive methods | A54 |

## Human-Animal Interactions

| Annika Lange | A comparison of reactions to different stroking styles during gentle human-cattle interactions | A55 |
|---|---|---|
| Stephanie Lürzel | Improving the cow-human relationship – influence of restraint during gentle interactions | A56 |
| Silvia Ivemeyer | Is cows' qualitatively assessed behaviour towards humans related to their general stress level? | A57 |
| Amanda Hubbard | Impact of handling frequency on drylot-housed heifer behavior | A58 |
| Sabine A. Meyer | Human-directed behaviour of calves in a Novel Human & an Unsolvable Problem Test | A59 |
| Congcong Li | Artificial mothering during early life could boost the activity, exploration & human affinity of dystocia dairy calves | A60 |
| Susanne Waiblinger | Effects of dam bonded rearing on dairy calves' reactions towards humans vanish later in life | A61 |
| Asja Ebinghaus | Validity aspects of behavioural measures to assess dairy cows´ responsiveness towards humans | A62 |
| Daisuke Kohari | Reproducibility, similarity & consistency of the flight responses of beef cattle | A63 |
| Kaleiah Schiller | Chute scoring as a potential method for assessing arousal state in ewes | A64 |
| Katrina Rosenberger | Dairy & dwarf goats differ in their preference for familiar & unfamiliar humans | A65 |
| Míriam Marcet-Rius | The provision of toys to pigs could improve the human-animal relationship: the use of an innovative test, the Strange-person | A66 |

| Céline Tallet | Are human voices used by pigs (*Sus scrofa domestica*) when developing their relationship with humans? | A67 |
|---|---|---|
| Ella Akin | Providing handling tools to move non-ambulatory pigs on-farm | A68 |
| Sofia Wilhelmsson | Animal handling & welfare-related behaviours in finishing pigs during transport loading & unloading | A69 |
| Cristina Santos Sotomaior | Comparison of a behavior test between does & bucks at different Brazilian rabbit farms | A70 |
| Rosaria Santoro | Beekeeper-honeybee interactions to measure personality at colony level: a pilot study of *Apis mellifera* in central Italy | A71 |
| Tayla Hammond | Tickling does not increase play in young male rats | A72 |
| Carlos Grau Paricio | Light spectrum for illuminating dark periods in reversed light cycles in laboratory rodent facilities | A73 |
| Helena Chaloupkova | Heart rate - the effect of the handler on the dog after a potentially stressful situation | A74 |
| Natalie Solheim Bernales | Small dogs display more aggressive behaviour than large dogs in social media videos | A75 |
| Anastasia Stellato | Efficacy of a 4-week training program for reducing pre-existing veterinary fear in companion dogs | A76 |
| Syamira Zaini | Cat pain & welfare: barriers to having good cat care management in Malaysian veterinary practices | A77 |
| Silvia Michela Mazzola | Does the personality of tigers influence the interaction with their keepers? | A78 |
| Anita Hashmi | The variable effect of visitor number & noise levels on behaviour in three zoo-housed primate species | A79 |
| Sandra Wirth | The influence of human interaction on guinea pigs: behavioural & thermographic changes during animal assisted therapy | A80 |
| Lena Lidfors | The use & well-being of cats in elder care in Sweden | A81 |
| Malin Larsson | Camels in animal-assisted interventions – survey of practical experiences with children, youth & adults | A82 |
| Karen Thodberg | Optimal dog visits for nursing home residents with varying cognitive abilities – an interdisciplinary study | A83 |
| Lynette A. Hart | Geography affects assistance dog availability in United States & Canada | A84 |
| Lina S.V. Roth | Horses & ponies differ in their human-related behaviour | A85 |
| Elske N. De Haas | Practice what you preach - the discrepancy in knowing & doing based on moral values to farm animals | A86 |
| Sarah Ison | North American stakeholder perceptions of the issues affecting the management, performance & well-being of pigs | A87 |
| Huw Nolan | What's in a name? The role of education & rhetoric in improving laying hen welfare | A88 |
| Yuki Koba | A food & agriculture course to raise awareness of animal welfare in university students majoring in pre-school education | A89 |

**Group B Posters**

Cognition & Welfare, Behaviour & Genetics, Social Environment, Addressing Future Trends in Animal Production, Free Communications
(Thursday 8th August - Friday 9th August)

## Cognition & Welfare

| | | |
|---|---|---|
| Jessica E. Monk | Can an attention bias test discriminate positive & negative emotional states in sheep? | B90 |
| Thomas D. Parsons | Judgement bias of group housed gestating sows predicted by personality traits, but not physical measures of welfare | B91 |
| Benjamin Lecorps | Pessimistic dairy calves are more vulnerable to pain-induced anhedonia | B92 |
| Miguel Somarriba | The effects of a composite chronic stress treatment on fear responses & attention bias in beef cattle | B93 |
| Stephanie Buijs | Roll out the green carpet! Dairy cows prefer artificial grass laneways over stonedust-over-gravel ones, especially when lame | B94 |
| Lindsey Robbins | Evaluation of sow thermal preference | B95 |
| Laura Shewbridge Carter | When lying down, would cows rather have a preferable surface or unrestricted space? | B96 |
| Vlatko Ilieski | Spatial distribution & preferable zones for dairy cows in a free stall confined area | B97 |
| Valerie Monckton | The effect of soiling & ammonia reductant application on turkeys' perceived value of wood shavings | B98 |
| Jade Fountain | Operant learning is disrupted when opioid reward pathways are blocked in the laying hen | B99 |
| Thomas Ede | The burning ring of fire: place aversion as evidence of felt pain in calves disbudded using different types of pain control | B100 |
| Tellisa Kearton | Social transmission of learning of a virtual fencing system in sheep | B101 |
| Pawan Singh | Effect of management enrichment on cognitive performance of Murrah buffalo calves | B102 |
| Erica Feuerbacher | Using applied behavior analysis to identify & enhance positive reinforcers for domestic dogs | B103 |
| Yuna Sato | Do pigs understand time interval & adopt optimal foraging behaviour? | B104 |
| Sara Hintze | Doing nothing… Inactivity in fattening cattle | B105 |
| Yoko Sakuraba | Care & rehabilitation activities for a chimpanzee with cerebral palsy: a case study | B106 |
| Cheryl O'Connor | Assessing pasture-based dairy systems from the perspective of cow experiences | B107 |
| Elena Navarro | Intra- & inter- observer reliability of facial expression in sows around farrowing | B108 |
| Emma Tivey | Sex differences in the behavioural response to tickling in juvenile Wistar rats | B109 |
| Ashley Bigge | Heart rate variability patterns as indicators of stress & welfare in Leghorn hens: a comparative housing system study | B110 |
| Jashim Uddin | A sampling strategy for the determination of infra-red temperature of relevant external body surfaces of dairy cows | B111 |
| Anna Juffinger | Effects of injection of clove oil or isoeugenol on mechanical nociceptive threshold & behaviour during injection in calves | B112 |

## Behaviour & Genetics

| Christoph Winckler | Genetic parameters for maternal traits in sows recorded by farmers | B113 |
| Julia Malchow | Space requirements for fast & slow growing chickens on elevated grids | B114 |
| Abdelkareem Ahmed | A comparison of tonic immobility reactions between domesticated *Columba livia* & *Columba guinea* | B115 |

## Social Environment

| Nedup Dorji | Behaviour & sociability of yaks among different regions in Bhutan | B116 |
| Julia Johns | Effects of herd, housing & management conditions on horn-induced altercations in cows | B117 |
| Miroslav Radeski | The relations between the individual & social behaviour of dairy cows & their performance, health & medical history | B118 |
| Catherine McVey | Extracting herd dynamics from milking order data: preliminary insights comparing social network & manifold-based approaches | B119 |
| Annika Krause | The effect of coping style on behaviour & autonomic reactions of domestic pigs during dyadic encounters – A pilot study | B120 |
| Kathrin Büttner | Focussing on significant dyads in agonistic interactions & their impact on dominance indices in pigs | B121 |
| Daniela Alberghina | Maternal serum cortisol & serotonin response to a short separation from foal | B122 |
| Colline Poirier | Understanding stereotypic pacing: why is it so difficult? | B123 |
| Kristina Horback | Identification of seven call types produced by pre-weaning gilts during isolation | B124 |
| Paula Ramírez Huenchullán | The use of pig appeasing pheromone decreases negative social behavior & improves performance in post-weaning piglets | B125 |
| Chiara Mariti | The influence of sex on dog behaviour in an intraspecific Ainsworth Strange Situation Test | B126 |
| Gudrun Illmann | Sow response to sibling competition during nursing in 2 housing systems in early & late lactation | B127 |
| Sanjay Choudhary | Influence of bull biostimulation on puberty & estrus behaviour of Sahiwal breed (*Bos indicus*) heifers | B128 |
| Rodolfo Ungerfeld | Observation of ram-estrous ewe interaction enhances mating efficiency in subordinate but not dominant rams | B129 |
| Aline Freitas-De-Melo | The phase of oestrous cycle influences the stress response to social isolation in Corriedale ewes | B130 |

## Addressing Future Trends in Animal Production

| Melissa C. Cantor | Are feeding behaviors associated with recovery for dairy calves treated for Bovine Respiratory Disease (BRD)? | B131 |
| Oleksiy Guzhva | Smart ear tags for measuring pig activity: the first step towards an animal-oriented production environment | B132 |
| Eloise Fogarty | Sensor technology & its potential for objective welfare monitoring in pasture-based ruminant systems | B133 |

| Justin Mufford | Comparison of heat stress behaviour between different Canadian *Bos taurus* cattle breeds using unmanned aerial vehicles | B134 |
|---|---|---|
| Joao H.C. Costa | Voluntary heat stress abatement system for dairy cows: does it mitigate the effects of heat stress on physiology & behavior | B135 |
| Isabel Blanco Penedo | Are Freewalk systems offering more comfortable housing conditions for cattle? | B136 |
| Elise Shepley | A cow in motion: are we really providing 'exercise' to dairy cows? | B137 |
| Amanda Lee | Evaluating the effects of mean occupation rate & milk production on two automatic milking systems | B138 |
| Knut Egil Bøe | Virtual fences for goats on commercial farms, an animal welfare concern? | B139 |
| Lucille Bellegarde | Introduction of a new welfare labelling scheme in France | B140 |
| Takuya Washio | Development of a theoretical model to explain consumers' willingness to purchase animal welfare products in Japan | B141 |

## Free Communications

| Kees Van Reenen | The effect of floor type on lying preference, cleanliness & locomotory disorders in veal calves | B142 |
|---|---|---|
| Sarah J.J. Adcock | Pain response to injections in dairy heifers disbudded at 3 different ages | B143 |
| David Bell | Behavioural responses of dairy calves to temperature & wind | B144 |
| Jung Hwan Jeon | Effect of environmental temperature on rectal temperature & body-surface temperature in fattening cattle | B145 |
| Emma Ternman | REM sleep time in dairy cows changes during the lactation cycle | B146 |
| Madan Lal Kamboj | Development of a dairy cattle welfare assessment scale & evaluation of cattle welfare at commercial dairy farms in India | B147 |
| Daniel Enriquez-Hidalgo | Effect of SF6 equipment to estimate methane emissions on dairy cow behavior | B148 |
| Joan-Bryce Burla | Lying behaviour in dairy cows is associated with body size in relation to cubicle dimensions | B149 |
| Rielle Perttu | The effects of mesh fly leggings on number of flies & fly-avoidance behaviors of dairy cows housed on pasture | B150 |
| Mette S. Herskin | Transport of cull animals - are they fit for it? | B151 |
| Martyna Lagoda | A first look at the relationship between skin lesions & cortisol levels in stable groups of pregnant sows | B152 |
| Louisa Meira Gould | An examination of the behaviour & clinical condition of cull sows transported to slaughter | B153 |
| Armelle Prunier | Behaviour of piglets before & after tooth clipping, grinding or sham-grinding in the absence of social influences | B154 |
| Solveig Marie Stubsjøen | Associations between sheep welfare & housing systems in Norway | B155 |
| Alejandra Feld | First steps for the development of behavioural indicators in dairy sheep: ethogram & functional categorization | B156 |
| Beth Ventura | Effects of multimodal pain management strategies on acute pain behavior & physiology in disbudded neonatal goat kids | B157 |
| Grete H.M. Jørgensen | Using faecal glucocorticoid metabolites as a method for assessing physiological stress in reindeer | B158 |
| María D. Contreras-Aguilar | Changes in saliva analytes reflect acute stress level in horses | B159 |
| Irena Czycholl | Test-retest reliability of the Animal Welfare Indicators protocol for horses | B160 |

| Freddie Daw | A study into the impact of sweet itch on equine behavioural patterns | B161 |
| Alice Ruet | Welfare of horses living in individual boxes: methods of assessment & influencing factors | B162 |
| María Díez-León | Effects of cage size on the thermoregulatory behaviour of farmed American mink (*Neovison vison*) | B163 |
| Inger Hansen | Efficiency of sheep cadaver dogs | B164 |
| Vladimir Vecerek | Changes in behaviour of shelter dogs post adoption | B165 |
| Eva Voslarova | Factors affecting behaviour of dogs adopted from a shelter | B166 |
| Mirjana Đukić Stojčić | Associations between daytime perching behavior, body weight, age & keel bone deviation in laying hens | B167 |
| Manisha Kolakshyapati | Association of fearfulness at the end of lay with range visits at 18-22 weeks of age in commercial laying hens | B168 |
| Jutta Berk | Feather corticosterone: a new tool to measure stress in laying hens? | B169 |
| Nienke Van Staaveren | A cross-sectional study into factors associated with feather damage in laying hens in Canada | B170 |
| Ariane Stratmann | Influence of ramp provision during rear & lay on hen mobility within a commercial aviary system | B171 |
| Jakob Winter | Piling behaviour of laying hens in Switzerland: origin & contributing factors | B172 |
| Leonie Jacobs | Euthanasia - manual versus mechanical cervical dislocation for broilers | B173 |
| Xavier Averós | Use of routinely collected slaughterhouse data to assess welfare in broilers | B174 |
| Randi Oppermann Moe | Associations of lameness in broiler chickens to health & production measures | B175 |
| Inma Estévez | Testing by numbers: addressing the soundness of the transect method in commercial broilers | B176 |
| Marisa Erasmus | Changes in broiler chicken behavior & core body temperature during heat stress | B177 |
| Joanna Marchewka | Identifying welfare issues in turkey hen & tom flocks applying the transect walk method | B178 |
| Elisa Codecasa | The artificial blood feeding of *Aedes aegypti* mosquitos as an alternative to the use of live research animals | B179 |
| Moira Harris | The EU Platform on Animal Welfare & its subgroups – strengthening ISAE's voice in Europe | B180 |
| Nicki Cross | Above the minimum: Integrating positive welfare into animal welfare policy | B181 |
| Melissa Elischer | Building an informed audience: engaging youth in animal behavior & welfare education | B182 |
| Kathryn Proudfoot | Impact of frame reflection on veterinary student perspectives toward animal welfare & differing viewpoints | B183 |

| 1: Visualizing and analysis of individual-level data within large group systems | |
|---|---|
| Kongesal 2-3 | Chairs: Michael Toscano and Janice Siegford |

| | |
|---|---|
| Mike Toscano | Assessing individual movement and location patterns of laying hens |
| Janice Siegford | Comparison of patterns of substrate occupancy by individuals versus flocks of 4 strains of laying hens in an aviary |
| Joshua Peschel | Using synthetic video for automated individual animal behavior analysis |
| Magnus R. Campler | Some cows don't like to get wet: individual variability in voluntary soaker use in heat stressed dairy cows |
| Elske N. De Haas | Visualizing behaviour of individual laying hens in groups |

| 2: Future trends in the prevention of damaging behaviour | |
|---|---|
| Dræggen 3 | Chair: Andrew Janczak |

| | |
|---|---|
| Jen-Yun Chou | Using experimental data to evaluate the effectiveness of tail biting outbreak intervention protocols |
| Rachel Peden | Factors influencing farmer willingness to reduce aggression between pigs |
| Lisette Van Der Zande | Reduce damaging behaviour in laying hens & pigs by developing sensor technologies to inform breeding programs |
| Margrethe Brantsæter | Experience with laying hens that are not beak trimmed |
| Kristine Hov Martinsen | Experience with pigs that are not tail docked |
| Andrew Janczak | Films demonstrating production without tail docking of pigs or beak trimming of laying hens |

| 3: Novel indicators of fish welfare | |
|---|---|
| Dræggen 4 | Chair: Oliver Burman |

| | |
|---|---|
| Felicity Huntingford | Using thermal choices as indicators for fish welfare |
| Marco A. Vindas | The role of the serotonergic system as a welfare indicator in salmonids |
| Becca Franks | Developing measures of positive welfare for fish |
| Tanja Kleinhappel, Tom Pike, Oliver Burman | Using social behaviour as a welfare indicator in fish |

| 4: Engaging students in learning about production animal welfare assessment | |
|---|---|
| Bugaarden | Chair: Marta E. Alonso |

| | |
|---|---|
| Marta E. Alonso, Xavier Averós, Alejandra Feld, Melissa Elischer | Engaging students in learning about production animal welfare assessment |

| 5: Managing the dairy cow around the time of calving – can we do better? | |
| --- | --- |
| Dræggen 8 | Chair: Margit Bak Jensen |

| | |
| --- | --- |
| Maria Vilain Rørvang | What can we learn from studying the motivation underlying maternal behaviour of cattle? |
| Kathryn Proudfoot | Assessing the welfare of dairy cows before & after giving birth |
| Peter D. Krawczel | Alternative management strategies to help dairy cows achieve successful outcomes during the transition into lactation |
| Marie J. Haskell | Keeping cow & calf together: behaviour & welfare issues |
| Cynthia Verwer | State-of-the-art dam rearing of calves – sector-wide assessment of scientific & practical knowledge on dam-rearing systems |

| 6: Animal training – efficacy and welfare | |
| --- | --- |
| Kongesal 4-5 | Chairs: Dorte Bratbo Sørensen and Kristina Horback |

| | |
| --- | --- |
| Dorte Bratbo Sørensen | Using positive reinforcement to train laboratory pigs: benefits for animal welfare & research models |
| Kristina Horback | Creating an animal training handbook for livestock producers, veterinarians & researchers |
| Jonas Riise Johansen | Training animals to co-operate for veterinary procedures – a practical approach and demonstration |

# Table of contents

Mechanical spiking as a killing method in Nile tilapia (*Oreochromis niloticus*) –
a pilot study 75
*Daniel Santiago Rucinque, Paulo Roberto Pedroso Leme, Pedro Fontalva Ferreira and Elisabete Maria Macedo Viegas*

## Session 03. Pre- & Postnatal Effects

### Oral presentations

Association between maternal social success during late pregnancy and
offspring growth and behaviour in dairy cows 76
*Mayumi Fujiwara, Marie J. Haskell, Alastair Macrae and Kenny Rutherford*

The effect of gilt rearing strategy on offspring behaviour 77
*Phoebe Hartnett, Keelin O'Driscoll, Laura Boyle and Bridget Younge*

Lameness in sows during pregnancy impacts welfare outcomes in their offspring 78
*Marisol Parada, Leandro Sabei, Bruna Stanigher and Adroaldo José Zanella*

Oesophageal tube colostrum feeding suppresses oxytocin release and induces
vocalizations in healthy newborn dairy calves 79
*Carlos E. Hernandez, Bengt-Ove Rustas, Lena Lidfors, Charlotte Berg, Helena Röcklinsberg, Rupert M. Bruckmaier and Kerstin Svennersten-Sjaunja*

Stress responses resulting from commercial hatching of laying hen chicks 80
*Louise Hedlund, Rosemary Whittle and Per Jensen*

Misdirected oral behaviors in orphaned neonatal kittens 81
*Mikel Delgado, Isabelle Walcher, Karen Vernau, Melissa Bain and Tony Buffington*

Identification of fear behaviours shown by kittens in response to novel stimuli 82
*Courtney Graham, David Pearl and Lee Niel*

The role of age in the selection of police patrol dogs based on behaviour tests 83
*Kim I. Bjørnson, Judit Vas, Christine Olsen, Conor Goold and Ruth C. Newberry*

Calf-cow contact during rearing improves health status in dairy calves – but is
not a universal remedy 84
*Edna Hillmann, Rupert Bruckmaier and Cornelia Buchli*

Effects of a 3-month nursing period on fertility and milk yield of cows in their
first lactation 85
*Kerstin Barth*

**Poster presentations**

## Session 04. Behaviour & Nutrition

### Oral presentations

### Poster presentations

## Session 05. Cognition & Welfare

### Oral presentations

## Session 06. Environmental Enrichment

### Oral presentations

**Poster presentations**

## Session 07. Plenary - ISAE Creativity Award

### Oral presentations

## Session 08. Behaviour & Genetics

### Oral presentations

### Poster presentations

## Session 09. Human-Animal Interactions

### Oral presentations

## Session 10. Plenary - ISAE New Investigator Award

### Oral presentation

## Session 11. Social Environment

### Oral presentations

## Session 12. Addressing Future Trends in Animal Production

### Oral presentations

**Poster presentations**

## Session 13. Free Communications

### Oral presentations

## Workshop 1. Visualizing and analysis of individual-level data within large group systems

**Oral presentations**

## Workshop 2. Future trends in the prevention of damaging behaviour

**Oral presentations**

## Workshop 3. Novel indicators of fish welfare

### Oral presentations

## Workshop 4. Engaging students in learning about production animal welfare assessment

### Oral presentation

## Workshop 5. Managing the dairy cow around the time of calving – can we do better?

### Oral presentations

## Workshop 6. Animal training – efficacy and welfare

### Oral presentations

**Synergies between fundamental and applied behavioural science: lessons from a lifetime of fish watching**

*Felicity Huntingford*

*University of Glasgow, Institute of Biodiversity, Animal Health and Comparative Medicine, University Avenue, G12 8QQ, United Kingdom; felicity.huntingford@glasgow.ac.uk*

Among numerous strengths, David Wood-Gush's research was informed by a deep knowledge of behavioural biology, so that his results often contributed to fundamental understanding as well as having practical outcomes. His work therefore provides many examples of the synergy between fundamental and applied science. In this lecture in David's honour, I will explore this synergy based on my own experience of moving between fundamental and applied research. Taking a chronological approach, my doctoral research explored the relationship between risk-taking by sticklebacks in two apparently-different contexts (confronting conspecific rivals and potential predators), using consistent individual differences as a research tool rather than treating them as a nuisance. This work provided an early example of what would now be called different stress coping styles, or animal personalities. After that, following my fundamental interest in aggression and still using fish as subjects, I studied the costs of fighting and their influence on the nature and outcome of aggressive encounters. This led naturally to a series of applied projects aimed at using such information to reduce aggression among farmed fish. Briefly, with several colleagues, I explored practical ways of shifting the cost: benefit ratio to make fighting uneconomic to fish, as it is for fish farmers. Still pursuing fundamental questions, I studied stress coping style in several fish species and explored how proactive and reactive individuals perform in different environmental conditions. This line of research also led to collaborative applied projects, in this case on the effects of stress coping style on growth and welfare in farmed fish. These include, for example, a major impact on immunological and behavioural responses to disease. After exploring the synergy between pure and applied research in the context of these well-established themes, I will talk briefly about an area of current interest, illuminated by Wood-Gush's writing. This concerns the detailed and accurate knowledge of animal behaviour that exists among non-scientists such as farmers and fishers, whose livelihoods depend on being able to predict what animals are likely to do. In a sense, this is the ultimate applied ethology and I will give examples of how such knowledge can complement both applied research and fundamental understanding of behaviour.

## Identifying enriched housing conditions for zebrafish (*Danio rerio*) that vary along a scale of preference

*Michelle Lavery[1], Victoria Braithwaite[2], Noam Miller[3] and Georgia Mason[1]*
*[1]University of Guelph, Animal Biosciences, 50 Stone Rd. E, Guelph, ON N1G 2W1, Canada,*
*[2]Penn State University, Biology, 410 Forest Resources Building, University Park, PA 16802, USA,*
*[3]Wilfrid Laurier University, Psychology, Science Building, 75 University Ave. W, Waterloo, ON N2L 3C5, Canada; lavery.j.m@gmail.com*

Widely used in research, zebrafish (*Danio rerio*) typically live in small, barren tanks. Environmental enrichment (EE) can improve their welfare, but only a few EEs have been assessed. Furthermore, although in other species more highly preferred EEs are known to produce better welfare states, current work on which EEs zebrafish most prefer is limited to just two studies, both with limited group replication ($n_{groups}$=1-5, depending on the study). We therefore aimed to assess zebrafish preferences for 13 putative EEs. These were black walls or walls with an underwater image; sloped or flat aquarium gravel; plastic grass-like or overhanging plants; the ability to see fish in neighbouring tanks; and combinations of these treatments. Twenty replicate groups of 8-10 adult TU zebrafish were housed in 19 L tanks. Groups underwent 13 consecutive preference tests (typically 10 groups per test; with the other 10 also being tested if results were ambiguous) with putative EEs installed on opposite sides of the tank (randomised across groups). Habituation to new treatments lasted at least three days, immediately after which fish preference stability was evaluated. Once preferences were stable (which took four to twelve days), data were collected on each group's occupancy of each side at intervals of 10-30 minutes, over the following three days. These were analyzed with one-sample $t$ or Wilcoxon signed rank tests. Preferred EEs were never taken away once fish had experienced them; new EEs to be tested were simply added. Fish preferred both gravel morphologies over standard bare floors (sloped: 77.12%, $P_{t=15.25, df=9}$<0.001; flat: 77.16%, $P_{t=9.93, df=9}$<0.001); gravel over plants (58.82%, $P_{t=-3.57, df=19}$=0.001); plants and gravel over plants or gravel alone (vs plants alone: 67.77%, $P_{t=-7.64, df=19}$<0.001; vs gravel alone: 69.71%, $P_{t=10.99, df=9}$<0.001 [grass] and 66.92%, $P_{t=8.19, df=9}$<0.001 [overhanging]), and more (i.e. four) plants over fewer (i.e. two) (58.68%, $P_{V=105, df=19}$<0.001). They were indifferent to wall type (when in barren tanks: 55.77%, $p_{t=1.35, df=9}$=0.21 [black] and 45.30%, $P_{t=-1.75, df=9}$=0.11 [underwater scene]; if combined with four grass plants and gravel: 53.75%, $P_{V=12, df=9}$=0.23 [black] and 49.46%, $P_{t=-0.13, df=9}$=0.89 [underwater scene]); visual contact with neighbouring tanks (52.57%, $P_{t=1.13, df=9}$=0.28); and between plant types (50.00%), if they were present in equal numbers. Given this series of results, we can place EEs along a scale from least preferred to most: plants < gravel < two plants+gravel < four plants+gravel. As zebrafish research facilities continue to grow in number, these results will aid the effective implementation of enrichment, since researchers can now provide different degrees of preferred EE depending on their research question, resources, and ethics. Our next work will evaluate how these different degrees of EE affect welfare state when provided long-term.

## Knowledge gaps in global aquaculture welfare: scope of the problem, current research, and future directions

*Becca Franks, Christopher Ewell, Isabel Fife-Cook and Jennifer Jacquet*
*New York University, Environmental Studies, 285 Mercer St, New York City, 10003, USA;*
*beccafranks@gmail.com*

Aquaculture is expanding at an unprecedented rate worldwide. It is the fastest growing sector of the food industry and already involves more individual animals than all terrestrial farming combined. This expansion has attracted the interest of investors, producers, governments and NGOs, yet despite some work examining the ecological and public health impacts of aquaculture, the welfare needs of these animals has yet to be assessed at a global scale. With this systematic review, our aim was to apply a multimethod approach to create the first comprehensive picture of global aquaculture welfare and highlight the knowledge gaps that could signal welfare-risks. First, to generate information regarding the number of species and their 2016 usage rates (i.e. tonnes), we extracted data from the Food and Agriculture Organization of the United Nations. Combining these data with taxonomic information from FishBase allowed us to generate a phylogenetic map of the diversity of species and extrapolate the number of individuals used annually worldwide. Finally, we conducted a bibliometric study using Web of Science (Clarivate Analytics) to determine the quantity and content of scientific information on each species, paying particular attention to the presence of the behavioral, ecological, and welfare research. Our results show that in 2016, global farming produced at least 84 million tonnes of aquatic animals, belonging to at least 543 unique species from 5 distinct Phyla. Of this total, over 7 million tonnes of animals were of indeterminate genera and species. Without information on genera or species, the welfare of these animals must be considered at risk because ensuring welfare requires, at a minimum, information regarding species-typical behavior, ecology and biological function. Of the remaining approximately 77 million tonnes of animals, which represent at least 501 distinct species, we could not find a single paper on the welfare of 318 species (63%). Without research to support welfare assessments, the absence of information on these species means that the welfare of an additional 29 million tonnes of animals is also being neglected. Finally, fine-grain coding of the aquaculture literature indicates that even when welfare is studied, it is overwhelmingly framed as a production value related to biological health, with natural behavior and affective states effectively ignored. In sum, this systematic review indicates that aquaculture welfare is not keeping pace with aquaculture production, raising alarms regarding the wellbeing of the trillions of animals involved. Future work needs to focus on the behavior and affective states of aquatic animals in naturalistic settings, especially in heavily used, yet under-studied species like, for example, Grass carp, (*Ctenopharyngodon idellus*).

## Behavioural predictors of stress in groups of fish

*Tanja Kleinhappel, Tom Pike and Oliver Burman*
*University of Lincoln, School of Life Sciences, Joseph Banks Laboratories, Green Lane, LN6 7DL,*
*Lincoln, United Kingdom; tkleinhappel@lincoln.ac.uk*

In many standard laboratory assays involving social animals, subjects are tested individually. This excludes potentially valuable behaviours only exhibited when animals are tested in groups, such as changes in group structure or social interactions. However, although testing animals in groups can provide a powerful tool for behavioural monitoring, the conventional measures of social behaviour that are typically recorded, such as dyadic aggressive encounters, may fail to reveal the full wealth of available information. Here, we aimed to determine if metrics describing the behaviour of grouping animals could be used to predict stress, hypothesising that metrics describing group structure can be valuable additions to the more standard behavioural indicators. Using zebrafish (*Danio rerio*) as a model, we observed replicated shoals of 7 fish when initially exposed to a novel tank (i.e. a mildly stressful environment) and again, after a habituation time of 24 hours (i.e. a familiar non-stressful environment). We quantified various standard behavioural measures, such as fish being stationary, displaying erratic movements or aggressive behaviours, in combination with metrics describing group structure, such as proximity (e.g. nearest neighbour distances), social (e.g. shoaling density and subgroups) and spatial (e.g. tendency for thigmotaxis) metrics. Generalised linear mixed-effects models showed that all analysed metrics significantly predicted the likelihood of fish shoals being stressed or non-stressed (P<0.001). However, ranking the models by their Akaike Information Criterion (AIC; an estimator of relative quality of the statistical model) indicated that there was a substantial difference in relative model fit. The model with the lowest AIC had fish being stationary as a binary predictor variable, showing that with fewer stationary fish present in a shoal, groups are significantly more likely to be stressed. This measure was followed by different metrics describing group structure, indicating that they performed markedly better compared to the standard behavioural measures, such as aggressive behaviours between individuals. Critically, our results show that metrics describing grouping behaviour can significantly predict stress, and, in many cases, outperform standard behavioural metrics. This can have a substantial impact on the selection of indicators for assessing welfare in group-housed fish, especially as the use of group metric data could allow for the refinement of behavioural protocols carried out in a range of research areas, by providing sensitive and rich data in a more relevant social context.

## Increasing welfare through early stress exposure in aquaculture systems

*Marco A. Vindas[1], Angelico Madaro[2], Thomas W.K. Fraser[1], Erik Höglund[3], Rolf E. Olsen[2], Øyvind Øverli[1] and Tore S. Kristiansen[2]*
*[1]Norwegian University of Life Sciences, Food Safety and Infection Biology, Ullevålsveien 72, 0454 Oslo, Norway, [2]Institute of Marine Research, Nordnesgaten 50, 5005 Bergen, Norway, [3]Norwegian Institute for Water Research (NIVA), Gaustadalléen 21, 0349 Oslo, Norway; marco.vindas@nmbu.no*

Pathologies arising from continued activation of stress systems may represent a mismatch between evolutionary programming and current environments. Ongoing rapid domestication of Atlantic salmon implies that individuals are subjected to evolutionarily novel stressors encountered under conditions of artificial rearing, requiring new levels and directions of flexibility in physiological and behavioural coping mechanisms. Phenotypic plasticity to environmental changes is particularly evident at early life stages. We subjected young salmon (10-month-old parr) to an early stress treatment (unpredictable chronic stress, UCS, n=3 tanks/129 fish each), and compared their subsequent performance over several months and life stages with unstressed control fish (n=3 tanks/129 fish each). The UCS was given three times per day for 5 min over 23 days. Three randomly selected stressor types (from the following: hypoxia, low water level, cold shock, heat shock, aberrant noise, flashing light, chasing, air exposure) were applied daily to maintain unpredictability. After the UCS, fish were kept following standard aquaculture rearing conditions. This work was approved by the Norwegian Animal Research Authority (NARA), following the Norwegian laws and regulations controlling experiments and procedures on live animals in Norway. UCS fish had higher specific growth rates compared with unstressed controls after smoltification (P<0.001), a particularly challenging life stage, and after seawater transfer (P<0.001). Furthermore, when subjected to acute confinement stress at the end of the experiment, UCS fish had lower hypothalamic catecholaminergic (P<0.01) and brain stem serotonergic (P=0.03) responses to stress compared with controls. In addition, serotonergic activity was negatively correlated with final growth rate (ρ=-0.3, P=0.01), which implies that serotonin-responsive individuals had a growth disadvantage. A subdued monoaminergic response may be beneficial to farmed fish because individuals may be able to reallocate energy from stress responses into other life processes, such as growth. In conclusion, by investigating the impact of a mismatch between current and past environments, we may be able to devise practices that result in an overall increase in the stress resilience and welfare of salmon in aquaculture systems.

### Does sudden exposure to warm water cause pain in Atlantic salmon?

*Jonatan Nilsson[1], Lene Moltumyr[1], Angelico Madaro[1], Tore Sigmund Kristiansen[1], Siri Kristine Gåsnes[2], Cecilie Marie Mejdell[2], Kristine Gismervik[2] and Lars Helge Stien[1]*
*[1]Institute of Marine Research, P.O. Box 1870 Nordnes, 5817 Bergen, Norway, [2]Norwegian Veterinary Institute, P.O. Box 750, 0106 Oslo, Norway; jonatan@hi.no*

Thermal treatment, where farmed salmonids are exposed to heated water for ~30 s to remove sea lice, has become the most used delousing method in salmonid aquaculture. There have, however, been concerns raised about it being painful for the fish. Under permit number 15383, we studied the behavioural response of farmed Atlantic salmon (age: 16 months; weight: 234±52 g) acclimated to 8 °C when transferred individually to temperatures in the range 0-38 °C. The exposure tank was of the same type as the stock tank (1.5×1.5 m), and the fish were transferred with a dip net. The temperature was increased with 4 °C steps from 0 to 24 °C, with 2 °C from 24 to 36 °C, and with 1 °C from 36-38 °C. Six individuals per temperature was tested in the range 0-36 °C, and four individuals per temperature at 37-38 °C. Individuals were euthanized immediately after exposure. In total 86 fish were used. As the transfer to the test tank could induce abnormal behaviour during the first 30 s also in control fish (8 °C exposure) which could mask effects of temperature, longer exposure durations than normally used by industry during thermal delousing were applied, with maximum 5 min exposure. Furthermore, to document which exposure durations that are life threatening at different temperatures, the endpoint for removing fish from exposure was when they lost equilibrium and laying on the side, a sign of imminent death. The behaviour of the fish was video recorded and analysed for swimming speed and the frequency of behavioural actions: Direction change; Circling swimming; Head shake; Collision with tank wall; Surface break; Body bend. Below 28 °C, none of the fish reached endpoint within the 5-min maximum, at 28 °C four of five fish reached endpoint after a mean of 252 s (236-267 s), and from 30 °C all fish reached endpoint, faster the higher the temperature (Spearman rank correlation, $r_s$=1.00, P<0.001) with a mean of 86 s (66-103 s) at 38 °C. The initial swimming speed increased with temperature ($r_s$=0.74, P<0.001) and differed significantly from the control from 32 °C and warmer. While fish exposed to temperatures below 28 °C calmed down within the first 30 s, fish above 28 °C maintained a high swimming speed until the they approached the endpoint. Direction change occurred frequently during the first 2 min also in the control, but with a higher rate (Welsh two sample t-test, P<0.05) from 32 °C. Collisions, Head shake, Circling, and Surface break occurred at higher rates than control from 28-30 °C (all P<0.05). Collision and Head shake started to occur immediately after transferred to the exposure tank. Body bend only occurred from 30 °C and warmer, just before the fish reached the endpoint, but varied much between individuals. The results suggest that even brief exposure to temperatures above 28 °C is painful for salmon, and that the 'safety margins' between the exposure durations used during thermal delousing and lethal durations are relatively small.

**Identifying potential play behaviors in fish: a study using online video analysis**

*Isabel Fife-Cook and Becca Franks*
*New York University, Environmental Studies, 285 Mercer St, New York, NY 10003, USA;*
*ifc219@nyu.edu*

Play behavior has fascinated scholars for millennia: it is found across a diversity of species and yet, by definition, serves no obvious immediate goal or function. While the relationship between play and welfare is complex, documenting its presence (vs absence) and the type of play in a species is relevant to welfare in at least three ways. First, for animals that are known to play, a total absence of play behavior is generally agreed to indicate poor welfare. Thus, if a species can be shown to play, an absence of play behavior under certain housing conditions may signal welfare problems for those animals. Second, in many species, the type of play (e.g. locomotor vs solitary) changes with welfare. As such, characterizing play types and the circumstances under which they occur has the potential to contribute additional welfare indicators for that species. Finally, identifying the presence and studying the types of play and where they occur contributes more generally to the ongoing refinement of our understanding of the relationship between play, play-types, emotions, and welfare. To date, play has only been anecdotally described in some non-mammalian/non-avian species and has not been systematically evaluated in fish. As an initial assessment of the potential presence play behavior in fish and to identify promising areas for future fish play research, we generated a database of YouTube videos depicting fish engaging in potentially play-like behaviors and evaluated them according Burgardt's five-point criteria for play. Videos were found by performing a YouTube search using various iterations of the search term 'fish play' or descriptions of potential fish play behaviors such as 'fish jumps over stick' and 'fish plays with [object]'. This procedure yielded 40 user-posted videos, which were then coded along the following dimensions: species (when possible), location (captivity or the wild), training (did the behavior appear to be trained with food rewards or not), social context (isolated or in a group), duration of behavior, functionality, relaxed-state, and play type: social, locomotor, and object. Analyzing these data produced 27 videos depicting fish engaging in behavior consistent with Burghardt's definition of play and without any signs of training or reinforcement (17 species from 12 distinct phylogenetic families). The majority of videos depicting potential play behavior were categorized as locomotor play (67%), followed by object play (57%) and finally social play (35%). Air-bubble streams were a fruitful source of potential play videos: approximately 60% of locomotor play videos involved interaction with bubble-jets, both in captivity and in the wild. These results suggest that at least some species of fish likely possess the ability to play and thereby contribute to the growing evidence of the need to include fish in our sphere of moral concern. Moreover, this study points to several areas of inquiry for future fish play research, specifically: (1) locomotor play facilitated by bubble jets and/or air-stones, (2) object play facilitated by the provision of novel objects, (3) controlled investigations into the environments that produce and modulate play behavior, and (4) determining whether and when play may involve positive emotions in fish.

## Impact of sea lice management methods on Atlantic salmon welfare: experimental findings and perspectives

*Cécile Bienboire-Frosini[1], Cyril Delfosse[1,2], Pietro Asproni[1], Camille Chabaud[1,2], Violaine Méchin[1], Céline Lafont-Lecuelle[1], Alessandro Cozzi[1,2] and Patrick Pageat[1,2]*
[1]*IRSEA, Quartier Salignan, 84400 Apt, France,* [2]*IRSEA-ARC, Daugstad, 6392 Vikebukt, Norway; c.frosini@group-irsea.com*

Salmon lice (*Lepeophteirus salmonis*) is an ectoparasite of salmonids, causing severe damage to Atlantic salmon (*Salmo salar*) aquaculture in the Northern hemisphere. Sea lice infestation must be managed to a sustainable level by the farmers during the rearing process. Due to the development of lice resistance to chemical treatments, interest in new management methods is increasing but their impact on Atlantic salmon welfare is not much investigated. We present several studies evaluating some of these methods in respect of Atlantic salmon welfare, measuring stress parameters. Studies respected C2EA125's Ethics Committee approval and ISAE ethical guidelines. The first experimental study assessed the effect of Atlantic salmon handling on plasma cortisol: in Norway farms, sea lice count is mandatory and implies to sample fish from cages out of the water before manual inspection. We used 16 salmon smolts (8 handled; 8 controls) to reproduce this handling with a landing net and measured their cortisol levels: handled salmon showed significantly higher plasma cortisol levels than controls (median=190.7 vs 33.5 ng/ml, respectively; Wilcoxon two-sample test; $P<0.001$). Snorkel lice-barrier is an example of new control methods: sea-cages are deeply submerged to reduce salmon exposure to sea lice and equipped with a snorkel-like tube to allow salmon surface access to refill their swim bladder with air. However, salmon have limited and competitive access to the surface, which can represent stressful conditions. In the second experimental study, we modeled this situation using 16 salmon smolts divided equally between two tanks, one undergoing a water volume reduction by 1/3 and in which salmon were continuously prevented from rising to the surface, except for 2 h/day, during 5 days, before stress parameters were measured: the plasma cortisol level was not significantly different between control and confined salmon (median=4.2 vs 4.9 ng/ml, respectively; Wilcoxon two-sample test; $P>0.05$) while the neutrophil/lymphocyte ratio was significantly higher in the confined than in control salmons (median=0.26 vs 0.18, respectively; Wilcoxon two-sample test; $P=0.0014$), suggesting a long-term stress effect. Finally, a recent on-farm study explored the use of a salmon semiochemical to enhance fish condition and decrease lice infestation. Descriptive data showed that treated salmon displayed better zootechnical performances (increased fish growth and shorter production cycle), lower plasma cortisol, higher skin mucus production, and seemed to swim in shallower depths. Importantly, the lice infestation was controlled to acceptable levels. To conclude, since the salmon vulnerability to lice infestation is linked to its health and welfare (stressed fish are more subject to infestations), it is important to consider the overall impact of the management methods on both sea lice infestation levels and fish welfare. In this holistic approach, the salmon semiochemicals' use seems to be a promising method encompassing the regulatory sea lice control needs and the fish welfare aspects.

## Mechanical spiking as a killing method in Nile tilapia (*Oreochromis niloticus*) – a pilot study

*Daniel Santiago Rucinque, Paulo Roberto Pedroso Leme, Pedro Fontalva Ferreira and Elisabete Maria Macedo Viegas*
*University of São Paulo, Faculty of Animal Science and Food Engineering, Department of Animal Science, Av. Duque de Caxias Norte, 225, 13635-900, Brazil; dsrucinqueg@usp.br*

Hypothermia in ice and water is historically used in Brazil in the slaughter process in most fish slaughterhouses, although not recommended by OIE. The study aimed to assess mechanical spiking in the induction of unconsciousness as a killing method in Nile tilapia. All procedures were approved by the Ethics Committee on Animal Use at the University of São Paulo (4446150817). Nile tilapia (mean ± SD) (574.0±170.8 g; 30.9±3.1 cm) were divided into two groups. Fish in Lateral group (L) (n=20) were placed on the left lateral side, and the access of the captive bolt gun was on the right side of the head. Fish in Frontal group (F) (n=20) were placed in a V-shaped wooden structure, and the access of the captive bolt gun was frontal. Three fish previously euthanized were dissected in order to both practice the method and determine the exact point of the shot. Mechanical spiking was applied using a commercial gun for fish – Ikigun®. The number of attempts for an efficient penetration, immediate loss of consciousness and number of fish with recovery indicators were assessed. Loss of consciousness was determined through behavioural evaluation observing swimming, equilibrium, handling, painful stimulus, vestibulo-ocular reflex (VOR) and breathing. Immediately after the shot, fish were placed in a plastic tank with 200 L of water and observed for 20 min. When observing any indicator of recovery of consciousness, fish were placed in a tank with water and 200 mg $L^{-1}$ of benzocaine before bleeding. Ten fish of each group were refrigerated for *rigor mortis* index (RMI) evaluation at 0, 3, 6, 24, 48, 72 and 120 h after bleeding. Loss of consciousness, number of shot attempts and RMI did not follow a normality pattern; hence, for comparisons between treatments, the Mann-Whitney test was used. Fisher's exact test was used to compare proportions of fish performing specific behaviours among treatments: loss of consciousness and recovery. All L fish lost consciousness immediately after the shot, in contrast with 95% F fish (P=1.0, df=39). All fish that lost consciousness immediately after the shot did not swim, lost equilibrium, did not respond to painful stimuli or handling, did not show any VOR and ceased breathing. Attempts of an effective shot did not differ between groups (1.15±0.49 L; 1.05±0.22 F) (P=0.5536, df=39). One L fish (5%) showed indicators of recovery, different from 25% F fish, which showed signs of recovery (P=0.182, df=39). The progress of RMI was delayed in L fish (median, min-max) (41%, 27-75%) at 6 h post-slaughter, in contrast with F fish (59%, 40-100%) at the same period of time (P=0.0445, df=19). However, previous studies from our research team showed that all tilapia submitted to hypothermia reached full *rigor mortis* (100%) at 6 h. Therefore, mechanical spiking is effective to induce unconsciousness without recovery, delaying the onset of *rigor mortis* in Nile tilapia, and its use in an automatic system may be an alternative to slaughter.

## Association between maternal social success during late pregnancy and offspring growth and behaviour in dairy cows

*Mayumi Fujiwara[1], Marie J. Haskell[2], Alastair Macrae[1] and Kenny Rutherford[2]*
*[1]University of Edinburgh, Royal (Dick) School of Veterinary Studies, Easter Bush, Midlothian, EH25 9RG, United Kingdom, [2]Scotland's Rural College, West Mains Road, Edinburgh, EH9 3JG, United Kingdom; mayumi.fujiwara@ed.ac.uk*

Prepartum dairy cows are often kept in dynamic social groups at a high stocking density and hence are subject to social stress. Stressful maternal experiences can have a negative impact on foetal development, which could affect the offspring throughout its postnatal life. This study aimed to investigate possible associations between maternal social experiences during the dry period and offspring growth and behaviour. Holstein cows (n=48) were dried off 8-9 weeks before their expected calving date and allocated to either high (H) or low (L) stocking density groups from then until calving (H: 0.5 headlock or 0.3 m feed-face space + 1 cubicle or 6 m$^2$ lying area per cow; L: 1 headlock or 0.6 m feed-face space + at least 1.5 cubicle or 12 m$^2$ lying area per cow). The number of times a cow displaced other cows (win) and a cow was displaced by other cows (lose) at the feed-face during the first 100 min after feed delivery was observed for two days per week throughout the experimental period. Social success rates (SSR) were calculated for each cow by dividing the number of wins by the total number of wins and losses. Calves born to these cows (n=44) were weighed at birth, on d7 (when introduced to a group pen) and on d49 (weaning). Average daily gains (g/day) from birth to d7 and birth to d49 were calculated. Video observations were made by a single observer to collect latencies of behavioural reactions to the novel group environment on d7. Calves were trained to drink from an automatic milk feeder after introduction to a group pen, and the number of trainings required for each calf was recorded. Generalised linear models were used to investigate associations between maternal SSR and calf body weight, growth and training count, and Cox regression was used to analyse latencies of behaviours at the introduction to the group pen. Both models included maternal SSR, treatment and parity, and calf gender, breed and disease incidence. Although calves from higher SSR cows tended to grow faster (by 1.8±0.1 g/day) from birth to weaning ($F_{1,33.0}$=3.5, P=0.070), calf body weight was not associated with maternal SSR (P>0.1) at any stage. There was a significant interaction between SSR and treatment, with a positive association between maternal SSR and birth weight for L calves, but not for H calves ($F_{1,38.0}$=4.9, P=0.033). Calves from higher SSR cows: (1) required significantly fewer trainings to learn to use the automatic milk feeder ($F_{1,35.0}$=13.3, P<0.001); (2) had a shorter latency to perform locomotor play (Hazard Ratio: HR=0.97 [95% CI: 0.95-0.99], P=0.002); (3) took longer to initiate social contact with pen mates (HR=1.03 [1.00-1.10], P=0.011). Higher social success in cows during the dry period was significantly associated with calves' learning ability and behavioural reactions to a novel group environment. Maternal SSR was associated with the body weight of L calves only and had a small impact on growth, which requires further investigation.

## The effect of gilt rearing strategy on offspring behaviour

*Phoebe Hartnett[1,2], Keelin O'Driscoll[2], Laura Boyle[2] and Bridget Younge[1]*
*[1]University of Limerick, Department of Biological Sciences, University of Limerick, Limerick, V94 T9PX, Ireland, [2]Teagasc, Pig Development Department, Teagasc, Animal and Grassland Research and Innovation Centre, Moorepark, Fermoy, Co. Cork, P61 P302, Ireland; phoebe.hartnett@teagasc.ie*

In Ireland, replacement gilts are often reared with finisher pigs destined for slaughter, including entire males. This exposes them to high levels of mounting and aggression, which is stressful and injurious. Moreover, finisher diets do not supply the minerals appropriate to promoting limb health. These factors can have a negative impact on sow longevity. Furthermore, stress experienced by the mother can affect offspring development prenatally. This experiment investigated whether the responses of piglets to three behavioural tests (open arena (OA), human approach (HA), and human touch (HT)) were affected by the rearing strategy of their dam. Gilts were reared under four strategies in a 2×2 factorial arrangement (Female (FEM) or mixed sex (MIX) groups; control (CON) or mineral supplemented (MIN) diet). Four piglets (2 male, 2 female) at 3 weeks old (d17-d21) were selected on the basis of a back-test score taken at approximately 17 days old (d16-d20) from approx. 10 gilts per rearing strategy (n=163 piglets in total). In the OA test the piglets were placed in an empty pen which was divided into 9 squares for 3 min. Behaviours (Duration: freeze, explore, stand, run, walk; Incidence: escape, eliminate, scream and low grunt) were recorded directly using a handheld computer (Psion Workabout Pro™ 3). Location (squares entered) was recorded by video, and analysed using The Observer™. After the OA, the experimenter entered the pen and timed how long it took for the piglet to approach (HA test; max 1 min). The experimenter then attempted to touch the neck of the piglet and scored the piglets response from 1 (flee) to 3 (completely calm). Data were analysed in SAS using Mixed models, Wilcoxon, and Fishers exact tests. Male piglets tended to freeze for longer than female piglets (P=0.09), and female piglets spent longer walking than male pigs (P<0.05). Piglets from MIN gilts tended to explore more (P=0.06), spent less time standing still (P<0.01) and grunted less (P<0.05) than piglets from CON gilts. Piglets from gilts reared in MIX groups tended to scream more than piglets from gilts that were reared in FEM groups (P=0.08). Gilt rearing strategy did not affect the number of squares entered, or the number of piglets that approached during the HA test. However, piglets from MIXxCON gilts tended to approach faster than the other groups (P=0.08). Finally, piglets from MIN gilts reacted more fearfully than those from CON in the HT test (P<0.05). In summary, piglets from gilts fed a mineral supplemented diet and reared in single sex groups showed behaviour consistent with less stress in the OA test. However, piglets from gilts reared in the poorest conditions (i.e. mixed sex groups without mineral supplementation) appeared less fearful of humans. In conclusion, diet and group composition during the gilt rearing period can influence offspring behaviour and response to stressful situations.

## Lameness in sows during pregnancy impacts welfare outcomes in their offspring

*Marisol Parada[1,2], Leandro Sabei[1], Bruna Stanigher[1] and Adroaldo José Zanella[1]*
*[1]University of São Paulo, Faculty of Veterinary Medicine and Animal Science, Department of Preventive Veterinary Medicine and Animal Health, Avenida Duque de Caxias Norte, 225, Pirassununga, SP, 13635900, Brazil, [2]Università degli Studi di Teramo, Fac. di Medicina Veterinaria, Loc. Piano d'Accio, 64100, Teramo, Italy; mparadasarmiento@unite.it*

Lameness in sows is a painful and common condition, affecting between 30- 60% of sows, according to recent data. The stress and the pain experienced by lame sows causes unfavorable scenarios not only for pregnant animals, but also for their offspring, possibly due to glucocorticoid-mediated effects on fetal programming. The objectives of this study are to assess the behavioral, emotional and physiological effects of sow lameness in their offspring. This study was carried out in a Brazilian commercial pig farm, studying sows and piglets, with follow up experiments performed, with the same piglets, at the University of São Paulo, Brazil. The protocol was reviewed by the Ethics and Animal Use Committee (protocol number 9870211117). Gait score was assessed in 582 pregnant sows, every 15 days, over a period of 4 months, using a validated scoring system (0 to 5, being 0 a sow without lameness and 5 a sow with severe lameness). Out of the 582 sows, 30 animals, 15 sows without lameness (group A: score 0 or 1) and 15 sows with lameness (group B: score 3 to 5), were selected, for the follow up study monitoring their offspring. From each of these sows, three piglets were studied (90 piglets in total). Piglets were weaned at 28 days, transported to the University Campus, housed in pens with 9 individuals, divided in groups by sex, weight and lameness score of their mother (A or B). The following data were obtained from the piglets: body photographs to count skin lesions (three days); behavior in the open field and novel object tests were performed three days after weaning, and nociception measures were taken with an electronic von Frey aesthesiometer in four body parts: left and right plantar pad (LPP and RPP); and left and right leg (LL and RL). Data were analyzed using t-test or the Wilcoxon test at a significance level of 0.05 in the programming language R. 91.41% of the 582 sows had at least one event of lameness, 37.46% were classified as Group A, and 62.54% scored as group B. Sows with lameness had fewer days of pregnancy than sows without lameness ($P<0.0005$), 115 and 116 days respectively. Piglets from lame sows vocalized more than piglets from sows without lameness when they were subjected to the novel object test. Piglets from lame sows responded numerically to higher pressures in all body parts in the nociception test than piglets born from sows without lameness. This was confirmed by the means values: group A-LPP=893.44 g; group B-LPP=1,007.26 g; group A-RPP=916.28 g; group B-RPP=997.11 g; group A-LL=896.35 g; group B-LL=981.99 g; group A-RL=961.58 g; group B-RL=974.12 g; ($P>0.05$). Lameness appeared to modify the nociceptive threshold of the offspring, suggesting an increase in pain tolerance of piglets from lame sows. Additionally, lameness in sows decreased pregnancy length. The results confirm the impact of lameness in altering behavior and welfare outcomes in piglets.

## Oesophageal tube colostrum feeding suppresses oxytocin release and induces vocalizations in healthy newborn dairy calves

*Carlos E. Hernandez[1], Bengt-Ove Rustas[1], Lena Lidfors[2], Charlotte Berg[2], Helena Röcklinsberg[2], Rupert M. Bruckmaier[3] and Kerstin Svennersten-Sjaunja[1]*
*[1]Swedish University of Agricultural Sciences, Animal Nutrition and Management, 75323 Uppsala, Sweden, [2]Swedish University of Agricultural Sciences, Animal Environment and Health, 53223 Skara, Sweden, [3]University of Bern, Veterinary Physiology, 3001 Bern, Switzerland; carlos.hernandez@slu.se*

Calves are born agammaglobulinemic and therefore depend on the transfer of passive immunity from the maternal colostrum to be able to fight disease and ensure survival. Because not all calves have a strong motivation to suckle soon after birth, some farms routinely feed newborn calves using an oesophageal tube (OT) feeder. The OT allows farmers to feed large volumes of colostrum regardless of the calf's motivation to drink. However, OT feeding is an invasive force feeding method that prevents calves from fulfilling their natural motivation to suckle. The effects of using an OT on the physiological and behavioural responses in healthy newborn calves are unknown. For these reasons, we measured oxytocin, cortisol and behaviour in healthy newborn dairy calves. Female calves were randomly allocated to receive their 1st colostrum feeding (an amount equal to 8.5% birth weight) either using an OT (n=5) or a nipple-bottle (bottle; n=5), 4 h after birth. The 2nd feeding was offered in buckets fitted with rubber nipples to all calves. Blood samples were taken from jugular catheters at -10, 0, 1, 2, 3, 5, 10, 20, 30 and 40 min from the start of feeding during the first two feedings for oxytocin and cortisol determination. Behavioural responses were analysed from videos and included vocalizations, drops/falls to ground and struggles, during 1st feeding only. Colostrum intake and time feeding were recorded in both feedings. Data were compared using t-test for time feeding and Wilcoxon test for leg movements and struggles. Vocalisations could not be analysed due to low frequency of the behaviour. Colostrum intake, oxytocin and cortisol area under the curve (AUC) data were analysed using standard least squares analysis with feeding method and feed number as fixed effects and calf ID as random effect. Data are mean±SEM. During the 1st feeding, the duration of feeding was shorter (OT=239.9±11.0 vs bottle=821.0±130.3 s, P<0.05) and there was a tendency for higher colostrum intake (OT=3.3±0.2 vs bottle=2.8±0.2 kg, P=0.08) in calves fed with an OT. However, no such differences were observed during the 2nd feeding. The OT fed calves had lower oxytocin AUC during the 1st feeding than bottle fed calves (OT=626.4±144.2 vs bottle=1,131.5±133.2 pg/ml min, P<0.05). No such differences were observed during the 2nd feeding. Cortisol AUC was higher during the 1st feeding (1st feeding=822.3±55.5 vs 2nd feeding=607.0±55.5 ng/ml min, P<0.01). There were no differences in cortisol AUC levels between the feeding methods. During OT feeding, 4/5 of calves vocalized during OT feeding (average 17.8±8.6 vocalizations) whereas no calves vocalized during bottle feeding. There were no differences in struggle or drops/falls to the ground. In conclusion, oesophageal tube feeding suppressed oxytocin release and induced distress vocalizations during feeding, suggesting a transient negative impact on the calves' welfare.

## Stress responses resulting from commercial hatching of laying hen chicks

*Louise Hedlund, Rosemary Whittle and Per Jensen*
*Linköping University, IFM Biology, AVIAN Behavioural Physiology and Genomics Group,*
*Linköping University, 581 83, Sweden; louise.hedlund@liu.se*

In the commercial hatchery, laying hen chicks are exposed to many potentially stressful events during their first day of life. We investigated effects of this in a short- and long-term perspective on behaviour and stress physiology by comparing chicks that had gone through the commercial hatchery process (HC) with a matched control group (CC). Behaviour tests and corticosterone measurements were performed at 1st week of age and repeated five weeks later. Chickens from both groups were weekly weighed and at 20 weeks of age, feather and comb-wattle scoring was performed. All experimental protocols were approved by Linköping Council for Ethical Licensing of Animal Experiments. Chicks had a significant increase in corticosterone during the hatchery process, which implies they were exposed to stress (HC=9,287.1±765.9; CC=1,367.9±548.5; GzLM; n=20; $\chi^2$=75.95; P<0.001). During the first week of life, HC were more fearful to enter a novel arena (HC=562.0±147.42; CC=1,031.8±160.2; GzLM; n=56; $\chi^2$=5.43; P=0.020) and less active when in the arena (HC=0.09±0.02; CC=0.16±0.02; GzLM; n=56; $\chi^2$=5.84; P=0.016), indicating more fearfulness. Further, HC had a higher corticosterone reactivity during restraint at 1st (HC=5,392.7±1,250.3; CC=2,146.9±383.6; GzLM; n=42; $\chi^2$=6.73; P=0.009) as well as 5th week of age (HC=3,616.1±719.1; CC=1,337.3±730.2; GzLM; n=46; $\chi^2$=4.94; P=0.026). HC weighed more than CC up to 14 weeks of age (HC=1,294.5±21.4; CC=1,212.8±20.9; GzLM; n=168; $\chi^2$=8.96; P=0.003). At 20 weeks of age, both males (HC=1.03±0.14; CC=0.53±0.07; Kruskal-Wallis H=9,1; n=86; P=0.003) and females (HC=0.91±0.15; CC=0.45±0.11; Kruskal-Wallis H=5.4; n=71; P=0.020) HC had more tail feather damage than CC, and hatchery treated males had more injuries on combs and wattles (HC=2.31±0.02; CC=1.63±0.13; Kruskal-Wallis H=10.8; n=86; P=0.001). We conclude that processing at the commercial hatchery was a stressful event with short- and long-term effects on behaviour and stress reactivity.

## Misdirected oral behaviors in orphaned neonatal kittens

Mikel Delgado[1], Isabelle Walcher[2], Karen Vernau[1], Melissa Bain[1] and Tony Buffington[1]
[1]University of California at Davis, Medicine & Epidemiology, School of Veterinary Medicine, 1 Shields Ave, Davis, CA 95616, USA, [2]Sacramento County Bradshaw Animal Shelter, 3839 Bradshaw Rd, Sacramento, CA 95827, USA; mmdelgado@ucdavis.edu

Recent shelter reports from the United States suggest that the number of hand-reared, bottle-fed orphaned kittens is increasing as rescue groups improve their ability to care for specialized populations. Despite this increase, we have little understanding of the potential long-term health and welfare effects of hand-raising kittens. One problematic behavior of neonatal kittens is sucking on the bodies of littermates (cross-sucking). Separation from the mother appears to be necessary, but not sufficient, for this behavior to develop. Misdirected oral behaviors have been well-documented in early-weaned piglets and calves, but there is little documentation of this behavior in kittens. We conducted detailed video observations of 68 fostered orphaned kittens from 23 litters. All kittens were surrendered to local rescue groups as orphans between 1 to 10 days of age and were hand-reared and bottle-fed in foster homes, and adopted into homes once weaned and altered. Nine litters were affected by cross-sucking behavior, with 17 kittens sucking on littermates, and 17 kittens sucked on by littermates (11 kittens were both suckers and victims). We observed kittens sucking on littermates dozens of times per day (observed range of sucking bouts: 25-80 times/day), spending several minutes a day engaging in this behavior (observed range of 55.8 -134.9 minutes/day). In addition to behavioral observations, we conducted an internet survey of foster caretakers including questions about characteristics of kittens that were sucking or being sucked on, as well as their environment and feeding schedules. We obtained survey data for 328 litters (1,106 kittens) and all data analyses were conducted in SAS University Edition. Male kittens were at a greater risk than females of being a victim of sucking behavior (63% of victims; chi-square test: $X^2(2)=22.16$, P<0.001). The largest kitten was a victim in 60.4% of litters (binomial test: Z=-3.17, P<0.001), and the smallest kitten was a sucker in 58.9% of litters (binomial test: Z=-2.71, P=0.003). There was no effect of kitten sex on the likelihood of being a sucker. The anogenital region (53%) and the stomach (29%) were the most commonly reported sucked on areas. Eleven percent of kittens who were suckers or victims required some form of medical care including for genital injuries (19/31) and sores on the skin (14/31). Several management strategies were tried, including interrupting the behavior (59%), separation from littermates (43%), and placing clothing on the kittens (13%), but 74% of kittens returned to sucking when reunited with littermates. Our results suggest that cross-sucking behavior among orphaned neonatal kittens is a common problem, and further research into risk factors and prevention is warranted.

## Identification of fear behaviours shown by kittens in response to novel stimuli

*Courtney Graham, David Pearl and Lee Niel*
*University of Guelph, Department of Population Medicine, Ontario Veterinary College, 50 Stone Road E., N1G 2W1, Guelph ON, Canada; courtney.graham@uoguelph.ca*

Fear is a negative emotional state and can impact animal welfare. We have some knowledge of what fear looks like in adult cats, but fear behaviours in kittens have not been properly identified through controlled studies. Using kittens within the sensitive period for socialization, we assessed which kitten behaviours occur in response to fear of novel stimuli, as identified through avoidance. Foster kittens (6-7 weeks old; n=20) were tested in home. Initially, they were habituated as a litter (n=6) for 5 min to a soft-sided pet exercise pen (1.2 m$^2$ × 0.7 m h), and then removed. Next, each kitten was placed back into the pen individually for seven trials: three stimulus trials with a noisy and unpredictable novel stimulus alternated with four trials with no stimuli (control). Stimuli included a plush cat paired with an audio recording of meowing and hissing, a plush dog paired with an audio recording of barking and growling, and a hand vacuum pulsed on and off. Order of stimulus presentation was randomized for each kitten. Each presentation lasted 10 s and was followed by 30 s in the pen without the stimulus; total observation time was 40 s. All sessions were video-recorded, and behaviour during each trial (n=136; four trials removed due to recording error) was scored by a blinded observer using Noldus Observer 12 event logging software. Trials were categorized as 'fearful' if the kitten showed avoidance during a stimulus trial, and responses during these trials were compared to control trials. Multivariable linear, logistic, and negative binomial mixed models, with kitten and litter as random effects, were used to assess associations for behaviour durations, occurrence, and counts, respectively, and included type of trial, stimulus type, and trial order. Durations of tucked tail (P<0.001) and hissing (P<0.001) were greater in fearful trials compared to control trials, whereas duration of upright tail (P=0.049) was greater in control trials. The odds of showing arched back posture (P<0.001), piloerection (P<0.001), freezing (P<0.001), and flinching (P<0.001) were greater in fearful trials, whereas the odds of eating (P<0.001) were greater in control trials. Further, the rate of ears back (P<0.001) was greater in fearful trials. No significant differences were found for head lowered, hesitant walking, hiding behind or inside of a retreat area, lip licking, paw lifts, rears, or meowing. The durations of tucked tail and hissing were significantly greater with the dog and vacuum stimuli compared to the cat stimulus, and the duration of hissing was greater in the third stimulus presentation regardless of stimulus type. The odds of freezing and rate of ears back were significantly greater in the last control trial compared to all other trials, suggesting that the kittens may have been sensitized to the trials by the end of testing. Overall, these results indicate that arched back posture, piloerection, tucked tail, freezing, flinching, ears back, and hissing are indicative of fear in kittens in the presence of fear-provoking stimuli. These fear responses are similar to those reported for adult cats, suggesting that the behavioural repertoire for fear in kittens is fully developed by 7 weeks of age.

## The role of age in the selection of police patrol dogs based on behaviour tests

*Kim I. Bjørnson[1], Judit Vas[1], Christine Olsen[2], Conor Goold[3] and Ruth C. Newberry[1]*
[1]*Norwegian University of Life Sciences, Faculty of Biosciences, Department of Animal and Aquacultural Sciences, 1432 Ås, Norway,* [2]*Dyrebar Omsorg, Drøbakveien 50, 1430 Ås, Norway,* [3]*University of Leeds, Faculty of Biological Sciences, LS2 9JT Leeds, United Kingdom; kim.ivbjo@gmail.com*

Police patrol dogs face many challenging situations, and only a proportion of dogs are suitable for this work. It is desirable to identify suitable dogs as early as possible, allowing unsuited dogs to be released for other purposes and reducing the emotional cost of separating the dog and handler at a later age. However, the selection process may be less reliable when dogs are young and their personality is less established. We investigated the stability of dog behaviour in the different successive subtests of a standardized police dog temperament test conducted at two different ages, and overall outcome of each test (pass or fail, based on subjective expert evaluation by testers). We also examined the extent to which behaviour in the first test predicted the outcome of the second test. Tests were administered to 62 male German Shepherd Dogs at approximately 6 and 12 months of age (mean±SD: 6.14±0.50 vs 12.31±0.64 months). Tests comprised 63 behavioural variables assessed across 14 subtests designed to measure behavioural responses in different situations. Each variable was scored from 1 to 5, with higher scores representing more desirable responses, and the mean score for each subtest was calculated. A positive association was found between test outcomes at 6 and 12 months ($\chi^2$=14.78, P≤0.001), with 74.2% of dogs having the same outcome at both ages. An absence of high correlations between subtest mean scores ($r_s$≤0.65) was suggestive of low redundancy across subtests. Binary logistic regression models identified that the mean scores from 3 subtests at 6 months, and 4 at 12 months, were significant predictors of test outcomes at the age tested. Furthermore, 3 subtests at 6 months, assessing reaction to strangers, flight and recovery when startled, and motivation and success when searching for a hidden object in an indoor environment, respectively, were significant predictors of test outcomes at 12 months. A model using the 6-month results from these 3 subtests had 82.4% sensitivity and 75.0% specificity in correctly predicting the 12-month test outcome. We compared the overall mean score from the 3 subtests between dogs that A) passed at both ages (n=21), B) failed at 6, but passed at 12 months (n=13) or C) failed at both ages (n=25). Back-transformed least squares mean scores (±SD) adjusted for multiple comparisons were higher for dogs in category A (4.2±0.4) than B (3.9±0.4, z=2.62, P=0.024) or C (3.4±0.4, z=7.74, P≤0.001), and category B scores also exceeded category C scores (z=4.00, P≤0.001). Our results suggest that some subtests are more predictive of test outcomes than others, raising the possibility of using a shorter testing procedure. They also suggest that testing can be implemented at the earlier age to exclude low scoring dogs and accept high scoring dogs while leaving open the possibility of a second test when older for a relatively small subset of young dogs with ambiguous (intermediate) test results.

## Calf-cow contact during rearing improves health status in dairy calves – but is not a universal remedy

*Edna Hillmann[1,2], Rupert Bruckmaier[3] and Cornelia Buchli[1]*
*[1]ETH Zürich, Ethology and Animal Welfare Unit, Universitätstr. 2, 8092 Zurich, Switzerland,*
*[2]Humboldt-Universität zu Berlin, Division Animal Husbandry and Ethology, Unter den Linden 6, 10099 Berlin, Germany, [3]University of Berne, Veterinary Physiology, Bremgartenstrasse 109a, 3001 Berne, Switzerland; edna.hillmann@hu-berlin.de*

We investigated if rearing dairy calves with contact to a dam or foster cow affects their health status compared to rearing calves without cow-contact, as it is common in dairy farming. Previous studies have found indications for an improved health status in dam rearing systems as well as effects on social behaviour and stress reactivity. However, to our knowledge this is the first study that used an on-farm approach to reflect the large variability between housing systems and management routines found in practice. We monitored calf health on 39 dairy farms in Switzerland and Germany (19 with calf-cow contact, 20 without) by assessing the immune status (114 calves/34 farms) and clinical health during two farm visits (414 calves/39 farms). Data were analysed using linear mixed effects models in R. In addition, farmers recorded the health status of their calves during the first three months of age over one year using a pre-defined protocol (775 calves/34 farms). For analysis, we used a survival analysis and cox regression in R. The serum concentration of IgG was not affected by the rearing system, the concentration of total protein, however, was increased by cow-contact compared to no-contact (<1 d, 12 farms) or short-term (1-14 d, 7 farms) contact (P=0.027). Long-term (>15 d, 20 farms) cow-contact positively affected the body condition score (BCS) of calves (P=0.09). The effect of cow-contact on symptoms of diarrhea depended on the season. In spring, no calf reared without cow-contact was found to show diarrhea symptoms. In all other seasons, however, long-term cow-contact decreased the proportion of calves with diarrhea (P=0.051). Similarly, the effect of cow-contact on respiratory symptoms differed between seasons. A positive effect of long-term cow-contact on respiratory health was found only in autumn. Short-term cow-contact did not reduce probability of respiratory symptoms, and in winter, calves without cow-contact showed respiratory symptoms most (P=0.07). Mortality was not reduced by cow-contact (3% without, 4% with contact). In the long-term, calves with cow-contact were less affected by reduced general condition (P=0.007) and respiratory diseases (P=0.043). The number of calves treated by a veterinarian was not reduced by cow-contact. However, 5.4% of the calves with cow-contact were treated with antibiotics compared to 8.3% of calves without, however, this finding was not supported statistically (P=0.175). We conclude that rearing dairy calves with cow-contact indicates some advantages regarding calf health. However, findings were not always straightforward and were strongly depending on the season. Also, farms providing calf-cow contact might differ from farms without calf-cow contact in several aspects despite calf-cow contact, e.g. the farmer´s attitude towards calf welfare. Thus, providing dairy calves with contact to the dam or a foster cow might improve calf health, but it is not a universal remedy.

## Effects of a 3-month nursing period on fertility and milk yield of cows in their first lactation

*Kerstin Barth*

*J.H. von Thünen-Institute, Federal Research Institute for Rural Areas, Forestry and Fisheries, Institute of Organic Farming, Trenthorst 32, 23847 Westerau, Germany; kerstin.barth@thuenen.de*

Dairy cows nursing their calves are expected to show a prolonged anestrus although not as distinctive as in beef cows. Nevertheless, farmers fear negative impacts on fertility if calves are allowed to suckle their dams at the beginning of lactation. Besides the impact on machine milk yield, this is one reason not to introduce cow-calf contact into practical dairy farming. Thus, the aim of the study was to investigate if there is a significant effect on fertility criteria caused by suckling. Data on 108 dairy cows (breed: 67 German Holstein and 41 F1 hybrids of German Holstein × German Red Pied) that calved for the first time at the research station of the Thünen-Institute from January 2015 to October 2018 were included. All cows calved in a single calving pen, 67 (62%) were separated from their calf within 24 h (control cows), while 41 cows stayed with their calf in the pen for 7 days (nursing cows). Afterwards, calves were allowed to have contact to their dams via selection gates between the cow and calf areas. Suckling was possible for 24 h over 3 months, except during the two milking times when all cows were milked in a 2×4 tandem parlor. Calves were weaned on average at an age of 92 days (SD=3) using nose flaps (QuietWean®, Canada) and moved to another barn at an age of 108 days (SD=11). Estrus detection was done by activity measurement using a Rescounter (DairyPlan 21, GEA, Germany) applied to one of the cows' front legs and by visual observation. Artificial insemination was carried out by three herdspersons. Date of first insemination after calving, number of inseminations necessary to get the cow pregnant, calving interval as well as culling records were analyzed using the Wilcoxon rank-sum test. Mean milk yield was compared by Student's t-test. Nursing and control cows were inseminated at day 76 and 74 p.p. (median) respectively. The Wilcoxon rank-sum test showed no significant effect of treatment (W=906.5, Z=-1.43, P=0.154). Also, the number of necessary inseminations did not differ (median in both groups: 1, W=1,179, Z=0.662, P=0.511). Median calving interval was 372 and 375 days for nursing and control cows, respectively, showing no effect of calf contact (W=520, Z=-0,216, P=0.832). Compared to nursing cows (12 out of 41), more control cows (29 out of 67) were culled up to 2019. However, the median age at culling did not significantly differ between groups (nursing cows: 3.0, control cows: 3.6 years, W=205.5, Z=0.903, P=0.375). Infertility was the main culling reason in both groups (nursing cows: 6, control cows: 9 cases). The amount of milk gained by machine milking during the first 100 days of lactation was higher for control than nursing cows (mean difference: 1.196 kg, t=14.29, df=93, P<0.001, 95% CI: 1.029-1.362 kg). These preliminary results indicate that nursing during the first three months of lactation does not necessarily impair fertility of dams, but milk yield during that period is reduced.

## Behavioural and physiological effects of prenatal stress in different genetic lines of laying hens

*Mariana R.L.V. Peixoto and Tina M. Widowski*
*University of Guelph, Animal Biosciences, 50 Stone Rd E, N1G 2W1, Guelph, Canada;*
*twidowsk@uoguelph.ca*

Prenatal stress can have long-lasting effects on the behaviour and physiology of a hen. Stress vulnerability can be impacted by genetics, resulting in different responses across strains. We hypothesized that offspring response to prenatal stress will depend on the genetics of the breeder flock. For this, two stress models (Natural and Pharmacological) were tested in five genetic lines of breeder hens: two commercial brown (B1 & B2), two commercial white (W1 & W2) and a Pure Line White Leghorn (WL). To form the Parent Stock, fertilized eggs were incubated, hatched and housed identically in 4 flocks of 27 birds (24F:3M) per strain. Each strain was equally separated into two groups: 'Stress', where hens were subjected to a series of acute psychological stressors (e.g. physical restraint, transportation) for eight days before egg collection, and 'Control', which received routine husbandry. At three maternal ages, fertile eggs from both treatments were collected and additional eggs from Control were injected with corticosterone (10 ng/ml egg content)('CORT'). A 'Vehicle' treatment was included to account for effects of egg manipulation. Each maternal age comprised a replicate over time. Eggs were incubated, hatched, and offspring (n=1,919) brooded until 17 weeks (wks) under identical conditions. Animal use was approved by the University of Guelph Animal Care Committee. Behavioural analyses included tonic immobility at 9 wks (n=450) and a combined voluntary human approach (n=140) and novel object test (n=180) at 16 wks. HPA-axis response and recovery were measured at 13 wks (n=640), through analyses of plasma corticosterone concentration in baseline samples and after 10 and 20 minutes of physical restraint. The effects of the stress model, genetic line, sex and the interaction between stress model and genetic line were subjected to ANOVA using the Glimmix procedure in SAS. Random effects included pen, room and maternal age. Further pre-planned comparisons included stress model (Control vs Stress; Control vs CORT) and strain (brown vs white). Prenatal stress did not affect any of the traits measured (P>0.1052). Baseline levels of corticosterone did not differ among strains (P>0.0656), but white lines showed higher HPA-axis responsiveness (P<0.0128) and slower recovery (P<0.0087) than brown. Brown birds stayed longer in tonic immobility (F=40.95; P≤0.0001) and B2 spent less time close to human than W1 (P=0.0001) in the human approach test. Interestingly, time spent close to the novel object did not vary between W1, B1 and B2 (P>0.2204). Results suggest that the white lines used in this study are more reactive and less fearful than the brown lines. Moreover, brown hens seem to respond differently to humans and objects, being more afraid of the former. Although our original hypothesis was rejected, this study highlighted the variety of behavioural and physiological responses across different genetic lines of laying hens.

## Heat exposure of pregnant sows modulates behaviour and corticotrope axis responsiveness of their offspring after weaning

*Elodie Merlot, Caroline Constancis, Rémi Resmond, Ayra Maye Serviento, David Renaudeau, Armelle Prunier and Céline Tallet*
*INRA, Agrocampus Ouest, Pegase, Saint-Gilles, 35590, France; elodie.merlot@inra.fr*

The aim of this study was to determine, in swine, what are the consequences of heat exposure during most of gestation on the development of stress response systems in the offspring. For this purpose, pregnant sows were housed at thermoneutrality (TN, 18-24 °C, n=12) or in heated rooms (HS, 28-34 °C, n=12) from the 7th (G7) to the 106th day of gestation (G106), the gestation length being of 115 days. Half of the groups were primiparous and the other half consisted of multiparous sows. From G107 to the end of the lactation period, all sows and their progeny were housed at thermoneutrality in the same maternity rooms. After 28 days of lactation, 5 males and 5 females per litter were weaned and housed in pens of 10 littermates in thermoneutral rooms. At 65 days of age, they were mixed within treatments to form groups of 12 piglets from at least 4 different litters. Two pigs per gender and per litter were randomly selected for blood sampling during lactation (d26) and post-weaning periods (d37). One pig per gender and per litter was randomly selected for saliva sampling before (d64 at 09.00, 13.00 and 17.00) and after mixing (d65 at 13.00, 17.00 and d66 at 09.00). Six piglets per litter were chosen by chance and their individual behaviour (postures, social and explorative activity) was observed by scan sampling on the 7-8th and 15-16th days of lactation. Behaviour was then observed at pen level on the day of weaning and on the 3 following days. Behaviour of sows was not investigated. After a transient increase in rectal temperature during the first week of exposure to elevated temperature, HS sows acclimatized well. Heat did not affect their body weight gain during gestation nor the subsequent growth of their litter during lactation. In the progeny, total cortisol blood concentrations were similar in the two treatment groups at 26 and 37 days of age. Salivary cortisol concentrations were similar between treatments before the mixing (TN: 0.79 vs HS: 0.92±0.07 ng/ml, P>0.05), but higher in HS pigs after the mixing stress (TN: 1.6 vs HS: 2.2±0.1 ng/ml, P<0.001). Behaviour traits were not affected by treatment during lactation but there was an effect on the first two days after weaning. HS piglets spent less time involved in an activity (29 and 20% for HS vs 40 and 27% for TN on the 1st and 2nd day after weaning, P<0.05), and this was especially noticeable for the time spent walking and exploring the pen on the 1st day (18% for HS vs 24% for TN, P<0.05). Accordingly, HS pigs spent more time lying on those 2 days (P<0.05). During the 3 days of observation, HS pigs spent more time sleeping in groups of 3 or more pigs (56 vs 50%, P<0.01). To conclude, long lasting exposure of pregnant sows to high ambient temperature increased the response of the corticotrope axis after a social stress, and altered the behaviour of piglets during both resting and active phases after weaning. These preliminary results suggest that prenatal heat exposure could alter both endocrine and behavioural stress systems in pigs, reminding what can be observed after other kinds of prenatal stressors (social stress, restraint stress…).

## Trends in early life conditions of pigs and laying hens in order to prevent damaging behaviour: a GroupHouseNet update

*Elske N. De Haas[1,2], Irene Camerlink[1], Sandra Edwards[1], Armelle Prunier[1], Xavier Averos[1], Johannes Baumgartner[1], Boris Bilcik[1], Nadya Bozakova[1], Anja Brich Riber[1], Ivan Dimitrov[1], Inma Estevez[1], Valentina Ferrante[1], Lubor Kostal[1], Ragnar Leming[1], Martina Lichovnikova[1], Dimitar Nakov[1], Sezen Ozkan[1], Evangelia Sossidou[1], T. Bas Rodenburg[1,2] and Andrew M. Janczak[1]*

*[1]COST Action CA15134 GroupHouseNet, Horizon 2020 Framework Programme, Brussels, Belgium, [2]Utrecht University, Domplein 29, Utrecht, the Netherlands; andrew.janczak@nmbu.no*

Damaging behaviours in pigs and laying hens are among the most concerning current welfare issues. They are multifactorial in origin and difficult to stop once occurring. Prevention is crucial, and this is the focus of the COST Action project 'GroupHouseNet'. In laying hens, feather pecking (FP) can best be prevented at the hatchery. The parental stock's predisposition for FP can be assessed by epigenetic analysis, allowing careful stock selection. The fertilized eggs are then conventionally incubated in the dark, but studies show that light provision may reduce FP. Recent development of in-ovo sexing reduces chick handling, thus reducing stress. It further allows hatching with food and water provision, enabling on-farm hatching. Accessing feed right after hatching can be crucial for chicks to learn what to peck and eat. Artificial dark brooders, mimicking the mother hen's wings, reduce fear, stress and FP. Matching the rearing and laying system through an all-in-all-out system reduces stress by eliminating catching, handling, transport and changing environment. In practice, multiple all-in-all-out barns can provide continuous egg output. A sustainable and circular system of feeding chickens insects grown using chicken manure seems feasible and positive in preventing FP. The use of pasture and supplementing hens whole grains are also being researched as potential strategies. Together, these trends can help in preventing damaging FP in laying hens. For pigs, early life factors can also influence later damaging behaviour, such as aggression, tail and ear biting. Aggressive biting is mainly reduced by socialisation (i.e. co-mingling) before weaning. Here, piglets learn to establish dominance relationships early in life which reduces fight duration when older. Tail biting, which is largely unrelated to aggression, is increased by early life undernutrition, social stress due to competition and cross-fostering. These factors are all influenced by litter size at birth. Familiar odours may contribute to reducing biting when pigs are moved from one environment to another by alleviating the level of stress associated with novelty. Tail and ear injuries pre-weaning may also occur due to mycotoxins from feed or straw, and can result in the affected piglet being bitten and the development of biters. Paying attention to tail and ear injuries pre-weaning is therefore recommended. Even though the barren environment of the pigs represents the major risk for expression of damaging behaviour, the pre-weaning environment should be optimized to reduce the likelihood of this problem. Foraging possibilities are essential for both laying hens and pigs. Providing pigs and poultry with the requirements for expressing natural behaviour and for reducing stress is therefore necessary to prevent damaging behaviour.

## Identification of dominance in suckling piglets

*Friederike Katharina Warns and Martina Gerken*
*University of Goettingen, Department of Animal Science, Albrecht-Thaer-Weg 3, 37075 Goettingen, Germany; fwarns@uni-goettingen.de*

Tail biting in pigs is a major welfare concern under practical conditions. We investigated whether the behaviour of piglets can be used for early identification of animals with undesirable behaviour which might result in intensive manipulative behaviour during their subsequent rearing. In this study, we analyzed the agonistic interactions of piglets with their littermates during suckling. In total, 117 non-docked crossbred piglets (Piétrain × BHZP Victoria) originating from 9 litters were observed during the suckling period of 28 days in two replication groups. Each farrowing crate was filmed continuously during the whole suckling period. We analyzed two consecutive suckling bouts on 3 to 4 days in each experimental week, resulting in 30 nursings per sow in total. Weights of the piglets were measured at birth and after weaning. When analyzing individual suckling behaviour, we differentiated between 9 agonistic behaviours. We classified each agonistic behaviour which resulted in the access to a teat as dominant (the piglet pushed another piglet away from a teat with/without biting, or defended the teat it was suckling at while it was attacked by another piglet with/without biting) and each behaviour without teat success as submissive (the piglet could not push another piglet away from a teat with/without biting, or could not defend the teat it was suckling at while it was attacked by another piglet with/without biting, or was trapped between its littermates and therefore could not reach a teat at all). An average dominance index (ranging between 1 = dominant and -1 = submissive) was calculated for each piglet based on the ratio of the total frequency of its dominant or submissive behavioral patterns across 30 sucklings. Analysis of variance included the fixed effects of sex and replication group, and the random effect of the sow. Additionally, correlations between dominance index, birth and weaning weight were calculated. No significant differences were found for dominance index, birth and weaning weight between sexes and replication groups. However, the dominance index was positively correlated with weaning weight (r=0.47; P<0.001) and birth weight (r=0.33; P=0.012). Thus, heavier piglets had a higher dominance index than lighter piglets irrespective of sex and dam. In further studies, we will analyze whether this difference persists during the subsequent rearing period and could allow for early identification of intensive manipulative behaviour.

## Postnatal effects of colostrum quality and management, and hygiene practises, on immunity and mortality in Irish dairy calves

*John Barry[1,2], Eddie Bokkers[1], Donagh Berry[2], Imke De Boer[1], Jennifer McClure[3] and Emer Kennedy[2]*
[1]*Wageningen University and Research, Animal Production Systems, P.O. Box 338, 6700 AH, Wageningen, the Netherlands,* [2]*Teagasc, Animal and Grassland Research and Innovation Centre, Moorepark, Fermoy, Co. Cork, Ireland,* [3]*Irish Cattle Breeding Federation, Shinagh, Bandon, Co. Cork, Ireland; eddie.bokkers@wur.nl*

Pre-weaned dairy calves are particularly vulnerable to welfare issues associated with high susceptibility for diseases, which often result in high mortality. Important factors are colostrum quality (Immunoglobulin (Ig) G>50 mg/ml), and/or colostrum management practises, combined with hygiene practises. The aim of this study was to investigate postnatal effects of colostrum quality and management, and hygiene practises, on the welfare indicators passive immunity and mortality among Irish dairy calves. Forty seven spring-calving, pasture-based dairy herds were enrolled in the study. To investigate if colostrum quality and management, and hygiene practises change as the calving season progresses, each farm was visited twice. Calf managers were interviewed to collect information on colostrum management and hygiene practices. Physical assessment of hygiene practices were conducted on feeding implements using 3M[tm] Clean-Trace surface protein plus test kits. All calves >1 day and ≤6 days of age present at the time of visit were blood sampled to assess serum IgG concentrations. Six colostrum samples from freshly calved cows were also collected. Risk factors associated with colostrum and calf serum IgG concentrations and calf mortality rates (excluding stillbirths) were assessed using mixed models. Probability of mortality occurring when serum IgG concentrations were below the recommended threshold (10 mg/ml after 24 h) was analysed with a generalized linear model. Mean colostrum IgG concentration was 85.1 (SD=51.6) mg/ml, and 21% of samples were below the 50 mg/ml quality threshold. Mean calf serum IgG concentrations in visit 1 and 2 were 30.9 (SD=13.4) mg/ml and 27.1 (SD=14.0) mg/ml, respectively. Herd size (P<0.01) and age at blood sampling (P<0.01) were negatively associated with serum IgG concentrations; no effects of sex, visit, calving difficulty or birth week of year were identified. Hygiene scores tended to worsen from visit 1 to 2 (P=0.07), however no association was found with herd size and 28 day mortality. Most calf managers (49%) cleaned feeding equipment every second day while 36% cleaned once a week or less frequently. When providing colostrum, 60% of calves were fed within 2 hours of birth, however, on 87% of the farms colostrum quality was not tested prior to feeding. No associations were found between colostrum management and calf feeding practises in visits 1 and 2, for male and female calves, and the independent variables calf serum IgG concentrations, and 28 day mortality rate. This study identified large variation in colostrum quality, emphasising the need for assessment prior to feeding. Hygiene practises associated with calf rearing can be improved, particularly in the latter half of the calving season. Yet, colostrum quality and management in Irish dairy herds is generally good, with >90% of calves having adequate passive immunity. This should prevent the occurrence of health related welfare issues in pre-weaned dairy calves.

## Prevalence and early life risk factors for tail damage in Danish long-tailed piglets prior to weaning – preliminary results

*Franziska Hakansson and Björn Forkman*
*University of Copenhagen, Department of Veterinary and Animal Sciences, Groennegaardsvej 8, Frederiksberg-C, 1870, Denmark; fh@sund.ku.dk*

Tail-biting is a significant welfare and economic problem in modern pig production. The reasons for performing this behaviour are generally said to be multifactorial, with a wide range of risk factors described. Previous results on post-weaning pigs have shown that the individuals being tail bitten are not randomly selected, but that e.g. sex and weight are risk factors. Although tail damage is reported to occur before weaning, most studies investigate tail-biting and resulting damage in the weaning and finisher period. Knowledge about the general prevalence of tail damage in piglets prior to weaning in relation to individual piglet characteristics could improve aforementioned studies. Therefore, the objective of this study was to assess the prevalence of tail damage in piglets prior to weaning and to compare piglets with and without damage regarding selected risk factors. This observational study was carried out on a commercial conventional Danish farm between October '18 and January '19. Sows were loose housed and piglets were not tail docked. A total of 741 crossbreed (LY×D) piglets from 51 litters were individually marked within 10 h after birth and followed until weaning. Selection of sows was based on the date of farrowing and a minimum of 13 life-born piglets. Nursing sows and sows with clinically observable illness were excluded from the sample. Within 4 d *post-partum* (day of birth = day 0), litter size was reduced to on average 15 piglets. No cross-fostering was performed in experimental litters. Individual piglet characteristics were collected at birth (F0), one week of age (F7) and prior to weaning (pW). Severity of tail damage of individual piglets was assessed pW. Variables included for preliminary results were: sex, weight (F0, F7 & pW) and treatment of piglets (F7). At 4w of age, 580 piglets were weaned (332 f/409 m). Piglets mean weight was 1.5 kg (range: 0.8-2.6) at F0 and 8.4 kg (range: 2.9-15.1) at pW. The prevalence of tail damage was 40.2% (233/580), of which 29.2% of the piglets had mild damage (superficial damages), 8.8% had moderate damage (bite marks/scratches of small size) and 2.1% had severe damage (clearly visible wound) on the tail prior to weaning. A multivariable proportional odds logistic regression model with fixed effects and interaction of weight and sex showed a positive association between increasing pW weight (in kg) and severity of tail damage in male piglets (OR: 1.2, 95% CI: 1.1-1.4, P=0.003). Our study confirms the occurrence of tail damage already before weaning and supports the argument that the probability of being tail bitten is potentially related to individual piglets' characteristics. No effect of sex and treatment on the occurrence of tail damage was found. However, for male piglets, a higher pW weight increased the likelihood of having severe tail damage.

## A cross-sectional study of associations between herd-level calf mortality, welfare and management on Norwegian dairy farms

*Julie Føske Johnsen[1], Kristian Ellingsen-Dalskau[1], Cecilie Mejdell[1] and Ane Nødtvedt[2]*
*[1]Norwegian Veterinary Institute, Terrestrial Animal Health and Welfare, Ullevålsveien 68, 0106, Norway, [2]Norwegian University of Life Sciences, Production Animal Clinical Sciences, P.O. Box 8146 Dep, 0033 Oslo, Norway; julie.johnsen@vetinst.no*

Dairy calf mortality is variable among Norwegian herds. Our aim was to investigate the distribution of calf mortality and associations to results from a national audit (NA) on calf welfare and management factors. Management of preweaned dairy calves is important, as milk allowance and social housing is directly related to behaviour and quality of life. Our source population was 912 NA herds. Cross-sectional data were available from 470 of the NA herds: herd calf management factors were obtained through a questionnaire and the results of the NA performed by Norwegian Food Safety Authority veterinary consultants. During a farm visit, welfare scores (satisfactory or not satisfactory) were assigned according to an evaluation of 9 calf welfare criteria: calf care routines including sick calves, feeding, bedding, housing comfort, mortality, water access, calves >8 weeks housed in single pens and colostrum feeding routines. A welfare index was created for analysis purposes: welfare scores (unsatisfactory=0, satisfactory=1) were summed. Data on calf mortality were obtained from the central databases of the National Dairy Herd Recording System. Outcome of interest was counts of calves <6 months at each herd that succumbed for any reason during 2016. A (multivariable) zero inflated negative binomial model was applied to assess factors that would influence mortality counts. This model was fitted manually, testing factors that were identified as significant in a (univariable) stepwise regression screening process. The mortality rate was 7.9±0.08%. The mean herd calf mortality was 2.3, median was 2 and range from 0 to 14. The distribution of calf mortality did not follow a normal distribution due to many zeros. Most of the herds had a high welfare index (mean 8.4, median 9.0 with a range from 2-9). At the age of 3 wk, calves received on average 7.1±1.99 l milk daily, distributed over 3.2±1.99 feedings per day. Calves were housed in single pens for 2.6±2.51 wk. Preliminary analysis indicate following associations: calf mortality was lower in herds where calves had free access to water throughout the milk feeding period ($\beta$=-0.3, P=0.001); a higher welfare index was associated with a lower mortality ($\beta$=-0.1, P=0.003), and calf mortality was higher in freestall barns vs tiestall barns ($\beta$=0.2, P=0.044). Increasing herd size (mean herd size in this study was 27±19.4 cows) was found to be a main contributor to explain variation in calf mortality: when herd size increased, calf mortality decreased. ($\beta$=-0.1, P=0.016). Calf milk allowance, milk type, number of daily feedings or weeks housed in single pens were not found to be associated with calf mortality. The results indicate an association between the lack of free water access and calf mortality. Measures of poor calf welfare were associated with measures of poor calf health.

## The effect of maternal contact on behaviour patterns indicative of emotionality and social competence in young dairy calves

*Janja Sirovnik, Noemi Santo and Uta König Von Borstel*

*Justus-Liebig University of Giessen, Institute for Animal Breeding and Genetics; Section Animal Husbandry, Behaviour and Welfare, Leihgesterner Weg 52, 35392 Giessen, Germany; janja.sirovnik@agrar.uni-giessen.de*

The aim of the present study was to assess the influence of maternal contact in the first 14 days of life on calves' fearfulness and social competence. Specifically, we hypothesised that Mothered calves (n=12) would show fewer vocalisations, longer exploration and locomotion, and shorter immobility and to be more vigilant and submissive during confrontation with an adult cow than Control calves (n=12, separated from mothers within 24 h after birth). The first 14 days of life were chosen as infants' brain is most plastic at an early age, and contact to mothers might influence infants' personality development. Mothered calves were prevented from suckling, but had almost uninterrupted contact with their mothers for the first five days of life (separated only during milking). During 6-14 days of life, Mothered calves were separated from their mothers for 12 h daily. Calves from both treatments and the mothers from Mothered calves were housed in a single pen until the calves were 14 days old, when Mothered calves were completely separated from the mothers. After 14 days of age, calves of both treatments were housed together in groups of up to four animals without visual and tactile maternal contact. To avoid potential influences of mothers' reactions to calves' stress due to testing, the testing began after final separation of Mothered calves at 14 days of age. Each calf was tested in three tests at seven-day intervals in the following order: open-field (OF), novel object (NO), and confrontation (CO) test with an unfamiliar adult cow. Calves' behaviour was recorded for 15 min/test and included: frequencies of locomotion, vocalisation, elimination, and durations of grooming, exploration of the test arena, complete immobility (in all tests), and play (NO, CO), plus any social behaviour of both animals (CO). A Wilcoxon Rank Sum Test (Python, SciPy 3.6.5) was used to assess differences in the recorded parameters between Mothered and Control calves in each test. Both of our hypotheses were confirmed as Mothered calves showed less fear-related behaviour (i.e. fewer vocalisations: P=0.044 in OF and shorter immobility: P=0.037 in CO), and showed signs of a greater social wariness (i.e. greater frequency (P=0.024) and duration (P=0.008) of vigilant behaviour during confrontation with an unfamiliar cow) than Control calves. Other behaviours measured did not differ between treatments (P≥0.05). Our results show a positive influence of maternal contact by reduction of calves' fearfulness in non-social contexts and greater wariness in confrontation with unfamiliar cows, which might indicate improved social competence. Furthermore, in combination with earlier studies without confounding of test-type and age, our results show that confronting calves with unfamiliar cows (CO) leads to a sufficient emotional reaction to observe treatment effects, which is less likely in less stressful situations (OF and NO).

**Associations between the early life husbandry and post-weaning behaviour of lambs**

*Anna Trevarthen[1], Poppy Statham[1], Jennifer Matthews[1], Freddie Daw[2] and Sarah Lambton[1]*
*[1]University of Bristol, Bristol Vet School, Dolberry Building, Langford House, Bristol, BS40 5DU, United Kingdom, [2]University of London, Royal Veterinary College, Hawkshead Lane, North Mimms, Herts, AL9 7TA, United Kingdom; anna.trevarthen@bristol.ac.uk*

The environment and management procedures lambs experience on sheep farms vary considerably. Lambs are born into a range of systems from extensive uplands, to lowland pasture or indoor buildings. Even within farms, the early life experience of lambs can differ dramatically, and little is known about how procedures (e.g. tail-docking, castration and vaccination) are associated with individual lamb behaviour in the longer-term. We explored links between early life husbandry and the behaviour of lambs post-weaning. We visited 18 farms during lambing and identified a subset of focal lambs (randomly selected, between 15-35/farm) which were 0-72 hours old at the time of visit. For each lamb, we recorded background information about their birth, early life experiences (including the number of procedures e.g. tail docking they had undergone and would undergo) and health. We subsequently conducted a post-weaning visit, during which the behaviour of the focal lambs was monitored whilst a health inspection took place in a weigh crate. An individual novel object test was also conducted as an indicator of fearfulness. Responses to the object were categorised as fearful, interested or no observable response. Across farms, 345 focal lambs were observed on both occasions. Preliminary analyses suggest that several early life factors were associated with lamb behaviour in the novel object test post-weaning. Lambs that had undergone fewer early life procedures were more likely to show interest (M=3.77, SD=1.19, n=171) than be fearful (M=4.02, SD=1.26, n=46) or not respond (M=4.16, SD=1.11, n=115) to the object ($X^2_{(2)}$=7.87, P=0.020). Similarly, lambs that were weaned later were more likely to show interest in the object ($X^2_{(2)}$=17.86, P<0.001). Interestingly, lambs that were kept indoors for longer at the beginning of their lives were more likely to show a fearful response in the test ($X^2_{(2)}$=7.00, P=0.030). No significant effects of sex, litter size or housing type during lambing were observed (P>0.05). In addition to the early life associations, we found that lambs which showed a fearful or interested response to the object were also more reactive when leaving the weigh crate following the health inspection, compared to those showing no response to the object ($X^2_{(3)}$=11.73, P=0.008), suggesting it was a good measure of general behavioural reactivity. Lambs with poorer body condition post-weaning were more likely to display no response in the novel object test ($F_{2,272}$=7.65, P=0.001), and those with a higher post-weaning body mass were more likely to show a fearful reaction (M=37.2, SD=6.4, n=43), than show interest (M=35.7, SD=6.4, n=155) or no response (M=34.3, SD=6.5, n=113) ($F_{2,308}$=3.62, P=0.028). These findings suggest that some early life experiences are associated with measurable differences in fearfulness and reactivity in the longer-term. Reducing the number of procedures in early life and weaning lambs later may serve to reduce fearfulness in lambs post-weaning.

## Maternal behaviour of Markhors (*Capra falconeri*) – a life worth living *ex situ*

*Jenny Yngvesson[1], Ewa Wikberg[2], Daniel Asif[1] and Jenny Loberg[1,2]*
*[1]Swedish University of Agricultural Sciences, Animal Environment & Health, P.O. Box 234, 52323 Skara, Sweden, [2]Foundtion Nordens Ark, Åby Säteri, 456 93 Hunnebostrand, Sweden; muif0001@stud.slu.se*

Wild animal kept in zoos for conservation purposes face multiple welfare challenges. Parental behaviour is the key to success in conservation and parental behaviour is considered an enrichment for the individuals. Markhors (*Capra falconeri*), a threatened goat species from the Middle East, are important in their ecological niche, not least due to their role as prey for large carnivores. They live in high altitude, rocky habitats. These goats suffer both from reproductive failure and health issues in captivity and are very difficult to study in the wild due to the political situation in this region. Information about their maternal behaviour is sparse. Bleats are a way of social communication and the high pitch bleats indicate stress, hence important to register. Our aim was to study details of maternal behaviour in Markhors in order to produce advice to increase reproductive success, increase welfare *ex situ* and, in the long run, facilitate conservation efforts. The behaviour of 4 goats and their 5 kids was observed during the spring/summer of 2018 in the zoo Nordens Ark in south west Sweden. We used direct observations and videos, during eleven days of observation. The enclosure was divided into different habitats due to its properties; elevated with rocks and low with grass and herbs. All kids were born in the elevated rocky part of the enclosure. Baby-sitting behaviour was observed in two of the goats, caring for all kids, only in the elevated area. Nursing time was significantly longer in the elevated rocky part of the enclosure ($48.5\pm2.3$ sec) compared to the lower part ($17.8\pm1.8$ sec) (One way ANOVA $F=116.1$ $P\leq0.001$). The number of high pitch bleats by the goats were significantly more in the lower part of the enclosure (median 6/day) compared to the elevated rocky part (median 1.5/day) (Mann-Whitney test $W=33.5$ $P\leq0.05$). Of six kids born (one died before the study started) the other five survived and hence all goats weaned at least one kid each. Goats showed a more calm behaviour indicated by longer nursing bouts and fewer high pitch bleats in the rocky elevated area. Considering Markhor behavioural ecology and our observation of maternal care, we conclude that successful reproduction is facilitated by an enclosure with elevated rocky habitats.

## Pullet rearing affects long-term perch use by laying hens in enriched colony cages

*Allison N. Pullin[1], Mieko Temple[2], Darin C. Bennett[2], Christina B. Rufener[1], Richard A. Blatchford[1] and Maja M. Makagon[1]*

*[1]University of California, Davis, Animal Science, 1 Shields Ave, Davis, CA 95616, USA, [2]California Polytechnic State University, Animal Science, 1 Grand Ave, San Luis Obispo, CA 93407, USA; apullin@ucdavis.edu*

Previous research indicates that pullets reared in cage-free aviary (A) housing utilize vertical space more than pullets reared in conventional cages (C). However, this effect has only been assessed up to 23 weeks of age, so it is unclear if this difference persists throughout the lay cycle. The objective of this study was to evaluate if rearing environment has a long-term effect on perch use in enriched colony cages. Lohmann LSL-Lite hens were reared in either C or A until 19 weeks of age and then moved into enriched colony cages (25±2 birds/cage) containing two elevated perches of different heights (n=6 cages/treatment). Video recordings from 21, 26, 35, and 49 weeks of age were used to determine the percentage of hens utilizing the low perch (9 cm from cage floor) and the high perch (25 cm from cage floor) for two separate days at each age (5-minute instantaneous scan samples over 5 hours/day). Overall perch use data (low perch and high perch percentages combined) were analyzed using a linear mixed effect model in R software. The likelihood of hens using the low perch or high perch were analyzed using a generalized linear mixed effect model in R software. For overall perch use, there was an interaction effect of age and treatment (P=0.002). At 21 weeks, a similar percentage of hens used perches for both C and A (C: 8.2 [5.4, 11.5], A: 8.7 [5.8, 12.2], estimated mean [95% CI] %). By 49 weeks, the percentage of C hens perching reached 10.3 [7.0, 14.3] % compared to 18.6 [14.1, 23.9] % of A hens perching. The A hens were more likely to use the high perch compared to C hens (treatment: P=0.02). For the low perch specifically, the opposite was shown with the C hens being more likely to use the low perch compared to A hens (treatment: P=0.01). All hens were increasingly more likely to use the high perch over the lay cycle (age: P<0.001). Additionally, all hens were increasingly more likely to use the low perch during the night observations over the lay cycle and less likely to use it during the day observations (age × time: P<0.001). Our results suggest that rearing environments do affect long-term perch use of hens in enriched colony cages. Rearing in aviaries results in more hens perching over time, but particularly there is a greater use of the high perch. The treatment difference in using the high perch particularly suggests that rearing has a long-term effect on the use of vertical space, which could have implications for resource access and health.

## Dietary fibre content in the feed and its effect on feather pecking, performance and cecal microbiome composition

*Antonia Patt[1], Ingrid Halle[2], Anissa Dudde[1] and E. Tobias Krause[1]*
*[1]Institute of Animal Welfare and Animal Husbandry, Friedrich-Loeffler-Institut, Dörnbergstraße 25/27, 29223 Celle, Germany, [2]Institute of Animal Nutrition, Friedrich-Loeffler-Institut, Bundesallee 37, 38116 Braunschweig, Germany; antonia.patt@fli.de*

Feather pecking is a multifactorial problem with genetic and environmental, e.g. nutritional, risk factors. When addressing nutritional risk factors, several studies suggest that the occurrence of feather pecking is associated with dietary fibre content in the feed. Generally, a high fibre content is assumed to reduce the prevalence of feather pecking. Further, dietary fibre affects intestinal microbiome composition, which has recently been proposed to be linked to the development of feather pecking. In the present study, we investigated the effect of systematic varying content of dietary fibre on behaviour, plumage condition, body weight, laying performance and cecal microbiome composition of hens with intact beaks of a commercial strain (Lohmann Tradition). Three feeds with varying dietary fibre content of 3, 6 or 9% and identical amounts of metabolisable energy were used. Around the onset of laying in week 21 of life, 12 groups were randomly assigned to one of the three feeds (= 4 replicate groups per feed, 20 hens per group). Groups were fed exclusively with the assigned feed until the end of the laying period in week 72 of life. The number of laid eggs was recorded daily on group level. Behavioural observations to determine feather pecking events, plumage scorings and body weight measurements were conducted systematically multiple times throughout the experimental period. Further, we counted the number of free feathers in the littered area within a defined quadratic area (= 625 cm$^2$) weekly. Both the observed feather pecking activity and the number of free feathers in the littered area were corrected for group size. After week 65 of life, three hens per group were euthanized and cecal samples for microbiome sequencing were taken. Data was analysed using linear mixed-effects models. The frequency of observed severe feather pecking decreased with increasing fibre content (3% fibre: 0.78 pecks/3 0 min/hen, 6%: 0.31, 9%: 0.12, P<0.0001). The number of free feathers in the littered area increased with an increasing fibre content (3% fibre: 0.06 feathers/625 cm$^2$/hen, 6%: 0.09, 9%: 0.16, P=0.0074). At all four time points overall plumage score was higher with an increased fibre content, i.e. quality of plumage was better. Over the course of the four evaluations, plumage quality decreased irrespective of fibre content (fibre content × week of age: P<0.0001). Uniformity of weight was neither affected by the different fibre contents (3% fibre: 66% uniformity, 6%: 69%, 9%: 71%, P=0.56) nor by time (week of life, P=0.65). Laying performance varied over time, but was not affected by fibre content (3% fibre: 79% laying performance, 6%: 80%, 9%: 82%, P=0.1). Diversity of cecal microbiome composition tended to increase with increasing fibre content (3% fibre: 36 Simpson effective diversity index, 6%: 41, 9%: 48, P=0.09). Results support the assumption that an increasing fibre content reduces feather pecking prevalence. Further, fibre content tends to affect cecal microbiome composition while seemingly not reducing laying performance.

**Early life microbiota transplantation affects behaviour and peripheral serotonin in feather pecking selection lines**

*Jerine A.J. Van Der Eijk[1,2], Marc Naguib[2], Bas Kemp[1], Aart Lammers[1] and T. Bas Rodenburg[1,2,3]*
*[1]Wageningen University & Research, Adaptation Physiology Group, De Elst 1, 6708 WD Wageningen, the Netherlands, [2]Wageningen University & Research, Behavioural Ecology Group, De Elst 1, 6708 WD Wageningen, the Netherlands, [3]Utrecht University, Department of Animals in Science and Society, Yalelaan 2, 3584 CM Utrecht, the Netherlands; jerine.vandereijk@wur.nl*

Early life environmental factors have a profound impact on an animal's behavioural and physiological development. In animal husbandry, early life factors that interfere with the behavioural and physiological development could lead to the development of damaging behaviours. The gut microbiota could be such a factor as it influences behaviour, such as stress and anxiety, and physiology, such as the serotonergic system. Stress sensitivity, fearfulness and serotonergic system functioning are related to feather pecking (FP), a damaging behaviour in chickens which involves pecking and pulling out feathers of conspecifics. Furthermore, high (HFP) and low FP (LFP) lines differ in gut microbiota composition. Yet, it is unknown whether gut microbiota affects FP or behavioural and physiological characteristics related to FP. Therefore, HFP and LFP chicks orally received 100µL of a control, HFP or LFP microbiota treatment within 6 hrs post hatch and daily until 2 weeks of age (n=96 per group) using a pipette. FP behaviour was observed via direct observations at pen-level between 0-5, 9-10 and 14-15 weeks of age. Birds were further tested in a novel object test at 3 days and 5 weeks of age, a novel environment test at 1 week of age, an open field test at 13 weeks of age and a manual restraint test at 15 weeks of age after which whole blood was collected for serotonin analysis. We analysed treatment effects within lines using mixed models with treatment, batch, sex, observer and test time as fixed factors and pen within treatment as random factor or Kruskal-Wallis tests. Early life microbiota transplantation influenced behavioural responses and peripheral serotonin, but did not affect FP. HFP receiving HFP microbiota tended to approach a novel object sooner and more birds tended to approach than HFP receiving LFP microbiota at 3 days of age (P<0.1). HFP receiving HFP microbiota tended to vocalise sooner compared to HFP receiving control (P<0.1) in a novel environment. LFP receiving LFP microbiota stepped and vocalised sooner compared to LFP receiving control (P<0.05) in an open field. Similarly, LFP receiving LFP microbiota tended to vocalise sooner during manual restraint than LFP receiving control or HFP microbiota (P<0.1). LFP receiving HFP microbiota tended to have lower serotonin levels compared to LFP receiving control (P<0.1). Thus, early life microbiota transplantation had short-term effects (during treatment) in HFP birds and long-term effects (after treatment) in LFP birds. Previously, HFP birds had more active responses and lower serotonin levels compared to LFP birds. Thus, in this study HFP birds seemed to adopt behavioural characteristics of donor birds, while LFP birds seemed to adopt physiological characteristics (i.e. serotonin level) of donor birds. Interestingly, homologous microbiota transplantation resulted in more active responses, suggesting reduced fearfulness.

# Modulation of kynurenine pathway and behavior by probiotic bacteria in laying hens

*Claire Mindus[1], Nienke Van Staaveren[1], Haylee Champagne[1], Paul Forsythe[2], Johanna M. Gostner[3], Joergen B. Kjaer[4], Wolfgang Kunze[2], Dietmar Fuchs[3] and Alexandra Harlander[1]*
*[1]University of Guelph, Department of Animal Biosciences, Guelph, Canada, [2]McMaster Brain-Body Institute, St. Joseph's Healthcare, Hamilton, Canada, [3]Innsbruck Medical University, Division of Biological Chemistry- Biocenter, Innsbruck, Austria, [4]Friedrich-Loeffler-Institut, Federal Research Institute for Animal Health, Institute of Animal Welfare and Animal Husbandry, Celle, Germany; cmindus@uoguelph.ca*

Tryptophan (TRP) is an essential amino-acid predominantly converted in kynurenine (KYN). A minor, but essential pathway of TRP breakdown is the production of the neurotransmitter serotonin. Altered serotonin neurotransmission is associated with feather pecking (FP) behavior towards conspecifics in laying hens. Additionally, birds with a high tendency to peck showed a lower population of *Lactobaccillacae* in their gut compared to non-peckers. TRP metabolism modulation by lactic bacteria is reported in mammals, where specific *Lactobacillus* bacteria showed positive effects on anti-social behaviors. We investigated the impact of a *Lactobacillus* probiotic bacteria supplementation on plasma concentrations of TRP and KYN, and on FP behavior in adolescent laying hens. Eighty-six birds (19 weeks of age) were assigned to 6 probiotic groups (P) or 6 control groups (C) (7 birds / group). During 5 weeks, P birds orally received $5 \times 10^9$ *Lactobacillus* strain dissolved in 1 ml of drinking water and C birds received 1 ml of water through the same procedure every day. Pens were video-recorded 5 min per day (for 3 days before treatment and 9 days during treatment) and were analyzed by a blinded observer. Blood samples were collected before and after treatment. Generalized linear mixed models were used to assess the effect of the supplementation on concentrations of TRP, KYN and KYN/TRP ratio, and severe (SFP) feather pecking. This experiment was approved by the Animal Care Committee of the University of Guelph (AUP 3206). Birds already showed SFP at the start of the experiment, with a numerical higher proportion of P birds showing SFP compared to C birds (P=0.05). Contrary to expectation, SFP did not spread and increase over time in both groups during the 5 weeks of treatment (P=0.34). TRP breakdown was significantly increased over the 5 weeks regardless of the supplementation treatment as indicated by a decline in TRP levels by 9% and KYN levels by 25% (P<0.001). However, the TRP breakdown to KYN in P birds was significantly less pronounced (P<0.05). Theoretically, this could underline an increased TRP availability to the brain for serotonin production, which may explain, at least partly, the countering of the expected spread of SFP within the P groups. In general, daily supplementation with *Lactobacillus* was found to be associated with beneficial effects on the TRP breakdown in young laying hens. Still, these findings are of preliminary nature and need further investigation into the pathways involved.

## Investigating the gut-brain axis: effects of prebiotics on learning and memory in pigs

*Else Verbeek[1], Johan Dicksved[2], Jan Erik Lindberg[2] and Linda Keeling[1]*
*[1]Swedish University of Agricultural Sciences, Department of Animal Environment and Health, Box 7068, 750 07 Uppsala, Sweden, [2]Swedish University of Agricultural Sciences, Department of Animal Nutrition and Management, Box 7024, 750 07 Uppsala, Sweden; else.verbeek@slu.se*

Gut microbes communicate with the brain through a pathway called the gut-brain-axis and can modify homeostatic processes in the host. An unbalanced gut microbiota early in life has been shown to have long-term negative consequences for physiology, behaviour and cognition in rodent models. Farmed pigs are reared under controlled conditions in terms of hygiene and diet, which may limit exposure to microbes essential for the development of a healthy and well-balanced gut microbiota. We hypothesise that supplementing piglets with prebiotics (a substrate that is selectively utilized by host microbes conferring a health benefit) early in life will enhance learning and memory in a t-maze. Half the piglets in each litter were randomly assigned to receive a supplement (oat $\beta$-glucan, 40 mg/kg), 20 supplemented (SUPP) piglets) and the other half received a placebo (water, 24 control (CON) piglets) three times a week between 7 and 35 days of age. Feeding behaviour was assessed by scan sampling on 20, 34, 41, 48 and 55 days of age. After a brief habituation period around 55 days of age, piglets were randomly assigned to one reward arm in a t-maze to assess spatial discrimination learning. They received one session consisting of 10 trials per day (acquisition phase) until reaching the learning criterion of 80% correct responding on two consecutive days. After this, piglets were tested for reversal learning (reward in opposite arm, reversal phase). Data were analysed using mixed models in R (lme4 and lmerTest packages) fitting a treatment by session interaction and sex as fixed effects and piglet ID nested within litter as a random effect. All piglets passed the learning criteria. In the acquisition phase, learning significantly improved over the sessions ($F_{(2,78.8)}$=33.9, P<0,001), but no effects of either treatment or sex were found (mean±sem for CON: 61±5%, 77±4%, 86±4% and SUPP: 65±5, 79±4 and 91±3% correct trials per session, analysed for sessions 1, 2 and 3 respectively). In the reversal phase, learning also significantly improved over time ($F_{(4,151.2)}$=185.7, P<0.001) but the supplemented piglets performed worse than control pigs ($F_{(1,39.4)}$=5.16, P<0.05, mean±sem CON: 12±3, 26±3, 57±5, 76±5, 89±2% and SUPP: 9±2, 16±5, 23±6, 62±5, 79±5% correct trials, analysed for sessions 1-5, respectively). No treatment differences in feeding behaviour or weight gain were found. Supplementing piglets with $\beta$-glucan did not affect initial learning, but reduced reversal learning. This suggests that piglets supplemented with $\beta$-glucan may have more difficulty in modifying existing behavioural patterns and may be less flexible in their behaviour. This was not in keeping with our hypothesis and may suggest that $\beta$-glucan acts differently from other types of prebiotics. However, we cannot exclude that $\beta$-glucan led to changes in confounding factors, such as reduced feeding motivation, even if this should also have led to poorer learning in the acquisition phase.

## Can dietary magnesium improve pig welfare and performance?

*Emily V. Bushby[1], Helen Miller[1], Louise Dye[2], Kayleigh Almond[3] and Lisa M. Collins[1]*
*[1]Univerisity of Leeds, Faculty of Biological Sciences, Leeds, LS2 9JT, United Kingdom, [2]Univerisity of Leeds, Faculty of Medicine and Health, Leeds, LS2 9JT, United Kingdom, [3]Primary Diets, Melmerby, Ripon HG4 5HP, United Kingdom; bsevbu@leeds.ac.uk*

On commercial farms, pigs experience stressful events at various times in their lifecycle. To improve welfare and performance during these stressful events, enrichment, handling and dietary strategies may be implemented. One dietary approach may be including an increased level of magnesium in the diet, as this has previously been shown to have a calming effect in other animals, such as horses. Dietary supplementation with phytase has been shown to increase the availability of minerals, including magnesium. The aim of this study was to investigate how supplementary dietary magnesium, with or without phytase, may reduce aggressive or stress related behaviours before and after mixing, as assessed by lesion scoring. To do this 240 large white landrace cross piglets were weaned at approximately 4 weeks of age and allocated to pens of 5 balanced for weaning weight, sex and litter. Two focal pigs per pen of five were identified at weaning (44 females, 52 males) for lesion scoring. From 20 days post weaning four different experimental diets were fed until 13 weeks of age. These were A (control), B (0.15% supplemented magnesium phosphate), C (0.03% supplemented phytase) or D (0.15% magnesium phosphate and 0.03% phytase). At five weeks post-weaning the pens of five pigs were mixed, creating pens of ten. Pig and feed weights were recorded on the first day of the trial diets, the day before mixing and the end of the trial, to enable the calculation of performance parameters. Lesion scores were assessed on a weekly basis using the same five stage scoring system as Stevens, *et al.* Lesion scores were split into tail score, ear score (sum of both ears) and body score (sum of all main body areas: left and right flank, hindquarters, shoulders and back). Friedman tests were used to assess body score, ears score and tail score in relation to a diet and time point (before or after mixing). As expected there was a significant effect of test time (before or after mixing) (P≤0.01) on the number of body (mean before=5.1, after=12.8), ear (mean before=1.7, after=3.8) and tail lesions (mean before=0.21, after=1.03) at pig level, showing that the overall number of lesions increased after the pens were mixed. There was also significant difference between the B (mean=11.6) and C (mean=14.3) (P=0.009) and C and D (mean=12.2) (P=0.011) diets in terms of the number of body lesions, showing that pigs receiving diet C (phytase) had a higher number of lesions in comparison to the two magnesium diets after mixing. There was no significant difference between the diets for the number of ear and tail lesions. Two-way ANOVAs were used to analyse the performance data, which showed that dietary treatment did influence the average daily gain at pig level (P=0.01) but not individual pig weights or pen feed intake. Overall, magnesium reduced the number of body lesions in comparison to pigs receiving a phytase but not the control diet, suggesting that magnesium may have some welfare benefits for grower pigs.

## Dietary modification of behaviour in pigs

*Jeremy N. Marchant-Forde[1], Severine Parois[1,2], Jay Johnson[1] and Susan D. Eicher[1]*
*[1]USDA-ARS, LBRU, 270 S Russell St, West Lafayette IN 47907, USA, [2]INRA, PEGASE, Agrocampus Ouest, 35590 Saint-Gilles, France; jeremy.marchant-forde@ars.usda.gov*

Put simply, with pig nutrition management, we aim to maximize output for minimal input, both in terms of litter size and subsequent pig growth rate. Food is the most important resource for pigs, and pigs will use aggression to gain or protect food. Naturally, pigs are designed to have gradual exposure to solid feed and wean late, and eat numerous meals over the course of the day, of variable quality, spending a high proportion of activity in foraging behavior. Under many commercial conditions, pigs have abrupt exposure to solid feed and wean early, eat fewer, high quality meals and spend little time foraging. Feeding system design for pigs housed in groups will impact not only aggression but also other aspects of behavior. Feeding system design incorporates not only the feed delivery system, but also the food composition, which in turn may comprise of physical form, flavor and ingredients, the last of which will be the focus of this presentation. The impact of ingredients on behavior may be intentional and unintentional. For example, increasing tryptophan in the diet can influence serotonin concentrations and intentionally reduce aggression when pigs are mixed. Similar behavioral effects can be seen with increasing resistant starch in the diet. However, adding the beta-agonist ractopamine to the diet for known production benefits, can also unintentionally increase aggressiveness, reactivity to handling and hyperactivity. More recently, we have been examining the proposed gut-brain axis, whereby the gut microbiota communicates with the CNS – possibly through neural, endocrine and immune pathways – and thereby influencing brain function and behavior. There is also a suggested role for the gut microbiota in the regulation of anxiety, mood, cognition and pain. Recent studies examining alternatives to antimicrobials, such as amino-acids and probiotics, delivered as short- to medium-term supplementation, can affect behavioral and welfare responses to stressors and impact gastrointestinal microbial populations. L-glutamine appeared to confer similar benefits to, and thus could be a viable alternative to dietary antibiotics. A synbiotic supplement of Lactobacillus + fructo-oligosaccharide + Saccharomyces cerevisiae cell wall may confer memory advantages in 3 cognitive tasks, regardless of the nature of the reward and the memory request. Feeding a probiotic supplement to the sow prior to farrowing has shown indirect short- and longer-term behavioral effects on piglets, demonstrating the existence of a sow-piglet axis. The existence of a gut-brain axis offers a potential mechanism by which we can develop dietary therapeutic strategies to help pigs cope with stressful events, and potentially alter affective states both ameliorating the negative and enhancing the positive.

## Why drink milk replacer when you can suckle the sow – foraging strategy in sow-reared large litters with milk replacer

*Cecilie Kobek-Kjeldager[1], Vivi Aarestrup Moustsen[2] and Lene Juul Pedersen[1]*
*[1]Aarhus University, Animal Science, Blichers Allé 20, 8830, Denmark, [2]SEGES, The Danish Pig Research Center, Agro Food Park 15, 8200 Aarhus N, Denmark; cecilie.kobek-kjeldager@anis.au.dk*

Breeding for high prolific sows has led to sows giving birth to more piglets than she has teats. Rearing them with the sow by supplementing with milk replacer may be an option. We therefore investigated how access to milk replacer affected the foraging strategy and nursing behaviour of the piglets. In the present study, 98 sow-reared litters were investigated in a 2×2×2 factorial design: litter size d1 (14 vs 17 piglets), housing (crate vs loose) and milk replacer (+/-). Five nursing events and drinking behaviour for ~12 h was observed d7 and d21 postpartum. Piglets were categorized according to their foraging strategy d7 and d21. Of piglets with access to milk replacer around 30% frequently drank while also suckling the sow (MIXEDDRINKER). Almost 70% of the piglets rarely drank milk replacer but suckled the sow (SOWDRINKER). Only 2% of the piglets seemed to live solely of milk replacer (CUPDRINKER) or was rarely seen suckling or drinking milk replacer (LOWDRINKER). In a binomial logistic regression model, the four foraging strategies (pairwise comparison), the risk of teat fighting, fighting over the milk replacer and udder massage post milk letdown were analysed for an effect of the starting litter size d1, housing, milk replacer and piglet weight. The mean nursing interval per litter and the ADG (d7-14 and d21-28) of piglets were analysed in linear mixed models. The nursing interval was analysed for an effect of the factorial design and the growth rate for effect of the foraging strategy on d7 and d21, respectively. Heavier piglets were more likely to be MIXEDDRINKER compared to SOWDRINKER (P<0.001). MIXEDDRINKER had a higher growth rate than SOWDRINKER and much higher than the CUP- and LOWDRINKER (P<0.001). The risk of teat fighting was not reduced in litters with access to milk replacer (P>0.1), and was higher in litters starting with 17 compared to 14 piglets d1 (P=0.007) and higher with decreasing piglet weight (P=0.008). MIXED- and SOWDRINKER had similar risk of teat fighting but the risk was lower than for CUP- and LOWDRINKERS (P<0.001). Fighting at the milk replacer occurred in 8-9% of the drinking observations, but was not affected by litter size d1, housing or the weight of the piglet (P>0.1). The nursing interval was slightly shorter on d7 in litters with access to milk replacer compared to litters without (P=0.005). The risk of a piglet performing udder massage 10 and 20 min. after milk letdown as a sign of hunger was not found to be consistent across piglets and was not affected by any of the factors (P>0.1). The results of the study indicate that access to these types of milk replacers could not reduce teat fighting and competition at the udder, and that particularly smaller piglets were involved in teat fighting irrespective of access to milk replacer. Particularly larger piglets appeared more likely to exploit the milk replacer while still suckling the sow, and thus the milk replacer seemed to act as a supplement to larger piglets than as a replacement for sow milk for smaller piglets.

## Individual characteristics predict weaning age of dairy calves when weaned automatically based on solid feed intake

*Heather W. Neave, Joao H.C. Costa, Juliana B. Benetton, Daniel M. Weary and Marina A.G. Von Keyserlingk*
*University of British Columbia, Animal Welfare Program, 2357 Main Mall, Vancouver, BC, V6T 1Z4, Canada; hwneave@gmail.com*

Dairy calves are normally weaned at a set age but automated milk and grain feeders can be used to wean calves automatically based on individual grain intakes. Calves are highly variable in when they are able to complete weaning based on intake of grain (between 6 to 13 wk of age), suggesting that some calves are ready to be weaned from milk at earlier ages than others. However, it is not clear how best to predict which calves are suitable candidates for early weaning. The aim of this study was to identify early behavioural measures that are the best predictors of weaning age when weaned based on grain intake. Individual characteristics of calves (n=43) included calf vitality, suckle strength, milk intake at first meal and in first week, training required to drink from the automated milk feeder, and personality traits (behavioural responses to novel arena, human and object). We hypothesized that calves with high vitality and learning ability, exploratory personality trait, and early grain intake would wean earlier. Calves were housed in 6 groups from d 1 of age and received 12 l/d of milk until d 30 when milk was reduced by 25% relative to the individual's previous 3-d intake average. Milk was reduced by another 25% when the calf reached each of two grain intake targets: 225 and 675 g/d (3-d rolling averages), and calves were completely weaned at 1,300 g/d of grain, resulting in variable weaning ages. Automatically-recorded feeding behaviour during the first 30 d included age to first eat grain (40 g), total milk and grain intake, total rewarded and unrewarded visits, and drinking speed. A principal component analysis (PCA) condensed individual characteristics into 5 factors explaining 66% of the variation which we labeled: low vitality, fearful, strong drinker, slow learner, and exploratory-active. Multiple regression analysis with backward elimination identified which combination of variables (5 from PCA, and 6 automated feeding behaviour measures) were best able to predict weaning age of calves. Variables were retained in the model if P<0.15. Mean weaning age was 59 d (range 44-84 d of age). Five explanatory variables were retained in the model explaining 67% of the variance in weaning age: age to start eating grain (P<0.01), total pre-weaning grain intake (P<0.01), slow learner in the milk feeder (P<0.01), fearful (P=0.03) and total unrewarded visits (P=0.08). The regression equation was: *weaning age = 48.4 + 0.39 (age to start eating) – 6.98 (total pre-weaning grain intake) + 1.45 (slow learner in milk feeder) – 2.23 (fearful) – 0.039 (total pre-weaning unrewarded visits).* Characteristics measured in the first 30 d of life can identify calves that are able to wean relatively early (e.g. high early grain intake) and calves that take longer to complete weaning based on solid feed intake (e.g. slow to learn to use the milk feeder). Early grain intake and learning ability are relatively easy to monitor and could be used to inform individualized feeding and weaning management decisions, such as allocating milk away from calves ready to wean and toward calves requiring more time on milk before weaning.

## The effect of omega-3 enriched maternal diets on social isolation vocalisations in ISA Brown and Shaver White chicks

*Rosemary Whittle, Reza Akbari Moghaddam Kakhki, Elijah Kiarie and Tina Widowski*
*University of Guelph, 50 Stone Road E, Guelph N1G 2W1, Canada; rwhittle@uoguelph.ca*

In commercial poultry production, addition of long-chain polyunsaturated fatty acids such as omega-3s (n-3) into the diet of parental stock may alter the yolk composition of the hatching eggs produced. Developing embryos absorb ~95% of the phospholipid content of the egg yolk. Therefore, increased n-3 in the yolk may alter embryonic brain development, specifically in the hippocampus. Previous research into the effects of n-3 enriched hatching eggs has found longer tonic immobility and stronger reactions to novel environments. We explored whether social isolation elicited different vocalisation responses between laying pullets hatched from hens fed either n-3 enriched diets or a control diet. The chicks tested were part of a separate ongoing nutritional study using two strains, ISA Brown and Shaver White, and 3 maternal diets. The maternal diets were linseed enriched (LinPro), algae enriched (All-G) or a control diet. Social isolation tests were conducted at 4-6 days old, when 108 chicks were tested (54 ISA, 54 Shaver). Six cages per maternal treatment were used with one chick randomly selected from each cage per day. Social isolation tests in chicks have been validated as a pharmacological model for anxiety-like-states by Sufka *et al.* The chick was carried to the test room, placed in a lit padded box and the voice recording was initialised. The test began when the researcher said 'start', the timer was set, and the lid was put in place. The test lasted five minutes. The voice recordings were input into WavePad software. Each recording was divided into distress periods (DP) lasting 30 seconds. The total number of vocalisations in each DP was counted, the vocalisations were classified according to intensity (<18 dB, 18-12 dB, 12-6 dB, >6 dB). The data were analysed in R using ANOVA with maternal diet, DP and strain as fixed effects. This study found that offspring from n-3 fed mothers vocalised more frequently during social isolation than those from control fed mothers ($\chi^2$=6.194, P=0.045). ISA Browns vocalised more frequently than Shaver Whites ($\chi^2$=12.565, P=0.0003). However, there was no significant interaction between maternal diet and strain ($\chi^2$=1.806, P=0.405). The number of vocalisations per distress period differed significantly ($\chi^2$=32.007, P=5.501e$^{-10}$). There was no significant difference between maternal treatments for vocalisation intensities <18 dB ($\chi^2$=3.047, P=0.218), 18-12 dB ($\chi^2$=0.307, P=0.858), or 12-16 dB ($\chi^2$=1.341, P=0.511). However, those most likely to be distress vocalisations (>6 dB), showed a significant difference between maternal treatments ($\chi^2$=7.376, P=0.025). This study shows that maternal-fed n-3 affects the response of chicks to social isolation, increasing the frequency of vocalisations. These results are contiguous with previous research, suggesting that increased levels of n-3 available to developing embryos can influence fear responses.

## Characterization of plant eating in cats

*Benjamin L. Hart, Lynette A. Hart and Abigail P. Thigpen*
*University of California, Davis, School of Veterinary Medicine, 1 Shelds Ave, Davis, CA 95616, USA; blhart@ucdavis.edu*

After reporting the frequency and demographics of plant eating in dogs, we launched a similar study on cats, taking into account the need to exclude cats that are indoor without access to plants, and cats that are outdoors where owners cannot see if they are eating plants. Our web-based survey of cat owners, received 2,296 returns. The main inclusion criterion was that the respondent (owner) had to have been able to see the cat's behavior 3 or more hours a day. The resulting 1,021 returns revealed that 71% of cats had been seen eating plants at least 6 times, 61% over 10 times, and 11% never eating plants. Comparing cats seen eating plants at least 10 times with those never seen eating plants, there were no differences in age range, neuter status, source or number of cats in the household. Of cats seen eating plants at least 10 times, 67% were estimated to eat plants daily or weekly. When asked about how their cat seemed to feel prior to eating plants, 91% of respondents said their cat was almost always appeared normal, beforehand. Vomiting was a bit more common – 27% reported the cat frequently vomiting after eating plants. The prior study on plant eating by dogs had very similar findings with regard to frequency of plant eating, appearing normal beforehand, and vomiting 20-30% of the time afterwards. Among young cats, 3 years of age or less, 39% engaged in daily plant eating compared to 27% of cats 4 years or older (P<0.01). While percent of younger cats showing no signs of illness prior plant eating was similar to older cats, just 11% of the younger cats were observed to frequently vomit after eating plants compared to a significantly higher 30% of older cats (P<0.001). These findings on more frequent plant eating by younger animals and less likelihood of vomiting afterwards parallel similar findings on dogs. A common explanation given for plant eating in cats and dogs is that they feel ill beforehand and plant eating induces vomiting, and they may feel better. Our findings here do not support this explanation; nor do the findings support the hypothesis that young animals 'learn' plant eating from older ones. The explanation offered here is that regular plant eating by domestic carnivores is a reflection of an innate predisposition of regular plant eating by wild ancestors which is supported by numerous reports of wild carnivores eating plants, as shown mostly by the non-digestible grass and other plant parts seen in their scats. Studies on primates reveal that non-digestible plants purge the intestinal system of helminthic parasites. Given that virtually all wild carnivores carry an intestinal parasite load, regular, instinctive plant eating would have an adaptive role in maintaining a tolerable intestinal parasite load, whether or not the animal senses the parasites.

## The effect of outdoor stocking density and weather on the behavior of broiler chickens raised in mobile shelters on pasture

*Hannah Phillips and Bradley Heins*

*University of Minnesota, Animal Science, 380A Haecker, 1364 Eckles Avenue, St. Paul, MN 55108-6118, USA; phil1149@umn.edu*

Mobile shelters with pasture access are becoming popular for small-scale organic broiler production. The aim of this study was to examine the behavior, and the relationship between weather and behavior, of broilers raised in mobile shelters with two stocking densities of pasture access. Fifty straight-run Freedom Ranger broilers were used in three replicates (n=150) at the University of Minnesota West Central Research and Outreach Center (Morris, MN, USA) from June to October 2018. From 4 to 11 weeks of age, birds were housed in a covered mobile shelter divided into two equal-sized pens (0.27 $m^2$ per bird) with daily access to pasture and received 0.15 kg per bird of starter feed (20.0% CP) once daily. The shelter was moved every 2 to 8 days (5.6±0.2 days) depending on age to maintain vegetative cover greater than 50%. Birds were randomly allocated to one of the two pens corresponding to one of two treatment groups: (1) high (0.46 $m^2$ per bird); and (2) low (2.5 $m^2$ per bird) density outdoor pasture allowance. Ten focal birds per treatment (n=20 per replicate) were randomly designated for behavior observations, which were performed twice daily, four times per week, from 5 to 10 weeks of age. Counts of the number of birds outside were recorded during each observation. Behaviors were recorded continuously for one minute (Animal Behaviour Pro© app) on each focal bird, with states recorded as durations and events recorded as binary outcomes. Pen means averaged for each week were analyzed in the MIXED procedure of SAS with fixed effects of week, treatment, the interaction of treatment and week, and replicate within treatment as a random effect with repeated measures of week. Analysis of pasture use included heat index as a covariate, and the Pearson correlation (r) of pasture use and weather with week as a partial correlation. Average temperature recorded during the study was 21.6 °C (6.7 – 32.8 °C). No behavioral differences between treatments were observed. Sitting awake was the most observed behavioral state in high and low density treatment groups (47.3 and 42.8±3.5% time, respectively), followed by standing and sitting asleep. Preening was the most observed behavioral event in high and low density treatment groups (25.2 and 26.0±2.4% flock, respectively), followed by foraging (20.6 and 23.6±2.0% flock, respectively). Sitting and aggression behaviors decreased with age (P<0.05). There were no differences in pasture use between high and low treatment groups (33.0 and 38.5±3.5% flock, respectively), and pasture use was negatively correlated (P<0.01) with temperature (r=-0.62) and solar radiation (r=-0.49). The results of this study indicate similar behaviors of broilers raised with two levels of pasture allowance, and suggests that hot weather negatively affects the use of pasture space.

## Outdoor feeder increased range usage but not bone quality in commercial free-range laying hens

*Terence Zimazile Sibanda[1], Richard Flavel[1], Manisha Kolakshyapati[1], Santiago Ramirez-Cuevas[2], Mitch Welch[1], Derek Schneider[1] and Isabelle Ruhnke[1]*
*[1]University of New England, Environment and Rural Science, 60 Wadgick, 2350, Australia, [2]FCR Consulting Group, Brisbane QLD 4000 Australia, Brisbane QLD, 4000, Australia; tsibanda@myune.edu.au*

Free-range hens that access the range frequently are not only less likely to obtain their formulated feed intake, they are also assumed to experience higher energy loss due to increased locomotion and maintaining temperature homeostasis. Offering the same diet to all hens as commonly performed ignores the different individual hen requirements and may result in reduced hen welfare and production. The aim of this study was to compare the effects of 3 different feeding strategies on tibia bone health and range usage in hens that are known to range frequently. A total of 3,125 Lohmann brown hens were housed amongst their 36,875 flock companions on a commercial free-range farm. These hens were individually monitored using a custom-made radio frequency identification (RFID) system for their daily range usage from 18 to 22 weeks of age. At 22 weeks of age, the top 60% of range users were selected and randomly assigned to one of 3 groups consisting of 625 hens each allowing for comparable stocking density. Hens of group 1 were housed under standard commercial conditions. Hens of group 2 were housed under standard commercial conditions and in addition provided with a gravity-filled feeder on the range placed 25 m distance from the shed. Hens of group 3 were housed under standard commercial conditions but received a diet that was formulated to contain +10% metabolisable energy and +10% amino acid levels. Standard commercial conditions comparable to the 36,875 remaining hens of the flock were achieved by sectional partitioning of the commercial shed and the range, allowing hens to utilise exactly the same facilities that they were placed in at 16 weeks of age when the shed was populated. All in-shed diets were delivered via a feeding chain, the additional outdoor feeder was available *ad libitum* during pop hole opening times (9am-8pm). Individual range usage of all hens was continuously monitored until hens were 72 weeks of age. At this age, 64, 52, and 43 left tibiae were randomly collected from hens of group 1, 2, and 3, respectively and individually scanned using a micro CT scanner. Bone characteristic parameters and hen ranging times were analysed using a mixed restricted estimate of maximum likelihood (REML) model with individual hen as a random factor, and the groups were compared using a least square means student`s t test pairwise comparison. Hens of group 2 spent significantly more time on the range (1.15±0.024 h/hen/day) compared to hens of group 1 (0.89±0.027 h/hen/day) and hens of group 3 (0.82±0.138 h/hen/day), while total the number of range visits (2.47±0.028 visits/hen/day) was comparable between the groups (P>0.05). There was no significant different in bone weight, bone length, diaphyseal diameter, cortical mineral density, relative bone marrow, and blood vessels volume between any of the groups (P>0.05). Hens of group 3 had significantly higher relative cortical volume compared to hens of group 1 and 2 (P<0.005). In conclusion, offering feed on the range increased range usage but had no impact on bone health.

## Pilot study of grazing strategies of beef cattle in relation to woody elements percentage in a silvopastoral system

*Karen F. Mancera[1], Heliot Zarza[2] and Francisco Galindo[1]*
*[1]Universidad Nacional Autónoma de México, Facultad de Medicina Veterinaria y Zootecnia, Av. Insurgentes Sur s/n, Ciudad Universitaria, 04510 México City, Mexico, [2]Universidad Autónoma Metropolitana, Unidad Lerma, Departamento de Ciencias Ambientales, CBS, Av. de las Garzas No. 10, Col. La estación, Lerma de Villada, 52005 Estado de México, Mexico; dra.kelokumpu@gmail.com*

Monoculture systems are associated with loss of environmental services and poor animal welfare. Silvopastoral Systems (SPS) associating pasture and woody elements (WE; trees and shrubs) in complex grazing environments are an alternative. SPS increase biodiversity, environmental services, and animal welfare. To date, there is no research on how WE in SPS affect grazing, which could directly impact animal welfare and productivity. Bites are the first source of grazing variation. Bites comprise grazing prehensions (GP) and mastications (GM), rumination mastications (RM) and jaw movements for the deglutition of the rumination boli (RB). Grazing deterrents (branches, thorns and large stems) create interference and decrease intake and bite rates (bites/min). Nevertheless, bite rate can also increase when bite size is reduced to select high-quality forage, favouring nutrition over forage quantity. Thus, this pilot study analysed the effects of different WE percentages in SPS paddocks on cattle jaw movements to: (1) provide a methodological framework to clearly identify cattle grazing strategies in complex environments; and (2) identify research priorities for the study of cattle-paddock interactions in SPS. Cows (n=116) rotated through six paddocks with different WE % in Puebla, México during the dry season (March-May). WE % was estimated by land cover classification using GIS. Herd behaviour was evaluated using scan sampling every 15 m and proportions of total time were calculated. Concurrently, average number of GP, GM, RM and RB were measured using a single grazing recorder rotated each day between nine trained heifers in the herd, creating repeated measures. A Generalized Linear Model with the factors level of WE % (High=50%≥WE>40%; Medium=40%≥WE>30%; Low=30%≥WE>20%), cow (random variable) and total free area (total paddock area – WE covered area; covariate) was used. High WE % increased GM (P=0.026, $F_{2, 19}$=4.47), probably due to increased fiber content and greater selection for feed quality. In contrast and contrary to predictions, GP, RM and RB decreased in medium and low WE % (GP: P=0.008, $F_{2,19}$=6.35; GP: P=0.035, $F_{2, 19}$=4.02; RB: P=0.029, $F_{2, 19}$=4.31), possibly due to grass stems acting as grazing deterrents and reducing intake and jaw movements, as found in other studies. Due to our inability to control for variables such as vegetation height and climate, our ability to draw general conclusions is limited. However, this study provides a methodological framework to assess the effects of WE % on cattle grazing in SPS and indicates that WE could influence cattle nutritional state and food intake, ultimately affecting welfare indicators such as body condition. We expect our findings to motivate rigorous scientific research on cattle-paddock interactions in the future, which is imperative to extend the benefits of SPS.

## The relationship between motion score and reproductive behaviour in dairy cows managed in pasture and feedpad environments

*Andrew Fisher, Caroline Van Oostveen, Melanie Conley and Ellen Jongman*
*The University of Melbourne, Animal Welfare Science Centre and Faculty of Veterinary and Agricultural Sciences, 250 Princes Highway, Werribee Victoria 3030, Australia; adfisher@unimelb.edu.au*

The detection of oestrous behaviour is used to identify the optimal timing for artificial insemination in dairy cows, and can be particularly challenging in extensively managed dairy cows. Because of changing climatic conditions, more Australian dairy farmers have installed concrete 'feedpads' to enable the effective supplementation of grazed pasture, which may be limited in dry weather. This study was conducted to identify whether the use of accelerometer-based behaviour loggers can overcome the challenges of oestrous detection on extensive dairy farms that utilise both pasture and feedpad environments. Following approval by the Institutional Animal Ethics Committee, 32 Holstein-Friesian cows (4.6 years (SD 1.48); 559 kg (SD 52.4); 86 DIM (SD 18.1)), were fitted with IceTagTM loggers. Feedpad cows (n=16) were fed on a feedpad for 80 min after each of morning and afternoon milking. Pasture-only cows (n=16) were allowed to walk straight back to the paddock after each milking. The diets offered were isoenergetic in that feedpad cows received a mixed ration on the feedpad, whereas pasture-only cows received grain during milking and silage while at pasture. Both cow groups received the same daily pasture allocation in adjoining plots. Continuous behavioural observations using two observers were conducted for three, 3-h periods per day over a total of 7 days across paddock, dairy yard, feedpad and laneway environments, following an Ovsynch program using GnRH and prostaglandin injections to induce oestrous. Reproductive behaviours were summed to a synthetic heat score for each animal for each of three periods per day, and logger data was used to derive a motion index per hour. A Spearman Rank correlation test was performed to examine whether there was a correlation between the motion index and the heat score. A MANOVA was used to examine whether there was an effect of feedpad or pasture-only environments on the motion index. A total of 19 cows were observed in standing oestrous, with 30 cows showing sexually active behaviours. There was a significant (P<0.001) but weak correlation between motion index and heat score (r=0.19). Feedpad cows had a lower motion index per observation hour compared with pasture-only cows (934 vs 1,286; SE=89.4; P<0.01), but overall motion index was greater in cows that were observed to be in oestrus (1,487 vs 784; SE=132.0; P<0.001). The use of behaviour loggers may assist with oestrous detection for cows in a pasture system that includes feedpad feeding but the data may need to be focussed on the pasture-based periods to optimise its effectiveness.

## Do grazing management practices influence the behaviour of dairy cows at pasture?

*Robin E. Crossley[1,2], Emer Kennedy[1], Imke J.M. De Boer[2], Eddie A.M. Bokkers[2] and Muireann Conneely[1]*
[1] *Teagasc Moorepark, Animal & Grassland Research and Innovation Centre, Fermoy, Co. Cork, Ireland,* [2]*Wageningen University and Research, Animal Production Systems, 6708 WD Wageningen, the Netherlands; robin.crossley@teagasc.ie*

The Irish dairy industry has undergone rapid change and growth since dairy quotas were eliminated in 2015; herd sizes are expanding and new grazing management practices are being explored to optimize productivity. However, more intensive grazing management practices with higher stocking rate (SR) and lower pasture supply (PS) may lead to elevated grazing pressure, influencing the behaviour and welfare of grazing dairy cattle. No recent research has identified the impact of different spring grazing management practices on the behaviour of dairy cows at pasture. We hypothesised that the behaviour of dairy cows grazing full-time at a higher SR and lower PS would differ from cows at a lower SR and higher PS. As part of an on-going system study, 144 Holstein-Friesian (HF) and HF × Jersey cows in 6 groups (balanced for parity, breed and milk production), were assigned to one of 6 grazing plans in a 3×2 factorial design; 3 levels of spring PS, Low (L; ~700 kg DM/ha, representing current commercial practices), Standard (S; ~900 kg DM/ha), or High (H; ~1,100 kg DM/ha), and 2 grazing intensities, Moderate (MI; SR=2.75 cows/ha, 90:10% pasture: concentrate diet, n=22) and High Intensity (HI; SR=3.25 cows/ha, 80:20% pasture: concentrate diet, n=26). SRs of 2.75 and 3.25 cows/ha are substantially higher than the average mean on Irish dairy farms of approx. 2 cows/ha. Direct behavioural observations were recorded at pasture for 30 min/group, 3×/day (~08.15, 12.00 and 15.45 h), and 3 days/week over 3 weeks, totaling 13.5 observation hrs/group. Recorded behaviours were threats, chases, head-head or head-body butts, self- or allo-grooming, cohort-resting, vocalising, drinking, tongue-rolling, mounting and head-rubbing. Group was the experimental unit; data were summarised by day, and analysed using the non-parametric Kruskal-Wallis test in SAS as they were non-normally distributed. Results indicated an effect of treatment on the number of head-head butts (H=13.1, df=5, P=0.023) and the number of allo-grooming events (H=19.0, df=5, P=0.002). Pairwise comparisons revealed that L-HI cows demonstrated significantly more head-head butts compared to all other groups (P<0.05). Additionally, L-HI and S-HI cows displayed more allo-grooming events compared to all other groups except one another (P<0.02 for L-HI compared to L-MI, S-MI, H-MI and H-HI; P<0.04 for S-HI compared to L-MI, S-MI, H-MI and H-HI). In general, cows' behaviour did not differ greatly in response to the examined grazing management practices. However, behaviours observed in the L-HI and S-HI groups may indicate an attempt to compensate for greater competition for available grass; either through aggressive or frustrated behaviour expressed by increased head-head butts, or by replacing unsatisfied grazing behaviour with increased allo-grooming. This may suggest that high levels of grazing pressure could have negative implications for the behaviour and welfare of grazing dairy cows.

## Use of a feed frustration test to assess level of hunger in dairy cows subjected to various dry-off management routines

*Guilherme Amorim Franchi, Mette Herskin and Margit Bak Jensen*
*Aarhus University, Department of Animal Science, Blichers Allé 20, 8830, Tjele, Denmark; margitbak.jensen@anis.au.dk*

Dairy cows typically have their milk production artificially ceased around 60 d before calving. This process, termed drying-off, often comprises changes of diet and milking frequency. Evidence suggest drying-off challenges the welfare of high-yielding cows. For instance, a nutrient-reduced diet can impair energy balance and cause hunger even when offered *ad libitum*. This ongoing work proposes a feed frustration test, inducing motivation conflict, to assess isolated and combined effects of daily milking frequency (twice; once) and diet (lactation diet or energy-reduced lactation diet with forage:concentrate ratios of 51:49 and 65:35, respectively, *ad libitum*) on the degree of hunger. Biweekly, groups of 2-6 healthy loose-housed Holstein cows were allocated to one of four treatments from 7 d (D-7) prior to the last day of milking (DO). In total, 92 cows (mean±SD) 754±80 kg in body weight and yielding 25±6 kg of milk/d were included in this study. From D-7 to DO, each cow had access to one computerised feed bin providing access to the respective diet. Tests were conducted on D-6 and D-1. First, barn staff remotely locked the feed bins and initiated the second daily feed delivery (at approx. 10.30 h). Duration of feed delivery ranged from 15-76 min and constituted the routine phase (phase A). When the feed delivery ended, all feed bins were remotely unlocked, except the experimental ones, which remained locked for additional 35 min, constituting the extra feed deprivation phase (phase B). During B, experimental cows could hear other feed bins opening and see cows in neighbouring pens feeding. During both phases, an experimenter standing in front of the feed bins approx. 4 m away recorded attempts to feed (i.e. focal cow puts the head over any locked feed bin, placing her collar past the feed bin gate) and high-pitched vocalisations (≥3 s). Following unlocking of experimental feed bins the latency to feed was recorded for 5 min. Data analyses were performed in R for each day separately. Attempts to feed in A were treated as counts/min due to a wide time range, while attempts to feed in B were treated as counts. Due to a low representation of vocalisations in the data set, this variable was analysed as binary (yes; no). Latency to feed was also analysed as binary (yes; no). All variables were analysed using GLMM including diet and milking frequency as fixed effects and block as random effect. On D-6, no significant differences were found between treatments, except the odds of vocalising being higher in cows on energy-reduced diet than normal diet [OR (95% CI)] [A: 5.4 (2.1 to 15.5), $\chi^2$=11.4, DF=1, P<0.001; B: 3.3 (1.4 to 8.2), $\chi^2$=7.1, DF=1, P<0.01]. On D-1, cows on energy-reduced diet attempted to feed 50% more in B [OR (95% CI)] [1.5 (1.08 to 2.09), $\chi^2$=5.93, DF=1, P<0.05], were more likely to vocalise [A: 8.4 (2.5 to 38.7), $\chi^2$=10.1, DF=1, P<0.01; B: 4.5 (1.9 to 11.5), $\chi^2$=10.9, DF=1, P<0.001] and to feed within 5 min [2.8 (1 to 8.6), $\chi^2$=3.84, DF=1, P=0.049] than cows on normal diet. These preliminary results suggest that during gradual dry-off hunger results from the negative dietary shift, while no mitigating effects of reducing the milking frequency were evident.

## The effect of step-down milk feeding and daily feeding frequency on behavior of male dairy calves

*Margit Bak Jensen, Annedorte Jensen and Mogens Vestergaard*
*Aarhus University, Department of Animal Science, Blichers Allé 20, 8830 Tjele, Denmark;*
*margitbak.jensen@anis.au.dk*

Dairy calves are often fed a fixed milk allowance the first 6 weeks, while an initial high allowance followed by a step down may better correspond to the calves' nutritional needs. The portion size is typically constant, while during natural suckling, the daily number of milk meals declines with age. Adopting a step down strategy without restrictions on meal frequency may stimulate dairy calves' ingestion of solid food. The present experiment investigated the effects of these factors on calves' feeding behaviour. Sixty-four male Holstein calves, purchased from conventional dairy farms at approx. 12 d of age, were housed in groups of eight and fed milk replacer via an automatic milk feeder (Urban Calf Mom, Germany). After a 2 d acclimation, where all calves were fed 6.5 L/d, the 14 day-old calves were assigned to one of two milk-feeding strategies (applied at group level) for the following 28 d experimental period, either Conventional (6.5 L/d throughout) or Stepwise (8 l/d until d 28 and 5 l/d thereafter). Within each group of eight, four calves were randomly allocated to each of two milk-feeding frequencies, either Restricted portion size (max. 2.3 L/portion) or Unrestricted portion size. Concentrates, hay and water were available *ad libitum*. The feeding behavior of the calves was recorded via video during 24 h on d 26 and d 40 of age and analyzed with a mixed model including the fixed effects of milk feeding strategy, frequency, day and their interactions. Initial body weight was included as a covariate, while pen and calf were included as random effects. Irrespective of treatment, calves spent more time sucking the teat on d 26 than d 40 (26.7 v. 20.1 (SE 2.41) min/d; $F_{1,57}=19.34$, P<0.001). However, on d 26 Unrestricted calves spent more time in the milk feeder than Restricted calves (50.5, 80.4, 51.6 and 56.7 min/d (SE 6.00) for Restricted/d 26, Unrestricted/d 26, Restricted/d 40, Unrestricted/d 40, respectively; $F_{1,57}=7.99$, P<0.01). Unrestricted calves also spent more time eating concentrate than Restricted calves (back transformed mean (95% CL): 21.9 (17.3; 26.9 9) and 27.6 (22.4; 33.2) min/d; $F_{1,50}=4.83$, P<0.05), and more time eating hay than Restricted calves (30.0 (16.4; 39.8) and 38.4 (28.8; 49.4) min/d; $F_{1,50}=4.39$, P<0.05). On the other hand, Stepwise calves spent more time eating concentrate than Conventional calves on d 40 (20.3 (14.4; 27.2), 16.9 (11.6; 23.2), 27.2 (20.3; 35.2) and 36.1 (28.2; 45.2) min/d for Conventional/d 26, Stepwise/d 26, Conventional/d 40, Stepwise/d 40, respectively; $F_{1,57}=7.87$, P<0.01). The higher appetite for concentrates in Stepwise fed calves on d 40 is likely due to the lower milk allowance at this point. However, higher appetite for concentrates and hay by calves fed milk in unrestricted portions may not be explained by milk intake. Rather, the longer time spent feeding in these calves is likely due to the larger milk meals, which appears to increase their appetite for milk (as indicated by the longer time spent in the milk feeder) as well as solid feeds. Thus, no restriction on milk portions may help calves transition from milk to solids before weaning off milk.

## Experience with hay influences the development of oral behaviors in preweaned dairy calves

*Blair C. Downey[1], Margit B. Jensen[2] and Cassandra B. Tucker[1]*
*[1]University of California, Center for Animal Welfare, Department of Animal Science, 1 Shields Avenue, Davis, CA, 95616, USA, [2]Aarhus University, Department of Animal Science, Blichers Allé 20, 8830 Tjele, Denmark; bdowney@ucdavis.edu*

Dairy calves are motivated to access fibrous feeds and will consume grass by 3 d of age in pasture-based settings. However, many early calf diets do not include forage. Abnormal behaviors, such as non-nutritive oral manipulation and tongue rolling, are seen in calves and may be the result of redirected natural feeding behaviors. We assessed how early experience with hay influenced the development of all oral behaviors. Twenty-two Holstein heifer calves were housed individually on sand bedding and fed a diet of starter grain and milk replacer (5.7-8.4 L/d step-up) via a bottle (C; ADG=0.85±0.17 kg) or given additional access to hay (H; ADG=0.88±0.15 kg) from birth through 50±1 d of age, when step-down weaning began. At wks 2, 4, and 6, all oral behaviors were directly recorded using 1-0 sampling at 1-min intervals for 24 h. We measured ruminating, eating, drinking water, sucking milk, grooming, and panting. Non-nutritive oral manipulation (licking, chewing, sucking non-feed items), tongue rolling, and repeatedly (≥2×) extending the tongue out of the mouth or up to the nose ('flicks') were also recorded. We analyzed the proportion of intervals calves performed each behavior using a general linear mixed model (glmmTMB in R) with week and treatment as fixed effects, and calf as a random effect. All calves were ruminating by wk 2 (mean proportion of intervals ± SE; 0.202±0.016), and H calves spent more of the observations ruminating compared to C calves throughout (H: 0.241±0.021, C: 0.155±0.015, $P<0.001$). H calves spent fewer observations eating starter than C calves (H: 0.037±0.003, C: 0.046±0.003, $P=0.028$), likely because they had an additional food source to consume (H: 0.038±0.003 eating hay). There was no treatment effect of intervals spent sucking milk (0.009±0.004, $P=0.983$). All calves engaged in self-grooming, but H calves performed it less than C calves (H: 0.120±0.006, C: 0.139±0.006 $P=0.025$). Tongue flicks and non-nutritive oral manipulation were consistently performed for a lower proportion of daily observations by H calves than C calves (tongue flicks H: 0.139±0.01, C: 0.175±0.011, $P=0.016$; non-nutritive oral manipulation H: 0.167±0.009, C: 0.209±0.01, $P=0.002$). Tongue rolling and panting were rare. Overall, calves with access to hay were more likely to engage in behaviors associated with food acquisition and processing, such as rumination, and were less likely to show non-nutritive oral manipulation and tongue flicks than calves without it. Hay likely provides an outlet for motivation to perform natural feeding behavior, leading to lower motivation to redirect oral behavior to other stimuli.

## Influence of water delivery method on dairy calf performance and inter-sucking after weaning

*Michal Uhrincat[1], Jan Broucek[1], Peter Kisac[1], Anton Hanus[1] and Miloslav Soch[2]*
*[1]NPPC, Research Institute of Animal Production Nitra, Hlohovecka 2, 951 41 Luzianky, Slovak Republic, [2]University of South Bohemia, Branisovska 1645/31a, 370 05 Ceske Budejovice, Czech Republic; uhrincat@vuzv.sk*

The aim of this work was to evaluate effects of water delivery method on the average daily gain, feed intake, health condition, and social behaviour of calves. Calves (n=62) were reared in hutches from the 2nd day to weaning at the age of 8 weeks. Calves received colostrum and mother's milk ad libitum 3 times a day from a bucket with nipple from the second to fourth day. From the fifth day they received 6 kg of milk replacer per day divided into 2 portions in 12 h intervals. From the second day until weaning the calves were offered concentrate mixture and alfalfa hay ad libitum. All calves were randomly divided according to the water delivery treatment into 3 groups – nipple sucking from bucket (N), drinking from bucket (B), and without delivery water (WW). Blood samples for analysis of white (leukocytes, basophiles, monocytes, and neutrophils) and red blood cells (haematocrit, haemoglobin, erythrocytes) were taken every week. After weaning at the average age of 56 days the calves were moved into the experimental barn. They were kept in loose housing pens in age and sex balanced groups. Calves were observed for two 24 h periods (first and seventh day) after moving into the barn. Other ethological observations were performed until the age of 6 months (once a week between 8:00 and 20:00). Behavioural data were obtained by video observations and electronic measurements. The social activity of each of the animal was recorded using following categories: licking/sucking barn equipment (tongue or mouth touching an object); self-licking/sucking (tongue or mouth touching the own body); cross-sucking (sucking on any body part of another calf); willing to be sucked. The study was approved by the ethics committee of the Research Institute of Animal Production Nitra. The data were analysed using a General Linear Model ANOVA. We found no significant difference among groups in average daily gain. Daily gain was highest in group N (N 0.46±0.13 kg, B 0.43±0.12 kg, WW 0.43±0.10 kg). N group of calves drank up more water to the weaning than B group (69.39±66.91 kg vs 50.72±51.95 kg), and group N had the highest intake of starter mixture (N 14.43±8.82 kg, B 11.30±5.45 kg, WW 13.31±6.86 kg). The highest alfalfa hay consumption was found in group WW (N 21.34±6.91 kg, B 22.26±7.52 kg, WW 23.59±8.76 kg). No calf became sick or died. There were no significant changes in blood parameters. We did not find significant differences between groups in the inter-sucking after weaning. The willingness to be sucked was significantly higher in calves from group N (N 10.50±0.77, B 5.10±0.80, WW 5.25±0.80, P<0.001). Water restriction did not result in any significant effect on growth, feed consumption, health condition, or social behaviour in experimental calves. However, we cannot recommend that water be delivered only as part of the milk replacer on the basis of this experiment alone.

## Feeding behavior of fattening bulls fed with an automatic feeding system

*Laura Schneider, Birgit Spindler and Nicole Kemper*
*University of Veterinary Medicine Hannover, Foundation, Institute for Animal Hygiene, Animal Welfare and Farm Animal Behavior, Bischofsholer Damm 15, 30173 Hannover, Germany; laura.schneider@tiho-hannover.de*

Automatic feeding of cattle offers multiple advantages in comparison to conventional feeding. While it is gaining more and more importance in dairy farming, there is still a lack of experience and scientific knowledge on its use in fattening cattle. Therefore, the objective of this study was to describe the behavior of bulls fed with an automatic feeding system. The study was conducted on 56 Simmental bulls on a commercial farm in Lower Saxony, Germany, with funding by the Ministry for Science and Culture of Lower Saxony. The bulls were housed in four groups of 14 animals in straw-bedded pens with a space allowance of 4 m$^2$ per head and 4.85 m manger space per pen. They were automatically fed six times a day a total mixed ration (feed delivery: 6:00, 9:30, 13:30, 17:30, 21:30, 22:30). Behavioral observations were performed during three observation periods (OP) at on average 11, 14 and 16 months of age via video recordings. The activity was recorded using scan sampling assuming each behavior to persist for the entire interval. The number of bulls feeding and lying was counted for 48 h per OP in intervals of 2 min from 4:00 to 23:30 and 10 min during the night. On individual level the behavior of all bulls was scanned in intervals of 10 min from 4:00 to 22:30 on three following days per OP. The feeding activity was widely spread from around 7:30 to 0:00 with averaged percentages of animals feeding never exceeding 20% per pen (data averaged for three 48 h-OP and four groups of 14 bulls each). Low percentages of bulls feeding as well as maximum percentages of more than 90% bulls lying occurred during the night and early morning hours. Two further lying periods with lower maximum percentages of 60 to 70% bulls lying were observed in the afternoon and early evening. The most frequent feeding condition showed one bull feeding alone (on average 28.0±2.2% of observation time) followed by two to three bulls feeding at the same time (16.6±2.0%; 6.7±1.9%). Simultaneous feeding of more than three bulls occurred rarely (3.0±1.4%). During 45.6±4.0% of observation time no bulls were feeding. Each bull spent on average 9.9±1.8% of the 18 h-period of individual observation feeding and 62.7±4.3% lying. During the observed feed deliveries (n=54) each bull was observed both feeding as the first one after feed delivery and as one of the latter (position 8 to 13). Automatic feeding with six feed deliveries per day ensures constant feed availability at any time of the day. This was confirmed by a widely spread feeding activity with low numbers of bulls feeding simultaneously at any time. Based on common feeding place recommendations of 75 cm per bull at the end of fattening, a manger space of 4.85 m should permit at least six bulls feeding simultaneously. Thus, the bulls´ preference to feed alone or in groups of two to three is unlikely to be caused by limited manger space. As the order of bulls feeding varied all bulls had similar access to fresh feed. In contrast to the feeding behavior the lying behavior indicates a high level of behavioral synchronization in conformance with the natural behavior of gregarious animals.

## The ins and outs of abomasal damage in veal calves

*Jacinta Bus[1], Norbert Stockhofer[2] and Laura Webb[1]*
*[1]Wageningen University & Research, Animal Production Systems group, P.O. Box 338, 6700 AH Wageningen, the Netherlands, [2]Wageningen Bioveterinary Research, P.O. Box 65, 8200 AB Lelystad, the Netherlands; laura.webb@wur.nl*

Of all cattle production systems, veal calves are most severely affected by abomasal damage, with current prevalence at slaughter ranging from 70 to 93% of animals affected. Though most damage is found in the pyloric region of the abomasum, fundic lesions are also found. Despite past research into the aetiology of abomasal damage and despite many risk factors being put forward, no agreement on the causal factors of abomasal damage in veal calves has yet been achieved. The aim of this review was to integrate and analyse available information on the aetiology of, and possible risk factors for, abomasal damage in veal calves. The review describes various proposed pathways through which risk factors may contribute to damage formation. We conclude that the aetiology of abomasal damage is most likely multifactorial, with diet being a main contributor. Pyloric lesions, the most common type of damage in veal calves, are likely the result of large and infrequent milk and solid feed meals, while fundic lesions may be caused by stress, though the evidence for this is inconclusive. Providing calves with multiple smaller milk and solid feed meals (or ad libitum provision) may decrease abomasal damage. There is also some indication that heavier calves, which may have higher access to feed resources because they are stronger, may consume more milk replacer and solid feed faster and thereby be more prone to developing abomasal lesions. In conclusion veal calf management, and in particular nutritional management, and the individual feeding behaviour of calves seem to be at the heart of the problem. Further research is required to understand the exact pathway(s) by which milk replacer causes abomasal damage in veal calves, and to verify the proposition that larger, more dominant calves are more prone to abomasal damage due to more rapid intake of feed. It is also crucial that we determine how calf welfare is impacted by the damage, and the level of pain these lesions may be causing.

## Individual differences in feeding behaviour of dairy goats

*Marjorie Cellier, Christine Duvaux-Ponter, Ophélie Dhumez, Pierre Blavy and Birte L. Nielsen*
*UMR Modélisation Systémique Appliquée aux Ruminants, INRA, AgroParisTech, Université Paris-Saclay, 16 rue Claude Bernard, 75005 Paris, France; marjorie.cellier@agroparistech.fr*

Detailed recording of ruminant feeding behaviour is becoming more common using automated feeders and electronic ID-tags, making it possible to phenotype individuals and monitor changes in feeding pattern over time, which may be indicative of health problems. However, most studies have concentrated on dairy cattle and calves. Here, we investigate if two commonly used breeds of dairy goats (Saanen: n=8; Alpine: n=8) differ in their feeding patterns. The goats (10-11 months of age in their second trimester of their first gestation) were housed in mixed-breed groups of 4 animals based on their feeding behaviour pre-weaning, where milk feeding frequency (MFF) differed between breed: Alpine High and Saanen Low had similar feeding frequencies when being milk fed (18-24 visits/day), whereas Alpine Low ($\leq$15 visits/day) and Saanen High (>32 visits/day) occupied the two extremes. Each pen had 4 feed troughs, and each goat (one of each breed × MFF combination per pen) had access to one trough only via an electronic ear-tag. The feed was a total mixed ration and available ad libitum, with troughs being replenished twice a day. Each trough was placed on a weigh-scale with weights recorded every 2 s. The goats were kept in these groups for 17 days, with data from the last 14 days used in the analyses. For each goat, data consisted of the weight of the trough (g) for each period when the trough weight remained constant, together with the timestamps (to the nearest second) for the beginning and end of this period of constant weight. After merging visits into meals (8 min inter-meal-interval), we calculated six meal pattern variables for each goat: frequency, size, and duration of meals, feeding rate, daily feeding time, and daily feed intake (see units for these variables below). A preliminary analysis was carried out on these data together with live weight, fitting pen as well as breed and pre-weaning type and their interaction as fixed factors in the model. A significant interaction was found between breed and pre-weaning type, with the two extreme MFF types, Alpine Low and Saanen High goats, having more meals (16.6±0.85 vs 12.1±0.85 meals/day; $F_{1,9}$=28.1; P<0.001) and smaller meals (175±18 vs 227±18 g/meal; $F_{1,9}$=8.3; P=0.018) of shorter duration (20±1.8 vs 26±1.8 min/meal; $F_{1,9}$=11.3; P=0.008) than Alpine High and Saanen Low goats, which had similar MFF. This difference in feeding pattern, however, resulted in no significant differences among the goats in feeding rate (overall mean: 10±0.4 g/min), daily feeding time (5 h 16±12 min/day), and daily feed intake (2.8±0.08 kg/day). No significant differences between breeds or pre-weaning feeding frequency were found for live weight. It is not yet clear why more extreme pre-weaning feeding behaviour, i.e. the high MFF in Saanen, and the low MFF in Alpine, would give rise to smaller, more frequent meals when adult. More data are currently being gathered to investigate these relationships in more detail.

## Foraging in the farrowing room to stimulate feeding behaviour

*Anouschka Middelkoop, Bas Kemp and J. Elizabeth Bolhuis*
*Wageningen University & Research, Adaptation Physiology Group, De Elst 1, 6708 WD Wageningen, the Netherlands; anouschka.middelkoop@wur.nl*

Timely intake of solid feed is essential to ease the nutritional change from sow's milk to feed at weaning and thereby to reduce weaning stress. A significant percentage of piglets, however, do not or hardly consume feed until weaning. Reducing sensory-specific satiety and stimulating exploratory behaviour towards the feed(er) may enhance pre-weaning feed intake. We therefore studied the effect of feed variety, feed presentation and their interaction on feed exploration, eating behaviour and growth of suckling piglets. Feed was provided *ad libitum* from d4 in two feeders (four feeding spaces each) per pen. In a 2×2 arrangement, piglets received either one feed item (creep feed) as a monotonous diet (MO) or received four feed items simultaneously (creep feed, celery, peanuts in shell and cereal honey loops) as a diverse diet (DV) and the feed was either presented without substrate (CON) or hidden in sand (SUB) in one of two feeders to stimulate foraging behaviour. Feed exploration and eating was observed live at d11, 18 and 27 using 2-min instantaneous scan sampling for 6 h/d. To study presentation preferences within SUB, it was noted at which feeder piglets were exploring/eating. Observations were also used to determine 'eaters' i.e. piglets scored eating at least once. Piglets were weighed at d4, 26 and 28 (at weaning). Data were analysed in mixed models with a random pen effect. No interactions between feed variety and feed presentation were found. SUB-piglets tended to spent more time on exploration in the feeder at d11 (0.5 vs 0.3% of time) and tended to eat more often at d18 than CON-piglets (4.2 vs 3.2% of time; P<0.10). Within SUB, piglets preferred to visit the feeder with sand to explore, as they spent at least two times more on exploration, but not eating, in this feeder at all observation days (P<0.05) versus the feeder without sand. DV-piglets spent at least two and a half times more time on feed exploration and eating than MO-piglets throughout lactation (P<0.0009 at all observation days). Eating the common creep feed, however, was seen two times more in MO than in DV-piglets at d18 (P<0.01) and d27 (P<0.05). DV enhanced the percentage of eaters/litter (d11: 71%, d18: 95%, d27: 99%) at all observation days (P<0.05) compared to MO (d11: 45%, d18: 71%, d27: 81%). DV-piglets also tended to grow faster in the last two days prior to weaning compared to MO-piglets (28 g/d, P<0.10). In conclusion, a diverse feeding regime for suckling piglets stimulated feed exploration, eating and the % of eaters from an early age onwards, and enhanced their growth towards weaning. Feed presentation in a foraging-stimulating context only subtly stimulated exploratory behaviour and eating. As such, the data suggests that dietary diversity is a promising feeding strategy in getting piglets to eat during lactation and may therefore benefit welfare and performance around weaning, while provision of substrate in the feeder stimulates natural foraging behaviour.

## Effect of flavour variety on productive performance of fattening pigs

*Jaime Figueroa and Daniela Frias*
*Pontificia Universidad Católica de Chile, Ciencias Animales, Av. Vicuña Mackena 4860, 7820436 Santiago, Chile; figueroa.jaime@uc.cl*

Diets formulated in pig production include several ingredients that satisfy animals' needs according to their specific requirements. Nevertheless, sensory proprieties of diets are very similar during the life of the animal and sensory specific satiety (SSS) could negatively affect pigs' intake and welfare. The hypothesis of the present experiment was that flavour variety between meals would increase fattening pig consumption and productive parameters. A total of 60 male and female pigs (84 days old, 39.2±4.0 kg) were weighed, individually identified by plastic ear tags and allocated to 12 pens (5 pigs/pen). Animals were fed a commercial unflavoured diet during 24 days (Control Group; 6 pens) or with the same diet with a daily inclusion of artificial flavours (Flavour Group; 6 pens). Lemon, coffee and cherry flavours (1.5 g/kg; Floramatic, Santiago, Chile) were rotated daily. Also the unflavoured diet was included for the diet rotation, reaching a total of 6 repetitions of each diet during the complete experimental period. Animals and feed waste were weighed at days 3,10,17 and 24 of the trial to estimate pig average daily gain (ADG), feed intake (ADFI) and feed to gain ratio (F:G). Data were analysed using the GLM procedure of the statistical software SAS (SAS Inst. Inc., Cary, NC). No differences were found between groups in relation to feed intake (P>0.4) or body weight (P>0.3) in any of the periods analysed. In relation to ADG, the Control Group tended to present a higher ADG than the Flavour Group (848 vs 592 g, SEM 104.6; P=0.089) in the first period analysed (0-3 days). However, during the second period (3-10 days), the Flavour Group presented a higher ADG (1,058 vs 1,297 g, SEM 40.3; P<0.001). No differences were found in the other periods analysed in relation to ADG. Finally, F:G ratio presented differences between groups only during the second period analysed (3-10 days), where Flavour Group presented a better feed conversion than the Control Group (2.4 vs 1.9, SEM 0.58; P=0.004). Flavour neophobia could have prevented the reduction of SSS during the first days of exposure, resulting in higher ADG in animals that continued with the unflavoured commercial diet. However, variety in the flavours given daily to fattening pigs could have increased their performance when the animals had experience with those flavours. Research is needed to determine if this effect could be maintained through the complete growing period.

## Influence of satiety on the motivation of stall-housed gestating sows to exit their stall

*Mariia Tokareva[1], Jennifer Brown[2], Alexa Woodward[1], Edmond Pajor[3] and Yolande Seddon[1]*
*[1]University of Saskatchewan, Western College of Veterinary Medicine, 52 Campus Drive, S7N 5B4, Saskatoon, SK, Canada, [2]Prairie Swine Centre Inc., 2105 8th Street East, S7H 5N9, Saskatoon, SK, Canada, [3]University of Calgary, 3330 Hospital Drive NW, T2N 4N1, Calgary, AB, Canada; mariia.tokareva@usask.ca*

The Canadian Code of Practice for the Care and Handling of Pigs permits the operation of existing stall barns after July 2024, provided that bred female pigs are given opportunities for greater freedom of movement. Previous research identified that stall-housed female pigs are motivated to leave the stall; with gilts equally motivated to leave the stall as to access a feed reward, whereas sows' motivation to exit the stall was less than that for a feed reward. Considering this, the objective of the present study was to determine whether the motivation of stall-housed sows to exit their stall is influenced by their satiety level. Stall-housed sows (n=42, parity 2-5) were studied in seven replicates (6 sows/rep), starting one week post breeding. Sow ad-libitum (ad-lib) high fibre feed (HF – soaked timothy cubes) intake levels were determined three days prior to training. Sows were assigned to one of three treatments: control (C), fed a standard gestation ration; moderately satiated (0.5 HF), receiving 50% of their ad-lib HF intake once per day additional to their standard ration, three hours prior to testing; and fully satiated (ad-lib HF), given unlimited access to HF additional to their ration. Sows were trained to use an operant panel containing two buttons: (1) active button (AB – push counts resulted in three minutes of free movement in alleyway); (2) dummy button (DB – push counts not rewarded). Animals were tested on an ascending fixed ratio (FR), with the required number of AB presses during a 30-minute testing session increasing by 50% each consecutive day. If a sow failed to reach the required FR within a daily testing period, no reward was given and the animal received a second opportunity to reach the required FR the following day. If a sow failed to reach the required FR for a second day, testing ended and the sow was off trial. Data were analysed by GLM (SAS 9.4) to determine the influence of treatment on the highest price paid (HPP, maximum AB push counts in one session) and latency to make the first AB push, in the session that the HPP was reached. Control sows showed a greater HPP than ad-lib HF sows, with sows fed 0.5 HF being intermediate (C: 94.43±18.21, 0.5 HF: 66.43±15.88, ad-lib HF: 59.14±23.65, F=3.38, P=0.044, mean HPP±SEM). The latency to press the AB was not influenced by treatment. Results suggest that stall-housed sows show a level of motivation to exit their stall for greater freedom of movement, and this motivation is influenced by their level of satiety. This also implies that provision of additional HF feed may be an option to increase the welfare of stall-housed sows, if periodic provision of greater freedom of movement is not viable.

## Feeding behavior of Iberian pigs in a multi-feeder system under high environmental temperatures

*Angela Cristina De Oliveira[1], Salma Asmar[2], Norbert Batllori[2], Iamz Vera[2], Uriel Valencia[2], Leandro Costa[1] and Antoni Dalmau[2]*
*[1]Pontifícia Universidade Católica do Paraná (PUCPR), Curitiba, Paraná, Brazil, 80215-901, Brazil, [2]Institut de Recerca i Tecnologías Agroalimentàries (IRTA), Monells, Girona, Spain, 17121, Spain; batista.leandro@pucpr.br*

Pigs during warm conditions can reduce the amount of feed consumed. One possible action to reduce this impact is to provide more fat, what increases energy and reduce heat production. In addition, in Spain, where this study was performed, it is usual that Iberian pig are fed restricted from 50 to 110 kg. The aim of the present study was to check the effect of this feed restriction using two types of diets (less and more energetic by replacing a 5% of starch (ST) for 5% of sunflower oil (E)) on the competition for feed during meals in Iberian pigs from 50 to 110 kg subjected to high temperatures (from 25 to 32 °C). Thirty-six male pigs were housed in pens (6 pigs/pen) and randomly distributed according to the initial BW in three-controlled rooms under heat-stress conditions. Three pens were fed with the ST diet and three with E diet. In parallel, other 6 pens were studied under an ad libitum regime to assess consumptions. The food given to the pigs was 90% of the amount consumed for ad libitum in the previous week. During the observations, total time used for feeding and time until the first displacement from the feeder, position of the animals, or number of displacements were recorded. The pig's hierarchies were evaluated based on agonistic interactions (focal sampling), in which it was given +1 point when an animal won a social interaction, -1 when the animal lost and 0 if there was no clear winner or loser. This classification was used inside each pen to classify each pig from position 1 to position 6 in hierarchy. The statistical analyses were done using proc mixed and genmod of SAS. The procedure was approved by the ethical committee of IRTA and by the Spanish authorities according to the EU regulations. The six pigs had six different positions for eating. In mean values animals repeated the same feeder a 36% of the times, and this ranged from 80 to 22%. When checked in relation to the hierarchy an effect was found ($P<0.05$), the pigs considered the first in the hierarchy showing a mean repetition in the feeder of 41.8%, the second one 39.1%, the third one 36.6%, the fourth one 35.1%, the fifth one 33.3% and the sixth one (the most subordinate) 32.4%. In relation to the stops, it was observed that pigs fed with E diet had lower ($P<0.05$) time eating (1,278.2±7.60 seconds) than ST diet (1,350.0±7.37 seconds) and had the first stop as well earlier ($P<0.05$) in E diet (773.4±12.20 seconds) than in ST (855.8±seconds). No differences were found between treatments in number of displacements/disturbances during the meal (mean of 3.77 per meal and animal). In conclusion, although a more energetic diet can have advantages in terms of performance, a lower total time eating is not reducing the competition between animals, that will appear earlier, probably when the most faster pigs have arrived to a certain point of satiety or level of available food in the feeder. Pigs have preferred positions for eating and a sign of dominance could be the percentage of repetitions in a specific feeder.

## Effect of long-term feeding on home range size and colony growth of free-roaming cats at a popular tourist site in Japan

*Hajime Tanida[1], Aira Seo[1] and Yuki Koba[2]*
*[1]Hiroshima University, Graduate School of Integrated Sciences for Life, 1-4-4, Kagamiyama, Higashi-Hiroshima, 7398528, Japan, [2]Teikyo University of Science, Faculty of Education & Human Sciences, 2-2-1, Senjusakuragi, Adachi-ku, Tokyo, 1200045, Japan; htanida@hiroshima-u.ac.jp*

The presence of free-roaming cats in an urban setting largely depends on food given by volunteer feeders. The uptown area of old-town Onomichi city, Japan, a popular tourist site with many historical temples and shrines, has been recently recognized as a town of cats. However, the cats, who are fed constantly by volunteer feeders and tourists, have become a public nuisance owing to their spraying and defecating in alleys, gardens, and parks. Cat feces can contain bacteria, viruses, and parasites that infect humans and pets. It has been predicted that feeding not only encourages the cats to settle down in a small area but also promotes colony growth. Therefore, the objective of our study was to examine the effect of long-term feeding by volunteer feeders and tourists on the home range size and colony growth of free-roaming cats in the uptown area of Onomichi city, Japan. The uptown area (21.4 ha) is located on the hillside of Mt. Senkoji. Because most of the area is a pedestrian zone, the cats can move around safely without the dangers typically associated with traffic. There were 4 main feeding sites (FS). Route censuses were conducted to study the home range and behavior of cats fed at the FS for 3 years from May 2011 to April 2014. When the cats were found on a route of 2.61 km, which covers most of the uptown area including the 4 FS, they were photographed, and their information was recorded on a 'cat identification card.' In total, 36 cats, 19 of whom were neutered, were identified at 4 FS in 3 years. At the beginning of the 1st year, the number of cats fed at FS1, FS2, FS3, and FS4 were 2, 3, 3, and 1, respectively. Regular feeders were found only at FS2, where a couple fed the cats once a day. Feeders at other FS were mainly tourists on a day trip who irregularly gave the cats a variety of foods including human foods. The number of cats fed at FS1, FS2, and FS3 increased significantly ($P<0.05$: Kruskal-Wallis test) to 4, 10 and 5 respectively in the 2nd year. In the 3rd year, however, the cats at FS1 and FS4 decreased ($P<0.01$: Kruskal-Wallis test) to 1 and 0, respectively. Only at FS2, the number of cats increased significantly ($P<0.01$: Kruskal-Wallis test) to a total of 15 free-roaming cats who were fed at the end of the 3rd year. The home range of the cats fed at FS2 during the 3rd year was significantly ($P<0.05$: Mann-Whitney's U test) lower than that during the 2nd year. This study reveals the possible effects of consistent feeding by residents versus sporadic feeding by occasional tourists or passers-by. In conclusion, our observations suggest that in this urban area, regular feedings at a feeding site increased the number of free-roaming cats that settled down at the site, whereas cats at a feeding site frequented by irregular feeders may disappear within a few years.

## Domestic hens: affective impact of long-term preferred and non-preferred living conditions

*Elizabeth Paul[1], Anna Trevarthen[1], Gina Caplen[1], Suzanne Held[1], Mike Mendl[1], William Browne[1] and Christine Nicol[2]*
[1]*University of Bristol, Veterinary Science, Langford House, Langford, Bristol, BS40 5DU, United Kingdom,* [2]*University of London, Royal Veterinary College, Hawkshead Lane, Hatfield, Herts, AL9 7TA, United Kingdom; e.paul@bristol.ac.uk*

While many studies have investigated domestic chickens' relative preferences for a range of living conditions, including features of pens, nest boxes, floors, stocking density, lighting and perches, little is known about the long-term impact of these on the birds' on-going affective states. In the present study, we designed three types of hen housing: Condition P comprised elements known to be preferred by chickens (floor space of 0.75 $m^2$ per hen, enclosed nest boxes, deep litter substrate, access to multiple perches, etc.); Condition NP comprised elements known to be non-preferred by chickens (floor space of 0.37 $m^2$ per bird, open nest boxes, wire flooring, single low-level perch, etc.); Condition I was intermediate between P and NP. Methods: Sixty point-of-lay hens (*Gallus gallus domesticus*) were housed in groups of 5 birds per pen, for 10 weeks in Condition I. During this time, 24 birds (2 per pen) were trained in a go/no-go judgement bias task. All birds were then moved to their long-term living condition for 24 weeks, in pens of either Condition P or NP. Behaviour tests to assess affect, including judgement bias tests, were administered to all judgement-bias trained birds at the end of the 10-week period during which they were in Condition I and again at the end of the 24-week period when they were in P or NP (n=24). Preference tests were also conducted at this time point: On seven occasions, birds were placed in a T-maze and given the choice to either return to the condition they had been living in for 24 weeks (P or NP) or to the condition they had lived in for the first ten weeks of the study (I). Results: When tested in the T-mazes, birds that had been kept in Condition P chose to return to that condition more often and more quickly, while those that had been kept in Condition NP chose to return less often and less quickly ($t_{(1,22)}$=2.50, P<0.05; $t_{=1,22)}$=-2.70, P<0.05). No significant effects of condition (P vs NP) were observed in the judgment bias tests: birds that had been living in Condition P (n=12) did not show a relative increase in the proportion of ambiguous probes pecked, and those living in Condition NP did not show a relative decrease. Instead, these birds showed consistent, trait like responses to probes (i.e. some birds remained 'optimists' while others remained 'pessimists' across the 24-week period). However, while birds experiencing P living conditions for the 24-week experimental period showed strong consistency across time (Pearson correlation between proportion of probes pecked after 10 weeks in Condition I and after 24 weeks in Condition P: r=0.781, P<0.005), hens experiencing the NP conditions did not (Pearson's r=0.491, n.s.). We conclude that judgement biases in hens can be robust across time; that demonstrating a preference does not necessarily coincide with such measures; and that living in less (as opposed to more) preferred conditions may reduce affective trait stability.

## Exploring attentional bias towards emotional faces in chimpanzees using the dot probe task

Duncan Andrew Wilson and Masaki Tomonaga

Primate Research Institute, Kyoto University, Language and Intelligence Section, Department of Cognitive Science, 41-2 Kanrin, Inuyama, Aichi, 484-8506, Japan; wilson.duncan.7a@kyoto-u.ac.jp

Recently, cognitive measures from human psychological research have been used to assess emotional states and welfare in animals. The dot-probe task has been used extensively to examine the relationship between attentional bias and emotion in humans and is especially suitable for use with non-human primates, as it requires no instruction and minimal training. This was the first study to investigate whether the dot-probe task can measure attentional biases towards emotional faces in chimpanzees. Eight adult chimpanzees at the Primate Research Institute, Kyoto University, voluntarily participated in a series of touchscreen dot-probe tasks. To begin each trial the chimpanzees touched a fixation cue in the center of the screen. This was followed by the simultaneous presentation of two stimuli for 150 ms. In the main experiment, conspecific threatening faces (12 scream faces or bared-teeth faces) were paired with non-threatening faces (12 neutral faces or scrambled faces). The stimuli and fixation cue then disappeared and a black dot (probe) appeared randomly in place of either the threatening face (congruent trials) or the non-threatening face (incongruent trials). Response times (ms) to touch the dot were recorded. It is assumed that if attention is biased towards one stimulus type (threatening faces or non-threatening faces) response times to detect the dot located in the same spatial location as that stimulus type will be relatively faster. A total of 48 trials ×12 sessions were completed. We predicted faster response times to touch the dot appearing after threatening faces relative to non-threatening faces. Generalized Linear Mixed Models were used to analyse the relationship between response times and congruency for each stimuli pair comparison. Fixed effects were congruency and stimuli pair and random effects were chimpanzee and session. Response times were significantly faster to touch the dots replacing bared-teeth faces (congruent, $M=418$ ms) than scrambled bared-teeth faces (incongruent, $M=430$ ms), ($SE=0.01$, $Z=2.34$, $P=0.019$) and scream faces (congruent, $M=420$ ms) than scrambled scream faces (incongruent, $M=440$ ms), ($SE=0.01$, $Z=3.28$, $P=0.001$). However, no significant difference in response times was found to touch the dots replacing bared-teeth faces (congruent, $M=428$ ms) versus neutral faces (incongruent, $M=433$ ms), ($SE=0.01$, $Z=0.61$, $P=0.544$), or scream faces (congruent, $M=421$ ms) versus neutral faces (incongruent, $M=432$ ms), ($SE=0.01$, $Z=1.83$, $P=0.067$). Therefore, we found no evidence that the touchscreen dot-probe task can measure attentional biases specifically towards threatening faces in our chimpanzees. Methodological limitations of using the task to measure emotional attention in non-human primates including emotional state, stimulus threat intensity, stimulus presentation duration and manual responding, as well as procedural improvements will be discussed.

## An automated and self-initiated judgement bias task based on natural investigative behaviour

*Michael Mendl[1], Samantha Jones[1], Vikki Neville[1], Laura Higgs[1], Emma Robinson[2], Peter Dayan[3] and Elizabeth Paul[1]*
*[1]University of Bristol, Bristol Veterinary School, Langford House, Langford, BS40 5DU, United Kingdom, [2]University of Bristol, School of Physiology, Pharmacology and Neuroscience, Biomedical Sciences Building, Bristol, BS8 1TD, United Kingdom, [3]Max Planck Institute for Biological Cybernetics, Department of Computational Neuroscience, Max Planck Ring 8-14, 72076 Tubingen, Germany; mike.mendl@bris.ac.uk*

Scientific assessment of affective valence (positivity or negativity) in animals allows us to evaluate animal welfare and the effectiveness of 3Rs Refinements designed to improve wellbeing. Judgement bias tasks measure valence; however, task-training may be lengthy and/or require significant input from researchers. Here we develop an automated and self-initiated judgement bias task for rats which capitalises on their natural investigative behaviour. Rats insert their noses into a food trough recess to start trials. They then hear a tone (2 or 8 kHz) and learn either to 'stay' for 2 s to receive a food reward or to 'leave' the trough recess promptly to avoid an air-puff. Which contingency applies is signalled by two different tones. Judgement bias is measured by responses to intermediate ambiguous tones. We carried out two experiments to investigate this new task. In Experiment 1, 36 of 40 (90%) rats reached training criterion on the tone-discrimination task in a mean of 23.1 (sem: 1.14) sessions. Half the rats were partially-reinforced during training and they were more likely to 'stay' during ambiguous and negative tones than rats that were fully-reinforced (Likelihood-ratio test (LRT) of effect of removing predictor variable from model: Chi-square=17.71, df=1, P=0.001). When exposed to prior short-term positive affect manipulations (15 min gentle handling; enrichment), rats tended to show more 'stay' responses (LRT Chi-square=3.28, df=1, P=0.07) than when they were exposed to negative ones (15 min small box; isolation). In Experiment 2, all rats were partially-reinforced during training, and 11 of 12 (92%) rats reached criterion in 17.5 (sem: 0.65) sessions. Rats exposed to a prior short-term positive affect manipulation (16 food rewards in 15 min) tended to make more 'stay' responses (LRT Chi-square=3.75, df=1, P=0.053) than those exposed to a relatively negative one (1 food reward in 15 min). This task capitalises on natural investigative behaviour, can be learnt in fewer sessions than other automated variants, generates 4-5 self-initiated trials/min, yields generalised responses across ambiguous tones as expected, and can be tested repeatedly. Affect manipulations generate main effect trends in the predicted directions, albeit not quite significant at P<0.05, and not localised to ambiguous tones perhaps indicating that affect manipulations altered food and/or air-puff valuation which influenced responses to all tones. Further construct validation is thus required. We also find that reinforcement contingencies during training can affect responses to ambiguity. The task is likely to be readily translatable to other species and should facilitate more widespread uptake of judgement bias testing.

## Investigating animal affect and welfare using computational modelling

*Vikki Neville[1], Liz Paul[1], Peter Dayan[2], Iain Gilchrist[3] and Mike Mendl[1]*
*[1]University of Bristol, Bristol Veterinary School, Langford, BS40 5DU, United Kingdom, [2]Max Planck Institute for Biological Cybernetics, Tübingen, 72076, Germany, [3]University of Bristol, School of Psychological Science, Bristol, BS8 1TU, United Kingdom; vn15961@bristol.ac.uk*

Behaviour associated with poor welfare, such as 'pessimistic' decision-making, can arise from several different affect-induced shifts in cognitive function. For example, risk aversion can arise from an altered sensitivity to, or expectation of, rewards or punishers, and these processes can themselves be influenced by several environmental factors. By characterising the cognitive processes that generate behaviour, we can gain a better insight into the relationship between specific forms of adversity and indicators of welfare such as judgement bias. We aimed to use computational modelling to extract parameters relating to different aspects of cognitive processing from judgement bias decision-making data and to assess how these were influenced by reward experience, following the prediction that enhanced reward experience generates a positive affective state. To achieve this, we used an automated and self-initiated judgement bias task in which rats had to choose between a risky option which resulted in either an air-puff or apple juice, and a safe option which provided nothing. More specifically, rats initiated each trial by putting their nose in a trough which resulted in the immediate presentation of a tone, the frequency of which provided clear or ambiguous information about the potential outcome. Rats then either stayed in the trough for 2 s ('stay'=risky option) or removed their nose ('leave'=safe option). We manipulated reward experience by systematically varying the volume of juice in a sinusoidal manner (mean=1 ml, SD=0.3 ml). Rats were not water or food restricted as part of these studies And all rats were rehomed as pets at the end of the study. These experiments adhered to the ISAE and ASAB/ABS guidelines for the ethical use of animals in research. Following data collection, we modelled decision-making on the task (binary variable: 'stay' or 'leave') as a partially-observable Markov decision process with a two-dimensional state space describing each rat's perception of the tone and time left to make a decision. The model provided a good fit of the data (RMSEA=0.028). The computational analysis revealed that variation in risk aversion could be attributed to changes in prior beliefs about the likelihood of reward which was modulated by what an individual had learnt from previous outcomes in the test environment. Specifically, an individual's expectation that the trial would be rewarded prior to presentation of the tone was greater when they had learnt that they were in a high reward environment, assumed to generate positive affect, resulting in more 'optimistic' decision-making (dAIC=4.979, P<0.001). As such, these models inform our understanding of the relationship between the environment, affect, and decision-making. The parameters obtained using this approach may provide a more precise measure of welfare than the decision itself and hence provide a better estimate of the affective impact of poor or improved husbandry. Computational modelling can be a useful tool in the study of animal welfare.

## Effect of early life and current environmental enrichment and personality on attention bias in pigs

*Lu Luo[1], Inonge Reimert[1], Elske De Haas[2], Bas Kemp[1] and Elizabeth Bolhuis[1]*
[1]*Wageningen University & Research, Department of Animal Sciences, De Elst 1, 6708 WD, Wageningen, the Netherlands,* [2]*Utrecht University, Department of Animals in Science and Society, Yalelaan 1, 3584 CL, Utrecht, the Netherlands; lu.luo@wur.nl*

Animals may show increased attention towards threatening stimuli when they are in a negative affective state, i.e. attention bias. A barren, stimulus-poor housing environment can induce stress and potentially a negative mood in pigs. Apart from current housing conditions, however, also the early life environment and personality characteristics might influence affective state. In this study, we aimed to investigate the effects of early life and current housing conditions and personality (coping style) on attention bias in pigs. Pigs (n=128, 32 pens) housed in barren or enriched housing in early life (B1 vs E1), experienced either a switch in housing conditions at 7 wks of age or not (creating B1B2, B1E2, E1E2 and E1B2 treatments). They were classified using a backtest as 'high resister' (HR, proactive coping style) or 'low resister' (LR, reactive coping style) at 2 wks. Pigs were subjected to a 3-min attention bias test at 11 wks of age. Half of the pigs were exposed to a 10-sec potential threat (T) and the other half not (C) in a test room with food in the centre. Attention towards the (location of the) threat, vigilance, eating and vocalisations were recorded. Firstly, behaviours of T and C pigs over the test were compared. Secondly, for T pigs, effects of early life and current housing, coping style and their interactions on behaviour during and for 150 sec after the threat were tested. Mixed models with random pen effects were used, except for squealing for which a Fisher's exact test was used. T pigs spent more time on vigilance behaviour (T: 13.6±1.4, C: 6.8±1.0%, P<0.001), less time on eating (T: 15.0±1.8, C: 27.8±2.4%, P<0.001), were more likely to squeal (T: 22% C: 6% of pigs, P<0.05) than C pigs, and paid more attention to the location of the threat (T: 7.1±0.6, C: 0.5±0.1% of time, P<0.001) throughout the 3-min test, indicating that pigs did respond to the threat. During presence of the threat, HR pigs showed more vigilance (P<0.05), particularly in E2 housing (E2-HR: 39.9±6.6, E2-LR: 6.7±2.9, B2-HR: 19.4±5.9, and B2-LR: 12.1±4.4, interaction P<0.05). E1-HR pigs (55.4±6.5%) tended to pay more attention to the threat than E1-LR pigs (30.3±5.9%), with levels of B1-HR (46.4±6.8%) and B1-LR (48.3±7.6%) in between (interaction P<0.10). After presence of the threat, no effects of housing or coping style on vigilance, attention to location of the threat or eating were found. E2 pigs grunted more often than B2 pigs (9.6±1.7 vs 3.6±0.9 per min, P<0.01). E2 pigs were also more likely to squeal than B2 pigs (P<0.05), particularly the HR pigs (E2-HR: 50%, B2-HR: 0%, E2-LR: 21%, B2-LR: 17%, interaction P<0.10). In conclusion, housing affected vigilance in a personality-dependent manner during a short period of exposure to a potential threat. We found no strong effect of early life or current housing on attention bias after the threat, but current housing conditions and personality did affect vocalisations.

## Half full or half empty – comparing affective states of dairy calves

*Katarína Bučková[1,2], Sara Hintze[3], Marek Špinka[1,2], Zuzana Andrejchová[1], Radka Šárová[2] and Ágnes Moravcsíková[1]*
*[1]Czech University of Life Sciences, Ethology and Companion Animal Science, Kamýcká 129, 16500 Praha, Suchdol, Czech Republic, [2]Institute of Animal Science Prague, Ethology, Přátelství 815, 10400 Praha, Uhříněves, Czech Republic, [3]University of Natural Resources and Life Sciences, Sustainable Agricultural Systems, Gregor-Mendel-Strasse 33, 1180 Vienna, Austria; katarinabuckova1@gmail.com*

Most dairy calves are reared individually in their early ontogenesis in spite of the fact that such housing impairs their welfare. On the one hand, there is an increasing focus on alternative housing systems of calves, but on the other hand, individual housing is seen as more practical by many farmers than housing with the dam or conspecifics. A good compromise to meet demands of welfare and practice could be rearing calves in pairs. Studies comparing pair with individual housing found some indications of improved welfare, e.g. calves in pairs played more or coped better with stress after weaning. However, as calves naturally form small groups within a herd, it is not clear whether pair housing is sufficient to induce a more positive long-term affective state compared to individual housing. Therefore, we aimed to assess long-term affective states in 10 individually and 10 pair housed female calves through a Go/No-Go judgement bias task with active trial initiation. First, calves were shaped to touch a target and then to walk 3 m to receive a milk reward from a teat bucket at a positive location (left or right, balanced for housing while consistent for each individual) in an experimental arena. Calves were trained on a spatial Go/No-go task in which the location of the teat bucket signalled either reward (positive trial) or non-reward (negative trial). To fulfil the discrimination criterion, calves needed to show at least 13 go-responses (= reaching the bucket presented at the positive location) out of 16 positive trials and 13 no-go responses (= not reaching the bucket at the negative location) out of 16 negative trials in two consecutive sessions. In the following 4 test sessions, 3 ambiguous trials were interspersed between the 16 positive and 16 negative trials, with the bucket being presented in 3 locations in-between the positive and negative locations. Data were analysed using generalised mixed-effects models with treatment (individual (IND) and pair housing (PAIR)) included as fixed effect and trial type nested in session nested in calf as random effect. We found that IND and PAIR housed calves did not differ either in their reactions to the positive stimuli (percentage of Go, mean ± SE, IND: 99.2±0.4%, PAIR: 97.6±1.5%) or in their reactions to the negative stimuli (IND: 9.5±2.5%, PAIR: 13.9±2.6%), thus documenting that calves from both housing conditions had the same motivation for milk and had learned the task equally well. Nevertheless, PAIR calves showed more go-responses towards the ambiguous locations than IND calves ($x^2_1$=6.76, P=0.009), indicating that they judged the ambiguous cues more optimistically and were thus in a more positive affective state. We conclude that pair housing is sufficient to improve long-term affective state in calves compared to individual housing.

## Brain size and cognitive abilities in Red Junglefowl selected for divergent levels of fear of humans

*Rebecca Katajamaa and Per Jensen*
*Linköping University, IFM Biology, Linköping University, 58183, Sweden; rebecca.katajamaa@liu.se*

Tameness, or low fear of humans, could be a major driving factor in the development of the domesticated phenotype, a coherent set of traits that are common to domesticated species. We selected the ancestor of all domestic chickens, the Red Junglefowl, over ten generations for divergent levels of fear of humans, using a standardised behaviour test. In previous generations, birds from the low fear line were larger, socially dominant and had altered brain size proportions in adult birds. Here, we used birds from the ninth and tenth selected generation to investigate if changes in brain size were apparent already in juvenile birds, and tested their cognitive skills in a conditioned place preference test and in a fear extinction test. The brains of 18 birds from each line at age 5 weeks were dissected, divided into different brain regions and weighed. Brain size relative to body weight was smaller in the low fear line (Generalized linear models; $\chi^2$=55.3, P<0.001). The mesencephalon was smaller relative to brain mass in the low fear line ($\chi^2$=6.0, P<0.05) whereas there were no differences between the lines for the other brain regions. Fear extinction was tested in 20 birds from each line 8 days after hatch. The test consisted of a conditioning phase on day one where a blue flash of light was paired with an air puff and a test on day two where the behavioural response to 10 consecutive blue light flashes was recorded. A control group of 10 birds from each line went through only the test phase. There was a significant effect of line in the group that went through the conditioning phase on day one. Birds from the low fear line were less reactive on the test day compared to the conditioning day ($\chi^2$=40.0, P<0.001). This effect was not seen in the control group ($\chi^2$=0.510, P>0.05). There was a significant decline in the intensity of the behavioural response over the 10 light flashes in both conditioned and control group ($\chi^2$=83.1, P<0.001 and $\chi^2$=29.7, P<0.001). The conditioned place preference test was done on 15 high fear birds and 16 low fear birds 8 days after hatch. On the first day of the procedure, birds were conditioned on a mealworm reward in one of two patterned compartments. The conditioning was done with a 30 second exposure to the conditioned compartment, followed by a 30 second exposure to the unconditioned compartment. This was then repeated four times in one single session. On the second day, the birds were allowed to freely explore both compartments of the arena. We measured compartment preference as the total time the bird spent in each compartment. No differences were found in compartment preference between the lines ($\chi^2$=0.713, P>0.05). Our results show that selection for reduced fear of humans, mimicking early chicken domestication, produces correlated changes in brain size as well as cognitive abilities. This supports the idea that tameness may to some extent have driven the diverse suite of phenotypes associated with domestication.

## Thermal comfort in horses: their use of shelter and preference for wearing rugs

*Cecilie M. Mejdell[1], Grete H.M. Jørgensen[2] and Knut E. Bøe[3]*
*[1]Norwegian Veterinary Institute, Section for Animal Health and Welfare, P.O. Box 750 Sentrum, 0106 Oslo, Norway, [2]Norwegian Institute of Bioeconomy Research, Division of Food Production and Society, P.O. Box 34, 8860 Tjøtta, Norway, [3]Norwegian University of Life Sciences, Department of Animal and Aquacultural Sciences, P.O. Box 5003, 1432 Ås, Norway; cecilie.mejdell@vetinst.no*

Horse owners are uncertain about how to care for their horse in winter. Thus, we have studied thermoregulatory behaviour in horses under various weather conditions in Norway, with emphasis on the cold season. We have investigated the preferences for using man-made shelters and for wearing rugs. Our first study followed a group of ~40 young (2-4 yrs) Icelandic horses kept outdoors 24/7 with free access to a non-insulated shelter bedded with straw. Ambient temperatures ranged from -31 to + 7 °C, and wind speed from 0 to 15 ms-1. Horses made more use of the shelter as temperature dropped (P<0.01), and during rain and wind. They were lying down almost exclusively indoors, and lying position was not influenced by temperature. Most horses in Norway are stabled at night and are turned out in a paddock during daytime. In the second study, we provided horses access to two shelter compartments in their outdoor paddock; one with an infrared heater turned on and one without. The compartments with heating was varied. Horses of different breed types were included. They were tested one at a time and on several test days. The horse's location (outside, inside heated or unheated compartment) and behaviour were recorded. In the first experiment, we included horses which did not wear a rug (n=22), and in the second we tested horses which did have a rug on (n=18). Ambient temperatures on test days ranged from -8.7 to +11.0 °C and wind speed from 0.5 to 22.8 ms-1. Among horses which did not have a rug on, small warmblood horses spent less time outdoors (34%) compared to small coldbloods (80%) (P=0.01). Shelters were used more on days with precipitation and wind, and there was a shift from using the unheated compartment towards the heated compartment as weather changed from dry and calm to wet and windy. Hair coat quality had significant effect (P=0.004) on shelter use. Wearing a rug reduced the effect of inclement weather, but horses still made use of the shelter and especially so during rain and wind. Finally, we trained 23 horses of different breeds to touch symbol boards in order to communicate their preference for having, or not having, a rug on. During different weather conditions, and after being outside for 2 h, horses were given the choice between changing the rug status or to stay unchanged (n=23. Ambient temperatures at test days ranged from -16 to +23 °C and wind speed from 0 to 14 ms-1. Again, chilly rain combined with wind turned out to be the most challenging weather type causing all horses to ask for a rug. Very cold weather (<-10 °C) also made horses ask for a rug. Very few horses preferred to keep a rug on in dry, calm weather above +10 °C. In all studies, we found individual differences regarding how much time they spent in the shelter, or at which temperatures they preferred to wear a rug.

## Can horses watch and learn – the good, the bad, and the not so ugly evidence of social transmission in horses

*Maria Vilain Rørvang[1], Janne Winther Christensen[2], Jan Ladewig[3] and Andrew McLean[4]*
*[1]Swedish University of Agricultural Sciences, Biosystems and Technology, Sustainable Animal Systems, Sundsvägen 16, Box 103, 23053 Alnarp, Sweden, [2]Aarhus University, Animal Science, Blichers Allé 20, 8830 Tjele, Denmark, [3]University of Copenhagen, Veterinary and Animal Sciences, Grønnegårdsvej 8, 1870 Frederiksberg C, Denmark, [4]Equitation Science International, 3 Wonderland Ave, Victoria 3915, Australia; mariav.rorvang@slu.se*

There are several mechanisms through which the behaviour of animals can be altered from observing others. These mechanisms are fundamentally different in terms of cognitive complexity, and range from simple hard-wired processes (social facilitation) to social enhancement of individual learning (local/stimulus enhancement) to genuine learning by observation (imitation). As group living animals, horses are often assumed to be able to learn new behaviour from watching other horses. Studies on social transmission of information in horses include a variety of studies, some of which may miscalculate their mental abilities. This is likely to be due to the confounding effect of associative learning mechanisms when interpreting results. Assuming higher mental abilities in their absence can have detrimental welfare implications, e.g. when isolating stereotypical horses on the assumption that these behaviours can be learned though observation by neighbouring horses. In this literature review we: (1) critically reassessed the various mechanisms through which the behaviour of animals can be altered from observing others; (2) evaluated the biological basis for social learning to be adaptive in horses; and (3) reviewed the current body of literature on social transmission of behaviour in horses with the aim of clarifying whether horses possess the ability to learn through true social learning; and lastly (4) elucidated the practical potential for using this knowledge when handling horses. More than 70 studies on social transmission of information in various animal species were reviewed, including 21 studies specifically concerning horses. Studies of social transmission in other domesticated farm animal species (e.g. cattle and pigs), were included in order to clarify interspecies similarities and dissimilarities. The review suggests a novel differentiation between simple social transmission processes (social facilitation, local, and stimulus enhancement) and true social learning (goal emulation, imitation). Such differentiation is essential in order to avoid mistaking higher mental abilities for mere associative learning mechanisms. The review further highlights that social learning mechanisms are costly, and may have been an unaffordable luxury in horse evolution. Lastly, the review reveals that many studies on social learning in horses unfortunately lack proper control groups, and that many demonstrations are applied to the test horses making it questionable to argue that the horse learned from observing another horse rather than from local or stimulus enhancement cues combined with individual associative learning. Horses may be unable to utilize true social learning but appear sensitive to socially transmitted cues, even at a group level. This highlights opportunities for horse owners who may use calm horses to attenuate fearfulness in groups of young horses.

## Visual laterality in pigs and the emotional valence hypothesis

*Charlotte Goursot[1], Sandra Düpjan[1], Armin Tuchscherer[2], Birger Puppe[1,3] and Lisette M.C. Leliveld[1]*

[1]*Leibniz Institute for Farm Animal Biology, Institute of Behavioural Physiology, Wilhelm-Stahl-Allee 2, 18196, Germany,* [2]*Leibniz Institute for Farm Animal Biology, Institute of Genetics and Biometry, Wilhelm-Stahl-Allee 2, 18196, Germany,* [3]*Faculty of Agricultural and Environmental Sciences, University of Rostock, Behavioural Sciences, Justus-von-Liebig-Weg 6B, 18059, Germany; goursot.charlotte@gmail.com*

The observation of laterality –hemispheric asymmetries in structure and/or function– is a promising non-invasive approach to better understand animal affect since it can give insight into cerebral processes underlying the key components of emotions. The emotional valence hypothesis states that positive emotions are mostly processed by the left hemisphere, while negative emotions are mostly processed by the right hemisphere. Testing this hypothesis in farm animals could help to gain insight into their emotional valence which is relevant for animal welfare but challenging to assess. Whereas previous studies found evidence of opposite hemispheric dominances for perceiving food or a predator, a distinction between emotional arousal and valence is impossible when using such complex natural stimuli. The aim of this study was to test the emotional valence hypothesis in the context of visual laterality in pigs by using a paradigm of emotional conditioning. Ninety male piglets were either positively (using a food-reward, 45 subjects) or negatively (using a mild punishment, 45 subjects) conditioned to an object (a ball). Afterwards, the object was presented without the reinforcer. Each subject was tested 3 times under 3 different treatments: right or left eye covered with a patch causing a reduced visual input to the contralateral hemisphere; or binocular viewing as the control treatment for which a patch was fixed between the eyes. We measured the duration, frequency and latency of touching the object, latency to vocalise after the introduction of the object, duration of locomotion and of exploration in the arena and number of vocalisations. Using a heart rate measurement belt, we also analysed heart rate variability. All parameters were analysed by repeated measures analyses of variance and multiple pairwise Tukey-Kramer-tests. In the negatively conditioned group, covering the left eye caused increased vocalisation rates (t=-3.24; P=0.004) compared to seeing with both eyes. Due to the lack of other effects, this did not provide a clear support for the emotional valence hypothesis with regard to negative emotions. In contrast, in the positively conditioned group, covering the right eye (reduced input to the left hemisphere) caused longer latencies to touch the object (t=-2.96, P=0.010) and to vocalise after the introduction of the object (t=-3.24, P=0.004), a shorter duration of exploring the arena (t=2.71, P=0.020) and an increased vagal activity (t=-3.14, P=0.007) compared to binocular viewing. This suggests that reduced visual input to the left hemisphere possibly resulted in a reduced positive appraisal which supports the emotional valence hypothesis with regard to positive emotions. This study shows that investigating the lateralised processing of emotionally conditioned stimuli can provide insight into the mechanisms of positive appraisal in animals.

## Hemispheric specialisation for processing the communicative and emotional content of vocal signals in young pigs

*Lisette M.C. Leliveld[1], Sandra Düpjan[1], Armin Tuchscherer[2] and Birger Puppe[1,3]*
*[1]Leibniz Institute for Farm Animal Biology, Institute of Behavioural Physiology, Wilhelm-Stahl-Allee 2, 18196 Dummerstorf, Germany, [2]Leibniz Institute for Farm Animal Biology, Institute of Genetics and Biometry, Wilhelm-Stahl-Allee 2, 18196 Dummerstorf, Germany, [3]University of Rostock, Faculty of Agricultural and Environmental Sciences, Behavioural Sciences, Justus-von-Liebig-Weg 6B, 18059 Rostock, Germany; leliveld@fbn-dummerstorf.de*

Studying auditory lateralisation may help to better understand the mechanisms involved in vocal communication and emotional processing in farm animals. In humans the perception of speech is lateralised, with the left hemisphere dominant for processing the communicative content and the right hemisphere dominant for processing the emotional content. Whether other species show a similar division of tasks is still largely unknown. Therefore, we studied the lateralised processing of communicative and emotional relevance of pig calls. We used recordings of group-housed male pigs during 5 minutes of social isolation and during 5 minutes of physical restraint for the playback of calls of more communicative relevance and calls of more emotional relevance, respectively. To avoid pseudo-replication, we created 36 different playback stimuli (based on recordings from 36 subjects) per context, as well as of a non-biological control sound, and presented these through earphones to 72 other group-housed male pigs. The earphones were attached to the ears with adhesive plaster and did not limit the pig's behaviour. Thirty-six pigs herd the stimuli binaurally to test for biases in the first head turn during playback. The other 36 pigs herd the playback stimuli 3 times (binaurally, left ear only and right ear only). The behavioural (locomotion, lying, exploration, freezing, vocalizations and escape attempts) and cardiac response (heart rate and heart rate variability) were analysed and compared between the ear treatments, using generalised linear mixed model analyses. Multiple pairwise comparisons were made with the Tukey-Kramer test. We found a significant left head turn bias to restraint calls (binomial test: P=0.027, n=21). During playback of these calls to the left ear, the pigs also showed subtle increases in arousal, i.e. increased heart rate (t=3.01, P=0.010) and vocalisations (t=3.43, P=0.003) and decreased freezing (t=-2.55, P=0.032) compared to the minute before playback. Together these results suggest lateralized processing in the initial perception (head turn bias) and subsequent appraisal of calls of more emotional relevance. Conversely, we found no significant head turn bias in response to isolation calls, while we found indications of heightened attention during playback to the right ear, i.e. more increased freezing compared to the left ear (t=-2.38, P=0.018) and more decreased escape attempts compared to both ears (t=2.99, P=0.009). This suggests lateralized processing only in the appraisal of calls of more communicative relevance. These results imply different lateralized processing of the communicative and emotional relevance of pig calls. Studying lateralized processing may, therefore, provide insight into how information related to the welfare state of conspecifics is processed in pigs.

## Assessing the impact of lameness on the affective state of dairy cows using lateralisation testing

Sarah Kappel[1,2], Michael T. Mendl[1], David C. Barrett[1], Joanna C. Murrell[1] and Helen R. Whay[1]
[1]University of Bristol, Bristol Veterinary School, Langford House, Langford, BS40 5DU Bristol, United Kingdom, [2]University of Plymouth, School of Biological & Marine Science, University of Plymouth, Portland Square, Drake Circus, PL4 8AA Plymouth, United Kingdom; sarah.kappel@plymouth.ac.uk

Lameness is one of the major concerns in dairy farming as the condition is often associated with painful foot lesions. Similar to humans, cognitive functions in animals can be biased by underlying emotions and negative affect associated with pain can result in the pessimistic appraisal of ambiguous information. Moreover, it has been shown that in animals cerebral lateralisation pattern such as dominant right hemispheric processing during withdrawal-related (negative) emotional states are manifested in lateralised motoric expressions and sensory processing (e.g. left eye preference). For example, cows show a left eye preference for visual processing of novel stimuli. However, how the underlying cerebral mechanisms of emotional appraisal are influenced by negative affect (e.g. associated with lameness) are still unknown. This study tested whether lame cows differ in their sensory processing of novel visual stimuli compared to non-lame cows. A greater left eye bias in lame cows compared to non-lame cows was expected as a reflection of affect-induced negative appraisal of the unfamiliar objects. 186 lactating Holstein Friesian cows (3-6 years old) were individually presented with unfamiliar objects presented bilaterally (i.e. simultaneously in the left/right visual field) or centrally (i.e. binocular field) within a familiar area (each repeated on three consecutive days) and the cow's response regarding approach behaviour and choice of approach direction (left or right side) recorded. The relationship between cow behaviour and mobility score assessed before and after lateralisation testing was analysed using Pearson's Chi-square. Milk yield (high or low) but no other predispositions were considered. Although most cows had no lateralised viewing preference during approach, a bias to view the right rather than the left object was found for cows that showed a clear head turn at initial object presentation (binomial test, n=63, P<0.001). With the bilaterally placed balloons, there was a trend for cows appearing hesitant in approaching the objects by stopping at a distance to them, to then explore the left object rather than the right, suggesting a left-eye preference. In contrast, cows that approached the objects directly had a greater tendency to contact the right object ($\chi^2$=6.88 df=2, P=0.032). No significant lateralised preference in right or left eye use was found for the centrally-located object (P>0.05). Although no relationship between lameness and lateralised behaviour was found, observed trends suggest that lateralised behaviour in response to bilaterally located objects may reflect an immediate affective response to novelty. However, further study is needed to understand the impact of long-term affective states on hemispheric dominance and lateralised behaviour in cows.

## Development of a novel fear test in sheep – the startle response

*Hannah Salvin[1], Linda Cafe[1], Angela Lees[2], Stephen Morris[3] and Caroline Lee[2]*
*[1]Department of Primary Industries, Livestock Industries Centre, Armidale, NSW 2351, Australia, [2]CSIRO Agriculture and Food, Animal Behaviour and Welfare, FD McMaster Laboratory, Armidale, NSW 2350, Australia, [3]Department of Primary Industries, Wollongbar Primary Industries Institute, Wollongbar, NSW 2477, Australia; hannah.salvin@dpi.nsw.gov.au*

Fear is a negative emotion leading to chronic stress and reduced productivity, safety, and welfare. Current fear tests involve isolating the subject from conspecifics making it difficult to separate fear towards the test stimulus from the anxiety generated by social isolation. We therefore aimed to develop a fear test in sheep which identified variability in fear whilst addressing the confounding effects of social isolation. The startle response is a consistent universal fear response to sudden and intense stimulation, however, the magnitude of startle can be highly variable. In this study, 60 ewes were allocated into 3 social groups: isolated (I); surrounded by 6 life-sized images of sheep (P) or; 3 conspecifics penned on either side (total 6; L). Sheep were tested over 2 days in sets of 10 per treatment. The use of a tactile rather than an auditory or visual stimulus allowed a targeted effect limited to the test sheep, this and the rapid onset of startle helped limit any social facilitation of the startle response by the conspecifics. Linear models were used to describe variation in response to treatments. Significance of treatment effects was determined using F-ratio tests. We hypothesised that the presence of conspecifics would reduce anxiety in sheep during testing, whilst still allowing variation in fear response towards the startle stimulus. Sheep entered a 2.1×1.2 m test arena, with a bowl of feed at one end, and remained there for 4 minutes. Approximately 15 seconds(s) after the sheep commenced eating, a sudden blast of compressed air was delivered to the face. Tri-axial accelerometers (HOBO®, ±3 g) were attached to the foreleg and set to record at 100 Hz to determine startle magnitude. Video footage was used to determine startle behaviour as well as anxiety (eating and vigilance) behaviours for 15 s pre- and 120 s post-startle. Ethics approval was granted by CSIRO Chiswick AEC under ARA #18-10. One I sheep had a foot abscess and 4 sheep (2 I; 2 P) never ate so were therefore unable to have the startle stimulus applied and were excluded from analysis. One P sheep failed to record accelerometer data. As hypothesised, there was a significant treatment effect on pre-startle eating (P<0.001) and pre-startle vigilance (P<0.001) with I groups spending less time eating (mean ± SE; 37.4%±5.5 vs 71.2%±5.3 and 83.3%±5.1) and more time vigilant (38.9%±4.9 vs 18.2%±4.7 and 11.3%±4.5) than P and L groups respectively. A similar treatment pattern was seen in post-startle eating (P=0.087; I: 48%±5.9, P: 48%±5.5, L: 76.5%±5) and vigilance (P=0.088; I: 31.9%±4.4, P:34.2%±4.2, L: 12.3%±3.9). Mean startle magnitude did not differ significantly (P=0.22) between L (314.3 g±13.5), P (272.2 g±15.6) and I (241.9 g±18.2) groups. This startle test shows promise as a valid fear test which can utilise social companions to negate the confounding effects of isolation by allowing a more accurate assessment of pre- and post-stimulus behaviour.

## Pecking in Go/No-Go task: Motor impulsivity in feather pecking birds?

*Jennifer Heinsius[1], Nienke Van Staaveren[1], Isabelle Kwon[1], Angeli Li[1], Joergen B. Kjaer[2] and Alexandra Harlander[1]*
*[1]University of Guelph, Department of Animal Biosciences, N1G 2W1 Guelph, Canada, [2]Friedrich-Loeffler-Institut, Federal Research Institute for Animal Health, Institute of Animal Welfare and Animal Husbandry, 29223 Celle, Germany; aharland@uoguelph.ca*

Motor impulsivity in mammals is associated with a number of disruptive behavioural disorders including attention-deficit hyperactivity and hair pulling in humans, but whether these findings can be generalized to birds is undetermined. Repetitive feather pecking (FP) where birds peck and pull out feathers of conspecifics could reflect motor impulsivity through a lack of behavioural inhibition. We assessed motor impulsivity in female chickens (n=20) during a Go/No-Go task where birds had to peck a key when only a visual cue (Go) was presented or inhibit a peck when a visual cue paired with a signal tone (No-Go) was presented in an operant chamber as approved by the University of Guelph Animal Care Committee (AUP 3206). Birds came from a high FP line (HFP) or unselected control line (CON) and extreme phenotypes were selected based on behavioural observations during 20 min observation periods over three days. Birds with an average of >5 pecks directed to the feather cover of conspecifics were classified as peckers (P), while those with an average of <2.5 pecks were classified as low peckers (LP). Birds were trained to respond rapidly and accurately until they reached response stability in the Go/No-Go task (>75% success rate over two consecutive sessions). Following task acquisition, birds' ability to inhibit pecks was measured by recording the number of pre-cue responses and number of responses during Go cues (false alarms). Generalized linear mixed models were used to assess the effect of genotype, phenotype and its interaction on the response variables. The number of pre-cue responses (HFP: 2±0.4, CON: 3±0.6, P: 2±0.5, LP: 2±0.5) and number of false alarms (HFP: 0.8±0.3, CON: 0.9±0.3, P: 0.7±0.3 LP: 1.1±0.3) were not affected by birds' genotypic or phenotypic inclination to perform FP (P>0.05). This suggests that birds are capable of inhibiting their motor response, even those birds that perform a high level of FP. This indicates that the repetitive motor action of FP does not reflect impulsivity as measured by behavioural inhibition in the present Go/No-Go task.

## Pharmacological intervention of the reward system in the laying hen has an impact on anticipatory behaviour

*Peta Simone Taylor[1], Ben Wade[2], Meagan Craven[2], Adam Hamlin[3] and Tamsyn Crowley[2]*
*[1]University of New England, School of ERS, Faculty of SABL, University of New England, Armidale 2351, Australia, [2]Deakin University, Centre for Molecular Medicine Research, School of Medicine, Deakin University, Waurn Ponds 3216, Australia, [3]University of New England, School of Science and Technology, Faculty of SABL, University of New England, Armidale 2351, Australia; peta.taylor@une.edu.au*

Valid reliable indicators of positive affective state are a critical component of welfare assessments, however, these are currently lacking for laying hens. Pharmacological interventions that disrupt specific neural pathways have shown to be a useful tool when validating indicators of affective states. As such, we utilised and validated the μ-opioid receptor antagonist nalmefene to block the reward pathway in laying hens. Eighty Isa brown hens at 80-85 weeks of age were randomly allocated to either a control (C) or nalmefene (N) treatment group. Hens were dosed daily with 0.9% saline (C) or nalmefene dissolved in saline (N) at a dose rate of 0.4 mg/kg. Exactly 30 minutes after dosing, hens were presented with five live mealworms in a transparent closed food container at front of their cage. Hens could see the mealworms and could reach the container but could not access the mealworms for two minutes due to the closed lid. After two minutes, the lid was opened and hens were provided with access to the meal worms for five minutes. Latency to peck the container and number of pecks when the lid was closed were calculated as an indicator of anticipation of the mealworm (reward). Anticipation is an emotional state related to the seeking system that arises when the arrival of a reward (or punishment) is expected. As such, we hypothesised that if hens were not experiencing reward (due to the administration of Nalmefene) they would take longer to and peck less at the closed container. The anticipatory behaviour test was performed on day 1, 3 and 4 after dosing. More hens from the control group (40-100%) pecked the closed mealworm container and were quicker to do so than hens from the nalmefene treatment group (10-40%) on all days (day 1: C: 95.1±8.8 s, N: 108.1±8.3 s, $\chi^2_{(39,1)}$=4.3, P=0.038; day 3: C: 42.3±11.4 s, N: 110.1±7.1 s, $\chi^2_{(39,1)}$=17.15, P<0.001; day 4: C: 18.4±7.9 s, N: 104.1±8.7 s, $\chi^2_{(39,1)}$=29.8, P<0.001). The average number of pecks on the closed mealworm container increased over time by hens in the control group ($\chi^2_{(98,1)}$=7.7, P=0.005). Although treatment hens pecked the closed mealworm container slightly more over time, they pecked fewer times than the control hens at all time points ($\chi^2_{(98,1)}$=408.7, P<0.001). We provide evidence that nalmafene is an effective pharmacological intervention to block the positive affective reward state in laying hens.

# Can an attention bias test discriminate positive and negative emotional states in sheep?

Jessica E. Monk[1,2,3], Caroline Lee[1], Sue Belson[1], Ian G. Colditz[1] and Dana L.M. Campbell[1]

[1]CSIRO, Agriculture and Food, Armidale, NSW, 2350, Australia, [2]University of New England, School of Environmental and Rural Science, Armidale, NSW, 2350, Australia, [3]Sheep CRC, Armidale, NSW, 2350, Australia; jessica.monk@csiro.au

Emotional states are central to the assessment and improvement of animal welfare. Tests for attention biases can potentially allow for the rapid assessment of emotional states in livestock. An attention bias is where an animal attends to certain types of information more than others, which can depend on their emotional state. Based on evidence from the human literature and previous studies in sheep, we hypothesised that an attention bias test could discriminate between pharmacologically induced positive and negative emotional states in sheep, by measuring allocation of attention between threatening and positive stimuli, relative to control animals. Eighty 7-year-old Merino ewes were used in our study. The protocol and conduct of the study were approved by the CSIRO FD McMaster Laboratory Animal Ethics committee. To test our hypothesis, 4 groups of sheep received one of the following drugs 30 min prior to attention bias testing to temporarily induce contrasting emotional states; anxious (m-chlorophenylpiperazine), calm (diazepam), happy (morphine) and control (saline) ($n=20$ per treatment). Sheep were monitored at all times for adverse reactions to the drugs or signs of extreme distress during testing. The attention bias test was conducted in a 4×4.2 m arena with high opaque walls. A life-size photograph of a sheep was positioned on one wall of the arena (positive stimulus). A small window with a retractable opaque cover was positioned on the opposite wall, behind which a dog was standing quietly (threat). The dog was visible for 3 s after a single sheep entered the arena, then the window was covered and the dog was removed. Sheep then remained in the arena for 3 min while behaviors were recorded. The photograph was visible for the duration of testing. Some of the key behaviors included time looking towards the photo wall or the previous location of the dog (dog wall), duration of vigilance behavior and latency to sniff the photograph. In contrast with our hypothesis, no significant differences were found between treatment groups for duration of vigilance or looking behaviors, although anxious sheep tended to be more vigilant than control animals as expected (Analyses of Variance, $P<0.1$) and took significantly longer to sniff the photo than the other groups (survival analysis, $P<0.01$). Twenty-four of 80 animals were vigilant for the entire test duration, 11 of which were from the anxious group while only 2 were from the control group (Fishers Exact Test, $P=0.02$). This censoring of data may explain why no significant differences were detected between groups for vigilance behavior. Previous studies using the same attention bias test paradigm had not observed a high prevalence of censored data, however the current study used older sheep than had been used previously and tested ewes in an environment with a higher level of background noise. Due to these potential effects, it remains unclear whether the attention bias test can detect positive states in sheep.

## Judgement bias of group housed gestating sows predicted by personality traits, but not physical measures of welfare

*Kristina M. Horback[1] and Thomas D. Parsons[2]*
*[1]University of California, Davis, Department of Animal Science, One Shields Avenue, 95616 Davis, CA, USA, [2]University of Pennsylvania School of Veterinary Medicine, Clinical Studies-New Bolton Center, 382 West Street Road, 19348 Kennett Square, PA, USA; thd@vet.upenn.edu*

Judgment bias testing has emerged as a potential tool for assessing affective state in animals. Researchers infer an animal's affective state (from positive to negative) based on an animal's response to an ambiguous stimulus that is intermediate to both the rewarded and punished conditioned stimuli. Recent reports in multiple species question what factors influence performance in judgment bias testing. In order to better understand these issues in sows, 25 female swine (PIC 1050, Landrace-Yorkshire crossbreed) were observed at 5 weeks of age for response to human handling, and, one year later, for reaction toward a human handling her piglets 24 h after first farrowing. After weaning, each sow was recorded for 1 h during the introduction to a novel, large group pen, and evaluated for body condition score, lameness and skin lesions 3 days post-mixing. From the behavioral data, three personality traits were determined using PCA: aggressive, submissive, and, active/exploratory. To investigate the relationship between personality traits and affective state assessment, judgment bias testing was carried out during the second gestation period using a 'go/no-go' paradigm. Gestating sows were housed in the same dynamic group pen containing ~120 sows and fed via electronic sow feeding stations. The sow pen allowed for 2.4 $m^2$ per sow, and included a straw bedded area and outdoor access. Both positive and negative biases were observed despite all sows living in the same enriched housing. When approaching the ambiguous cue, 11 animals exhibited a latency <20 s (mean=9.0±1.2 s), or 'positive bias', whereas 9 animals took >50 s (mean=59.6±0.4 s), interpreted as a 'negative bias', and 4 animals lacked a clear bias by exhibiting latencies between 20 and 50 s. Linear mixed-effects and post hoc linear models were used to analyze the time to approach the stimulus as an outcome variable. Sows which were more aggressive approached the ambiguous (m=-13.4, $F_{(1,22)}$=11.49, P=0.003, n=24), but not the positive (m=-0.1, $F_{(1,22)}$=0.21, P=0.648, n=24), stimulus significantly faster than non-aggressive sows. There was no relationship between the other personality traits and the sows' judgment bias valence. Sows with physical welfare measures suggesting more 'positive welfare' (no lameness, and good body condition) did not exhibit significantly more positive judgement biases. Furthermore, the latency to the ambiguous cue also was not significantly related to an animal's skin lesion score (anterior: m=-0.5, $F_{(1,22)}$=1.54, P=0.699, n=24; side: m=-1.3, $F_{(1,22)}$=1.05, P=0.316, n=24; posterior: m=-0.3, $F_{(1,22)}$=0.03, P=0.861, n=24). Our data demonstrate that individual personality can impact a sow's performance in a judgement bias test. These results highlight the complexity of how emotional state and personality can shape an animal's perception and response to unknown environmental stimuli as well as the challenges of assessing affective states in animals.

## Pessimistic dairy calves are more vulnerable to pain-induced anhedonia

*Benjamin Lecorps, Emeline Nogues, Marina A.G. Von Keyserlingk and Daniel M. Weary*
*Faculty of Land and Food systems, University of British Columbia, Animal Welfare Program, 2357 Main Mall 248, Vancouver, BC, V6T 1Z4, Canada; benjamin.lecorps@gmail.com*

Few methods are available to assess the affective component of pain in animals. In humans and laboratory rodents, pain affects cognitive functions such as decision-making (i.e. judgment bias) and causes deficits in appreciation of rewards (i.e. anhedonia). Pain responses often vary between individuals, and this variation may partly be explained by personality differences. Little attention has been given to pain-induced changes in reward appreciation and individual differences in pain responses in farm animals. In this study, we first assessed dairy calves level of Pessimism using a judgment bias test. With approval from the UBC Animal Ethics Committee (A16-0310), animals were first trained to discriminate a positive and a negative cue (i.e. location) before being subjected to ambiguous ones (three intermediate locations). Tests were performed twice at 25 and 50 d old and the average latency to touch the ambiguous locations was used to assess individual differences in pessimism. We then tested whether calves (n=17) reduced their consumption of a sweet solution (as an indicator of anhedonia) over 5 days following hot-iron disbudding, a known painful procedure. Calves showed evidence of anhedonia on the first (Fisher-Pitman permutation test: Z=2.42, P=0.012) and second day (Z=2.54, P<0.001) after hot-iron disbudding. We also found that the more pessimistic calves expressed higher levels of pain-induced anhedonia on the first day (Spearman rank correlation test, $r_s$=-0.53, P=0.029). These results suggest that pain in farm animals can be assessed through changes on their perception of rewards, and that pessimistic individuals are more vulnerable to painful procedures.

## The effects of a composite chronic stress treatment on fear responses and attention bias in beef cattle

Miguel Somarriba[1,2], Wendy Lonis[3], Rainer Roehe[2], Alastair Macrae[1], Richard Dewhurst[2], Carol-Anne Duthie[2], Marie J. Haskell[2] and Simon P. Turner[2]
[1]University of Edinburgh, Royal (Dick) School of Veterinary Studies, Easter Bush Campus, EH25 9RG, United Kingdom, [2]Scotland's Rural College (SRUC), Animal and Veterinary Sciences Group, Roslin Institute Building, EH25 9RG, United Kingdom, [3]Agrocampus Ouest, 65 Rue de Saint-Brieuc, 35000 Rennes, France; miguel.somarribasoley@sruc.ac.uk

Although much research has studied the effects of acute stress in bovines, the effects of chronic stress on cattle behaviour has been poorly studied. Attention biases have been used to evidence negative affective states by quantifying attention to a novel threatening stimulus. Here we examined if a composite chronic stress treatment impacts the behaviour of beef cattle as measured through attention bias during a Fear Response Test. This study complied with UK Home Office regulations and ISAE ethical guidelines. Limousin (n=40) or Angus (n=40) cross steers were assigned to a chronic stress (S) or control treatment (C), each treatment with four replicate groups. After a baseline of 8 weeks at a floor space allowance of 8.72 m$^2$ per animal, S steers received a reduced space allowance (4.35 m$^2$ per animal) and were weekly subjected to regrouping, transport for 20 minutes and isolation for 10 minutes. At the end of the chronic stress period (8 weeks), C and S animals experienced a fear test in a 4.4×16.3 m arena subdivided into 12 quadrants to track movements inside the arena. Each steer was allowed to acclimate for 90 s, followed by the sudden appearance of a startling stimulus (a person with a high visibility jacket opening an umbrella once). The person remained static in the arena for an additional 90 s. Reaction to the startle was recorded as well as subsequent behaviour and activity (running, walking, standing, vigilance, sniffing, orientation and location changes). Behavioural and location data were analyzed in separate Principal Components Analyses (PCA). For the Behaviour Analysis PCA, PC1 was labelled as 'Inactive – Active' correlating with behaviours such as vigilance (r=0.77; P<0.0001) and walking duration (r=-0.75; P<0.0001); PC2 was labelled as 'Distance travelled pre-startle vs post-startle' and PC3 as 'Attention towards novelty'. These PCs explained 29.3, 18.6 and 10.5% of the variance respectively. Animal scores on PC1 were affected by their sire identity (eta2 (analog to R2) 0.36; P<0.05), while PC2 and PC3 were affected by treatment (eta2 0.16; 0.19; P<0.01) and pen (eta2 0.38; 0.34; P<0.05). In PC2, C animals travelled further post-startle than pre-startle (proportional increase 6.92 and 18.33 for S and C animals; P<0.05) whilst in PC3, S animals showed more attention towards the novel stimulus (P<0.01). The PCA developed to describe the location of animals relative to the startling stimulus identified two dimensions that explained 34.9 and 29.6% of the variance, but the S or C treatment did not influence the location of cattle relative to the stimulus after its appearance. These results indicate differences in behavioural responses and attention to a novel stimulus by animals subjected to a composite stress treatment, suggesting that they may have experienced a more negative affective state. Faecal cortisol and response to an ACTH challenge will be presented to help interpret the response of cattle to chronic stress.

## Roll out the green carpet – dairy cows prefer artificial grass laneways over stonedust-over-gravel ones, especially when lame

*Stephanie Buijs, Gillian Scoley and Deborah McConnell*
*Agri-Food and Biosciences Institute (AFBI), Sustainable Agri-Food Sciences Division, Large Park, BT26 6DR Hillsborough, United Kingdom; stephanie.buijs@afbini.gov.uk*

Dairy cows often walk long distances between the parlour and the pasture and it can therefore be expected that a more comfortable laneway surface (e.g. softer, less abrasive) has a considerable positive impact on cow welfare. This may be especially important for lame cows that generally have more tender hoofs. However, knowledge on which type of laneway surface is preferred by dairy cows is currently lacking, and cannot be extrapolated from studies on barn floor types as both the way the cow uses the surface and the available top-surface materials differ. We used a preference test to determine if dairy cows favoured a softer laneway top-surface of artificial grass over a standard laneway (stonedust-over-gravel). We expected a preference for artificial grass, which would be more pronounced in lame cows than in sound ones because of increased hoof tenderness. Dairy cows (66 pairs) were tested whilst returning to pasture after their afternoon milking, after having been habituated to the preference test setup for several days. The setup consisted of a laneway divided lengthwise into two 2.2 m wide lanes, each consisting of four 23 m stretches. The 1st and 3rd stretch of the left lane and the 2nd and 4th stretch of the right lane were covered in artificial grass, whilst the other stretches were left bare. At the start of each stretch the cows were forced to the middle of the laneway using sideways exclusion triangles, so they needed to make a new choice between the standard surface and the artificial grass at each of the four stretches (a wire prevented switching laneways within a stretch). Because individual testing was expected to reduce willingness to continue on to pasture at a normal speed, cows were tested in pairs. Only the lane choices of the lead cow of the pair were analysed as choices were likely dependent within a pair. We analysed if the cows preferred one surface type over the other (i.e. used it more often than expected by chance) using Wilcoxon signed rank tests. This was done separately for sound cows (mobility score <3 on a 1-5 scale, as determined by an experienced assessor as the cows left the parlour, 69% of all lead cows) and lame ones (mobility ≥3, 31% of all lead cows). In addition, the number of artificial grass stretches used was compared between lame and sound cows (two-sample Wilcoxon rank sum test). Both sound and lame cows preferred the artificial grass over the standard surface (P=0.004 and P=0.001, respectively). As expected, this preference was stronger in lame cows than in sound ones (median stretches of artificial grass used ± IQR for lame cows: 3±3-4, for sound cows: 2±2-3, P=0.001). In conclusion, especially lame dairy cows preferred a softer artificial grass walking surface over a standard one, suggesting that such surfaces reduce pain or discomfort when walking.

## Evaluation of sow thermal preference

*Lindsey Robbins[1], Angela Green-Miller[2], Donald Lay Jr[3], Allan Schinckel[1], Jay Johnson[3] and Brianna Gaskill[1]*
[1]*Purdue Univ., Animal Science, 270 S. Russell St., West Lafayette, IN 47907, USA,* [2]*Univ. of Illinois at Urbana-Champaign, Agricultural and Biological Engineering, 332 G Agricultural Engr. Sciences Bld., Urbana, IL 61801, USA,* [3]*USDA-ARS, Livestock Behavior Research Unit, 270 S. Russell St., West Lafayette, IN 47906, USA; lrobbin@purdue.edu*

Despite advances in livestock cooling system technologies and management strategies, heat stress continues to negatively impact swine well-being, resulting in greater morbidity and mortality. The Federation of Animal Science Societies (*Ag Guide*) states that the thermal comfort zone (TCZ) for any pig over 100 kg is between 10-25 °C. These guidelines are dated and do not reflect the increased heat generation in current genetic lines. We hypothesized that sows have a lower TCZ than what is stated by the *Ag Guide* and this will be altered by gestational stage and parity. The thermal preference of 21 sows was tested in a factorial design of parity (2, 3, and 5; n=2 per gestational stage) and gestational stage: open (not pregnant, n=7), mid-gestation (45-65 days pregnant, n=7) or late gestation (95-108 days pregnant, n=7). Sows were placed individually inside a thermal apparatus (12.20×1.52×1.85 m: l×w×h) with a constant thermal gradient ranging 10-31 °C and acclimated to the environment for 24 h. During the following 24 h testing period, sows were continuously videotaped. The sows' location and behavior (active, inactive, or other) were documented using instantaneous scan samples every 10 minutes. Data were analyzed using a GLM and Log10 transformed for normality. Tukey tests and Bonferroni corrected custom tests were used for *post-hoc* comparisons. Peak temperature preference was determined by the maximum amount of time spent at a specific temperature and preference range was calculated using the peak temperature ±SE. Parity did not affect thermal preference ($F_{2,339}$=2.66, P=0.07). Results demonstrated a two-way interaction between thermal preference and behavior ($F_{1,339}$=34.48, P<0.01) and thermal preference and gestational stage ($F_{2,339}$=7.43, P<0.01). While inactive, sows spent the most amount of time at 14.8 °C but their preference ranged from 14.0-15.8 °C. Although sows did not spend much time active, their preference peaked at 14.4 °C but ranged from 12.8-16.0 °C. Differences between peak temperatures were observed between sows in late (14.4 °C) and mid-gestation (15.0 °C; $F_{1,339}$=13.68, P<0.01) and between late gestation and open sows (14.8 °C; $F_{1,339}$=8.07, P<0.01). However, no difference was found between mid-gestation and open sows ($F_{1,339}$=0.97, P=0.33). Thermal preference for open and mid-gestation sows ranged between 13.6-16.4 °C. However, sows in late gestation had a preference range that was shifted to slightly cooler temperatures (13.0-15.6 °C). Furthermore, late gestation sows spent the least amount of time between 27-30 °C, compared to open and mid gestation sows (P<0.01). The thermal preference range for sows tested in this study fell within the *Ag Guide* recommendations (10-25 °C). However, sows in the present study preferred a much narrower range of temperatures (12.8-16.4 °C). This study indicates that the TCZ of sows is affected by gestational stage and behavior. Thus, there is a need to refine thermal recommendations based on gestational stage to reduce thermal stress and improve sow welfare.

## When lying down, would cows rather have a preferable surface or unrestricted space?

*Laura Shewbridge Carter[1], Mark Rutter[1] and Marie J. Haskell[2]*
*[1]Harper Adams University, Shropshire, TF10 8NB, United Kingdom, [2]SRUC, West Mains Rd, Edinburgh, EH25 9RG, United Kingdom; lshewbridgecarter@harper-adams.ac.uk*

Dairy cows have a preference to access pasture when given the choice, which is not influenced by herbage mass. This suggests that cows value pasture access for lying down rather than for grazing, supported in part by evidence that cows have a higher motivation to access pasture at night, when they primarily lie down, than during the day. This study investigated cow preference for two different lying qualities that pasture might offer cows that may not be fulfilled by traditional cubicles; the surface for lying down and the extra space it offers for lying. Twenty-four Holstein dairy cows were used in the trial (July – Nov. 2018). Cows were allocated to one of six experimental periods (n=4×6) and housed in individual pens (6×15.2 m), during which they had visual and tactile contact with another cow. Each pen had three lying surfaces: deep-bedded sand (SA), mattress (M) and deep-bedded straw (ST). These surfaces were contained in boxes 2.4 m$^2$, which were designed to allow a cubicle to be fitted over the middle and removed again. Behaviour and location was recorded using video cameras. Each period lasted a total of 20 d over three stages; Stage 1 (8 d): cubicles were fitted over each surface. Cows were given 2 days training on each surface, to experience lying on all surface types, before having a 2 day free choice period with access to all surfaces. Stage 2 (9 d): 1 day to remove cubicles. Cows were then given 2 days training on each surface, followed by a 2 day choice period, asking the cows to choose their preferred surface to lie on when they had additional space. Stage 3 (3 d): a cubicle was fitted onto each cow's most preferred surface, with the other two surfaces remaining without a cubicle, and the cows given a final 3 day choice period. This created a trade-off between preferred surface with restricted space (C), second most preferred surface with additional space (S1) and third preferred surface with additional space (S2). REML was performed for the choice periods of each Stage, investigating differences in the percentage of time spent lying on each of the different surfaces, including time not lying. There was an effect of surface on lying times in Stage 1 (mean ± SEM; SA=5.5±0.5%, ST=27.5±1.1%, M=27.9±1.1%; P=0.006) and Stage 2 (7.8±0.6%, 42.5±1.3%, 14.3±0.8%; P=0.008). For Stage 3, data was available for nineteen cows and showed that cows spent significantly longer lying on their second most preferred surface with additional space (S1) than their preferred surface with restricted space (C) or their third preferred surface with additional space (S2) (P<0.05; predicted means for % lying: S1=34.81, C=13.44 and S2=13.05). 16 cows traded off lying on their preferred surface, showing a partial preference to lie on less preferred surfaces with additional space (S1+S2>50%), with 3 cows choosing to lie on their preferred surface and trading off additional space (C>50%). This result indicates that cows have a partial preference for more space when lying down compared to lying on a preferred surface, suggesting that cows value space above surface when lying down.

## Spatial distribution and preferable zones for dairy cows in a free stall confined area

*Vlatko Ilieski and Miroslav Radeski*
*Faculty of Veterinary Medicine, Animal Welfare Center, st. Lazar Pop Trajkov 5/7, 1000 Skopje, Macedonia; vilieski@fvm.ukim.edu.mk*

The spatial distribution of dairy cows in a confined area depends of the herd's social structure, individual preferences, the spatial structure and available resources. Thus, locations with more and/or better feed, comfortable lying areas and places with suitable surroundings within the stall might be more frequently occupied by the animals and considered as preferable. The objective of this study was to identify the favorable places in a confined area based on the individual and social behaviour of dairy cows. In a free stall confinement area of 900 m$^2$, 91 dairy cows were observed for 14 hours/day (07.00-21.00) on two consecutive days by using four cameras covering the whole area. For the subsequent video analysis, the cow catalogue was created with five photos of each individual cow (front, back, left and right side). The created ethogram was consisted of 16 different behaviours, such as state events – lying, moving, standing, feeding, brushing; and point events – headbutts, displacements, chasing, chasing up, licking, etc. Furthermore, the spatial distribution of cows was observed by dividing the whole confinement area into eight zones (four lying area with cubicles from ZL1 to ZL4 and four feeding areas from ZR1 to ZR4). The favorable zones in the area were identified using descriptive statistics and Pearson correlation coefficients (considered significant at P<0.001). Thus, in the feeding area, zone ZR3 received the longest stay of 1:55:59±1:18:31 hours (or 0.17±0.11 proportion of the observed time), while the least visited zone was the entrance zone ZR1 with 0:58:42±0:57:23 hours (0.09±0.09). In the lying area, the zones in the middle, ZL2 and ZL3, received the longest stay 1:32:34±1:35:00 hours (0.13±0.14) and 1:36:12±1:35:46 hours (0.14±0.14). Considering the correlation between behaviour and the spatial zones, the feeding duration was positively correlated with ZR3 and ZR4 (r=0.46, and r=0.47, respectively), and self-licking was positively correlated with ZR1 (r=0.34) relating this zone and this type of behaviour. Considering the interactions with other cows, the ZR4 was significantly positively correlated with the number of headbutts (r=0.31) and displacements received (r=0.44). These findings suggest that the zones in the middle of the confinement area were more preferable, and the salt cubes found only in the feedbank of ZR4 lead to most interactions between cows. Additionally, the correlations between different zones suggest that each individual stays in specific zones in the confined area. Applying this kind of observation and analysis of the animal's spatial distribution might locate the preferable places for the dairy cows and will point out the risky zones, i.e. zones that should be improved in the stall.

## The effect of soiling and ammonia reductant application on turkeys' perceived value of wood shavings

*Valerie Monckton, Nienke Van Staaveren, Peter McBride, Isabelle Kwon and Alexandra Harlander*
*University of Guelph, Department of Animal Biosciences, N1G 2W1 Guelph, Canada;*
*vmonckto@uoguelph.ca*

Modern farms house turkeys indoors in different conditions from those of their natural habitat. Litter substrate plays an important role in turkey housing, where infrequent changing of bedding (often wood shavings) results in the accumulation of excreta, as well as increased levels of litter moisture and ammonia gas, which adversely impacts turkey welfare. Moreover, it is unknown how this soiling alters the perceived value of litter substrate to turkeys. This study aimed to assess the effect of soiling and ammonia reductant application on turkeys' relative preference for wood shavings through a consumer-demand approach. Twenty-four 11-week-old turkey hens raised on wood shavings were housed in 6 pens (4 birds/pen) which consisted of a 'home' (H) and 'treatment' (T) compartment separated by a barrier with two unidirectional push-doors. The hens' relative preference for familiar, soiled wood shavings in H was compared to one of four substrates in T: unsoiled wood shavings (US), soiled wood shavings (SS), soiled wood shavings treated with an ammonia reductant (AS), and no litter (concrete floors covered in rubber mats, NL). Additionally, we compared the hens' relative preference for these substrates to a benchmark: feed (F) in T while the feed in H was blocked. To measure the hens' motivation to access the treatments, the doors that led to T randomly weighed 0, 20 or 40% of the hens' average weight, while T and H switched compartments with each treatment/door weight combination. Testing took place over 22 days. Birds' presence in each compartment was determined via video-recordings using instantaneous scan sampling every 30 min for an average of 14 hours per combination. The effect of treatment, door weight, and their interaction on the percentage of time birds spent in T was analysed using a generalized linear mixed model (PROC GLIMMIX, SAS V9.4). Turkeys pushed the doors, although door weight did not significantly affect where turkeys spent their time ($P>0.05$). Turkeys spent significantly more time in T with F ($64\%\pm9.0\%$, $P<0.001$) than with US ($21\%\pm13.3\%$), AS ($24\%\pm12.9\%$), or NL ($20\%\pm13.1\%$), while they spent the same amount of time in T when SS ($45\%\pm10.8\%$) were present. To the authors' knowledge, we present the first motivational test to assess how soiling and ammonia reductant application modifies the perceived value of wood shavings to turkeys. In conclusion, turkeys spent the same amount of time on SS and F, while they showed a lower relative preference for US, AS, and NL. It is possible that SS were not soiled enough compared to the H litter to influence turkeys' preference. Alternatively, there may be properties in SS that attract turkeys (e.g. foraging in more diverse litter); however, this will require further investigation.

## Operant learning is disrupted when opioid reward pathways are blocked in the laying hen

*Jade Fountain[1], Susan J. Hazel[2], Terry Ryan[3] and Peta S. Taylor[1]*
*[1]University of New England, School of Environmental and Rural Science, Elm Avenue, 2351 Armidale NSW, Australia, [2]University of Adelaide, School of Animal and Veterinary Sciences, North Terrace, 5005 Adelaide, Australia, [3]Legacy Canine Behavior & Training, Inc, Sequim, 98382, USA; jadekfountain@gmail.com*

There is limited research into mesolimbic and dopamine function specific to birds, indeed little is known about how chickens experience rewards and how reward might affect learning in chickens. A more thorough understanding of how laying hens experience reward in their environment can give insight into impact of meeting individual preferences that optimise health and wellbeing and which hold implications for welfare improvements. Implications could extend to using training as a means of providing enrichment and positive experience to hens. This study examined the effect that μ-opioid receptor antagonist nalmefene has on learning in laying hens. Thirty-eight laying hens were randomly assigned into either a pharmacological treatment group or a control group. Hens in the pharmacological group (n=18) were administered with μ-opioid antagonist (0.4 mg/kg nalmefene dissolved in 0.9% saline injected intramuscularly into the pectoral muscle) to disrupt the reward (dopamine) pathway. Hens in the control group (n=20) received 0.9% saline (0.5 ml/kg intramuscularly into the pectoral muscle). On day one of the trials no dosage was given, and all hens in the treatment and control group succeeded in passing a habituation task to expose the hens to the training table, cup and trainer. During the habituation task, hens were given an opportunity to follow a cup of meal worms at a slow speed, and then a fast speed across the length of the table. From day two onwards, hens were dosed thirty minutes prior to trials and trained on a series of three tasks using operant conditioning in which they were given three five minute sessions over three days to learn the tasks. The three tasks consisted of (1) peck on a target, (2) colour stimulus discrimination, and (3) pecking on cue presentation only. All tasks were trained by the same trainer blind to treatment and using a clicker training technique; correct responses were marked with a clicker in a procedure called 'shaping' to train hens to criteria on each task. All statistical analysis were conducted running a series of Generalised Linear Mixed Models (GLMM) with day, treatment and the interaction between day and treatment as fixed factors. By day four, more control hens had completed Task 1 and Task 2 than hens that received nalmefene ($F_{(1,17)}$=100, P<0.0001). In fact, no hens in the nalmefene treatment group achieved any task except the habituation task on any of the days in the trial. This study demonstrated hens treated with an opioid antagonist failed to learn any tasks during the trials. The results suggest that performing operant tasks is likely to be rewarding to laying hens, and when the ability to experience reward is blocked, they no longer choose to participate. Thus, operant training is a positive rewarding experience for laying hens.

## The burning ring of fire: place aversion as evidence of felt pain in calves disbudded using different types of pain control

*Thomas Ede, Marina A.G. Von Keyserlingk and Daniel M. Weary*
*University of British Columbia, 6947 Lougheed Highway, V0M1A1, Canada;*
*thomas.ede92@gmail.com*

Hot-iron disbudding is a common procedure on many dairy farms, and is known to cause calves inflammatory pain in the hours after the procedure. A variety of methods are available to mitigate this pain, but most research assessing the efficacy of these methods has relied on automatic or reflexive responses. The aim of this study was to investigate the affective component of post-operative pain, using a conditioned place aversion paradigm. The apparatus consisted of three pens adjacent to each other: two 'treatment' pens connected by a 'neutral' pen. Calves (n=31) were subjected to two disbudding procedures and their 6 h recovery (one bud in each treatment pen, 48 h apart, order balanced): one without post-operative pain control and the other with the use of a nonsteroidal anti-inflammatory drug (either Meloxicam, 0.5 mg/kg [n=16] or Ketoprofen, 3 mg/kg [n=15]). All disbudding procedures included sedation (Xylazine, 0.2 mg/kg) and a local cornual nerve block (Lidocaine 2%, 5 ml). Place conditioning was tested 48, 72 and 96 h after the last treatment by allowing calves to freely roam between the pens until they chose to lay down for 1 min. Time spent in each treatment pen was square root transformed and analysed with linear mixed models. The treatment associated with the pen in which calves chose to lay down was analyzed with chi-square tests. Calves spent more time and lay down more frequently in the pen where they received Meloxicam compared to the pen where they only received a local block (respectively, $t_{1,78}$=5.2, SE=2.0, P=0.01; $X^2$=6.8, P=0.009). Surprisingly, calves avoided the pen where they received Ketoprofen compared to the other treatment pen (time spent: $t_{1,73}$=-6.8, SE=1.8, P<0.001; lay down frequency: $X^2$=10.5, P=0.001). We hypothesized Ketoprofen may have some unintended side effects that result in increased aversion. These results support the use of meloxicam to control post-operative pain associated with hot-iron disbudding, and illustrate the value of place conditioning paradigms to assess the affective impact of pain in animals.

## Social transmission of learning of a virtual fencing system in sheep

*Tellisa Kearton[1,2], Danila Marini[1,2], Rick Llewellyn[3], Susan Belson[1] and Caroline Lee[1,2]*
[1]*CSIRO, Armidale, 2350, Australia,* [2]*University of New England, Armidale, 2350, Australia,* [3]*CSIRO, Glen Osmond, 5064, Australia; tellisa.kearton@csiro.au*

Virtual fencing (VF) is a technology which uses associative learning to train animals to respond to an audio tone in order to avoid receiving an electrical stimulus from a collar, and remain within a set area. Flock behaviour and social influences can play a role in learning and behaviour in relation to the VF. This pilot experiment (approved by CSIRO animal ethics committee ARA 17/24) aimed to investigate whether social grouping influences the effectiveness of a VF system in which varying proportions of the flock were subjected to a VF collar. Three groups of 9 merino ewes were randomly assigned to 3 VF collar treatment groups: (1) 33% of the group wore VF collars; (2) 66% VF; and (3) 100% VF. Dog training collars were used to apply the VF protocol. Testing was conducted in an 80×20 m field divided midway by a virtual fence. Testing occurred over 2 days for an 8 h period, allowing normal grazing behaviour across the field. Sheep were kept at pasture following testing. Behavioural responses of ewes receiving the audio cue and subsequent electrical stimulus (direct) and responses of peers (indirect) were recorded by video camera. The proportion of the flock responding either correctly (turn, stop walking) or incorrectly (continue walking, run forward), and the proportion of the indirect responses out of the total number were calculated for this study. Based on indirect interactions with the VF, and compared to the 100% group, ewes in the 33% group were more likely to continue walking forward when indirectly exposed to the VF electrical stimulus (Fisher's exact test, P=0.03). The 33% group tended to have fewer correct responses to the indirect audio cue (P=0.08). For the average numbers of ewes within each group indirectly interacting with the VF, the 66% group showed a higher number of peers involved in correct responses to indirect cues and stimuli than the 33 and 100% groups. In the 33% group (the only group which were not effectively contained by the fence), the majority of ewes were not subject to VF and were free to cross the fence line, incentivising the VF ewes to challenge the boundary, and reducing the number of ewes responding correctly to the cues. Differences between the 33 and 100% treatment groups for the proportion of correct responses to the indirect electrical stimuli and audio cues suggest that sheep being directly cued have some social influence on their flock members responding to indirect cues. When low numbers of a flock (33%) are collared there are few opportunities for learning. The 100% VF group ewes were more likely to respond to direct cues than indirect cues, while the 66% group showed a higher number of peers indirectly interacting with the fence. This preliminary study of group responses to learning a virtual fence provides some evidence of social facilitation of learning occurring in sheep not receiving cues directly. Further investigation of social transmission of learning and training in sheep is recommended.

## Effect of management enrichment on cognitive performance of Murrah buffalo calves

*Pawan Singh, Sudip Adhikar, Rajashree Rath and Madan Lal Kamboj*
*ICAR-National Dairy Research Institute, Livestock Production Management, Department of Livestock Production Management, ICAR-National Dairy Research Institute, 132001 Karnal, India; pawansinghdabas@gmail.com*

The purpose of management enrichment is to enhance the quality of animal care by providing environmental stimuli necessary for optimal psychological and physiological well-being. Intensive housing conditions and artificial rearing of dairy calves during early life under modern production systems are known to impair cognitive development of the animals which may affect their coping strategies with intensive life conditions and thus impact their subsequent performance. The aim of present study was therefore to investigate the effect of housing buffalo calves in individual calf boxes with or without social contact with con-specifics and provision of different environment enrichment stimuli on their cognitive development. For this study, 24 Murrah buffalo calves were selected at birth and randomly allotted to four treatment groups of 6 calves each (T0, T1, T2 and T3) for an experimental period of 3 months. The calves in T0 (control) were housed in individual calf boxes which permitted visual and auditory social contact with calves housed in adjoining boxes. The calves in T1 were housed in individual boxes and groomed by using a curry comb brush twice a day for a period of 10 min each as management enrichment. The calves in T2 were housed in individual calf boxes which were isolated from each other and deprived social contact. The T3 calves were housed individual calf boxes with a provision of rubber nipple in the pen for dry suckling as management enrichment. All the calves were allowed to suckle colostrum for 5 days after birth, after which the whole milk was fed twice a day as per standard calf feeding schedule using a calf feeding bottle with a crew nipple. The calves were trained in a Y maze for 12 sessions of initial learning with milk feeding bottle with milk hidden at white side and 12 sessions of reversal learning with milk bottle hidden at black side. The calves were treated as qualified for that learning in case they made more than 80% correct attempts in 3 consecutive training sessions. Data were analysed with SPSS Version 17 using one way ANOVA and means were compared with Man Whitney test. For satisfactory initial learning the mean number of training sessions required were 6.17±0.60, 5.67±0.42, 6.00±0.63 and 6.00±0.63 in T0, T1, T2 and T3, respectively, which did not differ significantly among the 4 groups. However, during reversal learning, calves with a provision of artificial nipples as enrichment (T3) took lesser ($P<0.05$) training sessions (7.17±0.30) as compared to T1 (9.17±0.30), T2 (9.33±0.61) and T0 (9.33±1.09) calves. When compared between type of learning, the calves took lesser ($P<0.001$) sessions during initial learning (5.79±0.21) as compared to during reversal learning (8.75±0.36). Comparing the two enriched groups, calves with artificial nipples performed task in less sessions (5.33±0.21&7.17±0.30) compared to groomed (5.67±0.42&9.17±0.30) in initial and reversal learning, respectively. It was concluded that provision of an artificial nipple as management enrichment inside the calf boxes improved cognitive performance of buffalo calves.

## Using applied behavior analysis to identify and enhance positive reinforcers for domestic dogs

*Erica Feuerbacher[1], Chelsea Stone[2] and Jonathan Friedel[3]*
*[1]Virginia Tech, Animal & Poultry Sciences, 3460 Litton-Reaves Hall (0306), 175 West Campus Drive, Blacksburg, VA 24061, USA, [2]Carroll College, Anthrozoology, 1601 N. Benton Ave., Helena, MT 59625, USA, [3]National Institute for Occupational Health and Safety, 1095 Willowdale Road (Mailstop 4050), Morgantown, WV 26505, USA; enf007@vt.edu*

With increasing use of positive reinforcement in animal training, identifying effective reinforcers and ways to enhance efficacy is vital to ensuring that owners, keepers, and trainers can bring about effective behavior change using humane methods. Applied behavior analysis techniques allow for identification of and comparison between different reinforcers including by using break points on progressive ratio (PR) schedules. Several aspects of reinforcers have been found to affect reinforcer efficacy in both laboratory and applied settings, including quality, magnitude, schedule, and delay to reinforcement. We assessed the effect of different ways of food delivery (one large treat all at once or four small treats delivered one-by-one, but the same overall magnitude) as well as different magnitudes of food using four owned dogs. Dogs had not been fed for approximately 4 hours. Dogs emitted nose touches to experimenter's hand on a PR schedule such that schedule requirements increased within sessions. After two schedule completions, the dog had to emit one more nose touch than the last schedule to earn reinforcement. Reinforcers were small, moist treats. We measured total number of responses emitted and break points (BP) (largest schedule completed). We utilized an alternating treatment single-subject design such that all dogs encountered all conditions across multiple sessions. Our data allow for demonstration of experimental control of our interventions, showing predictable and repeated changes in our dependent variable with concomitant changes in our independent variable across repeated measures within individual dogs. Each dog experienced between 17 and 40 sessions. At least two but no more than four sessions were conducted each day for each dog to minimize satiation. We analyzed results using behavioral economic concepts unit price, demand/work curves, and visual analysis. Two dogs were high responders (mean BP: 12.5 and 12.4) and two were low responders (mean BP: 5.3 and 5.1). Delivery method had a small, inconsistent effect on reinforcer efficacy (mean BP, 12.6, 12.0, 5.4 and 5.2 for one large treat and 13.0, 13.1, 5.2 and 5.0 for four small treats, respectively). However, magnitude did reliably increase reinforcer efficacy, with larger magnitude reinforcement producing break points nearly twice that of small magnitude (mean large BP 12.5, 12.4, 5.3, and 5.1; mean small BP 6.3, 5.0, 3.1, and 2.5, respectively). Nevertheless, unit price for small magnitude reinforcers was nearly double that for large magnitude, suggesting that larger magnitudes might not have as large an effect on behavior as predicted. While mixed results exist regarding the effect of magnitude of reinforcement, the lack of effect of magnitude might be due to procedural issues including discriminability of the operating contingency, and unpredicted interactions between magnitude and schedule of reinforcement are resolved when those two variables are combined into one variable (unit price).

# Do pigs understand time interval and adopt optimal foraging behaviour?

*Yuna Sato, Takeshi Yamanaka, Chinobu Okamoto and Shuichi Ito*
*Tokai University, School of Agriculture, Department of Animal Science, 9-1-1, Toroku, Higashi-ku, Kumamoto-shi, Kumamoto, 862-8652, Japan; yuuna_sato@animbehav-tokai.com*

According to many studies, pigs possess the ability to explore foods and develop foraging behaviours as an adaptation to changing food environments. It is believed that pigs use various abilities to obtain their food. We assessed whether pigs can adapt to changing food environments by understanding extended time intervals. Two domestic pigs were fed with small round cookies personally by the experimenter when they approached. After 5 s, the next feed was given, and this process was repeated once again. The time interval between the feed was increased 15 s after the subjects were fed their third cookie. The interval for feeding was doubled after every third cookie. When the pigs shifted to a different experimenter, the feeding interval was initialised to 5 s. Two conditions were set such that the distance between the two experimenters was either close (approximately 4 m) or far (approximately 17 m). An experimenter who fed the pigs represented the feeding patch. In this experiment, the reward was given by human hand, but there was no correct or incorrect answer. We predicted that the energy consumption would increase as the distance increased. If pigs understood the feeding interval, when the distance between the patches was close, the pigs would move to the other patch as soon as possible or at the shortest time interval possible. Conversely, when the distance was far, the pigs would remain for a longer time in each patch. Two sows (sow #1 and sow #2) were subjected to only one treatment per day for 20 min with both close and far distance. Each treatment was performed for 3 days. In the case when the distance was close, the mean number of times food was ingested at a feeding patch was 2.47±0.03 (Sow #1) and 3.12±0.29 (Sow #2). The number of shifts to the feeding patches during the test period was 74.01±0.02 (Sow #1) and 48±6.68 (Sow #2). When the distance between the patches increased, the mean amount of ingested food at the patch was 4.24±0.43 (Sow #1) and 5.74±1.29 (Sow #2). The number of shifts between patches during the test was 18.33±4.19 (Sow #1) and 11.67±5.19 (Sow #2). In this study, when the distance between the patches was close, the domestic pigs shifted during a short feeding interval. In contrast, when the distance between the patches was long, the pigs stayed in the same patch even when the feeding interval became longer. These results suggest that domestic pigs understand extended time interval in feeding and possess the ability to develop foraging behaviours adapted to such changing food environments.

## Doing nothing – inactivity in fattening cattle

*Sara Hintze[1], Freija Maulbetsch[1], Lucy Asher[2] and Christoph Winckler[1]*
*[1]University of Natural Resources and Life Sciences Vienna, Department of Sustainable Agricultural Systems, Gregor-Mendel-Strasse 33, 1180 Vienna, Austria, [2]Newcastle University, School of Natural and Environmental Sciences, King's Road, NE1 7RU Newcastle upon Tyne, United Kingdom; sara.hintze@boku.ac.at*

Large numbers of farm, lab and zoo animals are housed under barren and monotonous conditions, in which they lack the opportunity to engage with their environment, often resulting in high levels of inactivity. Long episodes of inactivity are potentially highly relevant for animal welfare since they might be accompanied by aversive states, including boredom, apathy or depression. However, being inactive is not necessarily a problem because positive inactive states like relaxation and post-consummatory satisfaction exist as well. Currently, we lack knowledge to reliably differentiate between positive and negative states associated with inactivity. To study what inactivity means for animal welfare, we first need to develop methods to identify different forms of inactivity. To this end, we developed an 'inactivity ethogram' for cattle describing the exact posture of an animal while lying or standing, its head and ear positions, tail movements, as well as its location (e.g. feeding or lying area) and the distance to its closest conspecific. In addition, active behaviours were included to capture transitions between active and inactive behaviour. The ethogram was applied to Austrian Fleckvieh heifers across intensive, semi-intensive and pasture-based husbandry systems. Three farms per husbandry system were visited twice; once in the morning and once in the afternoon to cover most of the daylight hours. During all visits 16 focal animals were continuously observed for 15 minutes each (96 heifers per husbandry system, 288 in total). The time the heifers spent inactive increased from pasture (56%) to the semi-intensive (75%) to the intensive system (86%), but it did not differ significantly between husbandry systems ($F_{2,6}$=2.26, P=0.19; linear mixed-effects model, fixed effect: 'husbandry system', random effects: 'farm visit' nested in 'farm'). Moreover, heifers across husbandry systems did not differ in the time they spent standing ($F_{2,6}$=1.79, P=0.25) or lying ($F_{2,6}$=0.80, P=0.49) while being inactive. When analysing the simultaneous occurrence of different combinations of standing/lying, head and ear postures, we found that across husbandry systems, the animals' heads were in an upright position for most of the time during both standing (52%) and lying (85%), with forward pointing ears being the most common ear posture during standing (50%) and backward pointing ears during lying (52%). Low ears, previously shown to be an indicator of positive low arousal states in dairy cows, were mostly recorded in heifers on pasture, independent of the animals' body or head positions (60%). Data are currently further analysed to detect: a) the sequence/transitions of behaviours/postures over time; and b) the simultaneous occurrence of different postures and positions. Our preliminary results indicate that despite a lack of statistically significant differences in the total time the heifers spent inactive, behavioural patterns of inactivity vary across husbandry systems. Our study forms a basis for future research to differentiate between positive and negative inactive states.

## Care and rehabilitation activities for a chimpanzee with cerebral palsy: a case study

*Yoko Sakuraba[1,2], Nobuhiro Yamada[3], Ichiro Takahashi[4], Fumito Kawakami[5], Jun'ichi Takashio[6], Hideko Takeshita[7], Misato Hayashi[8] and Masaki Tomonaga[8]*

*[1]Kyoto University, Wildlife Research Center, 2-24, Tanakasekidencho, Sakyo, Kyoto city, 606-8203, Kyoto, Japan, [2]Kyoto City Zoo, Okazaki Koen, Okazakihoshojicho, Sakyo, Kyoto city, 606-8333, Kyoto, Japan, [3]Noichi Zoological Park, 738, Otani, Noichicho, Konan city, 781-5233, Kochi, Japan, [4]Home nursing station Otasukeman, 1-13-26, Ushioshinmachi, Kochi city, 780-8008, Kochi, Japan, [5]Chubu University, 1200, Matsumotocho, Kasugai chity, 487-8501, Aichi, Japan, [6]Biwakogakuen Medical and Welfare Center, Kusatsu, 8-3-113, Sasayama, Kusatsu city, 525-0072, Shiga, Japan, [7]Otemon Gakuin University, 2-1-15, Nishiai, Ibaraki city, 567-8502, Osaka, Japan, [8]Kyoto University, Primate Research Institute, Kanrin, Inuyama city, 484-8506, Aichi, Japan; sakuraba.yk30@gmail.com*

Each animal should be provided with good welfare and the best environment they need, whereas disabled animals need more careful consideration and specific approaches to improve their welfare. Although such cases are few, studying and discussing the practical methods to support them is crucial to improving their welfare not only in the present but also in the future. A female chimpanzee *(Pan troglodytes)* with paralysis of the right-side body caused by cerebral palsy, named Milky, lives in Noichi Zoological Park. This presentation aims to introduce our activities for supporting and evaluating her physical state and behavioral development. Milky was born on July 14, 2013, to a mother under anesthesia due to a difficult delivery, and she has been cared by humans ever since. After birth, Milky was found to have developmental retardation, including serious paralysis of her right-side body caused by cerebral palsy. Therefore, caretakers provided her with many rehabilitation supplies in her daily living enclosure (e.g. mattresses, a slope, stable/unstable chairs, a mirror, balance balls, monkey bars). At 1 year of age, Milky began rehabilitation activities to improve her physical/cognitive development. These activities were administered collaboratively by her caretakers, a physical therapist, an occupational therapist, and researchers. For example, we performed a massage on her to extend her paralyzed body parts and encouraged her to climb on a handmade jungle gym during play sessions with therapists. To evaluate her physical states, we analyzed data collected from January 2017 by time sampling per 10 seconds of video recording before/during the rehabilitation activities. In results, these activities may have reduced her stress because she emitted laughter during the activities. The regression analysis of dorsiflexion of the right ankle joint, focusing on sitting position, to longitudinal data shows a weak and positive linear relationship, although the variation is large ($R^2$=0.197). This result may suggest that our support and rehabilitation activities have a positive effect on improving her physical state. However, this is just a case study, and it is necessary to compare the present behavioral data with that of no activity days to clarify whether these activities lead to good rehabilitation for her. Moreover, social interactions with other chimpanzees must be encouraged in the future to further promote her well-being.

## Assessing pasture-based dairy systems from the perspective of cow experiences

*Cheryl O'Connor and Jim Webster*
*AgResearch Ltd, Ruakura Research Centre, Private Bag 3123, Hamilton 3240, New Zealand; cheryl.oconnor@agresearch.co.nz*

Opportunities for positive experiences are needed to support 'a life worth living' for farm animals. Affective state is the amalgamation of experiences created from perceptions and sensations and is a cognitive interpretation and representation of the outside world. A number of scientific frameworks have been developed that emphasise the importance of affective state to welfare, and how it represents the ultimate domain for welfare assessment. A framework is needed to discuss affective state with farmers, in order to raise the profile of an animal's actual experiences. To ultimately have farm systems that support positive experiences, we must know how animals are interpreting those systems and farmers are ideally placed for this. Appraisal is the cognitive process of interpreting sensory information and incorporating an emotional aspect to it. The inherent flexibility in appraisal means an animal's affective state cannot be directly or accurately assessed from the resources available to the animal or situations they are exposed to. Experiences are shaped not only by external sensations such as; sight, sound, smell, physical activity and movement, satiation and hydration but also from internal appraisal such as; contentment, choice, stimulation and reward. The framework for discussion will allow anthropomorphic terminology to aid a common understanding, as used in QBA. We have conceptualized an example framework containing the main requirements to assess experiences of pasture-based dairy cows. External sensory experiences would include visual, sound, smell and social sensations from herd mates, freedom of movement, satiety (dependent on quality of pasture) and hydration (dependent on adequate water supply). Contentment experienced by cows should consider the degree of herd stability, positive social interactions, consistent routine and good stockmanship. In large herds negative aspects of social contact can arise and in most dairy systems contact between cows and their calves is low. The distance cows must walk in some pasture-based systems must be included as it could reduce the time available for a cow to choose preferred activities and there may be little novel stimulation or exploration beyond grazing of uniform pastures. Opportunity for rewarding emotions from successful shade seeking behaviour, exploration or interactions with calves or herd mates should be included. Using the framework highlights aspects of pasture dairy systems that may reduce positive experiences for cows such as limited time, shade/shelter or variability in the environment. Designing ways to increase variability, reduce time 'working' and improve social dynamics will increase the opportunity for choice, stimulation and rewarding experiences. An experience-based approach to review the cow's perspective of farm systems will help reveal where changes can be made to facilitate positive experiences in pasture systems, improve welfare and promote a life worth living.

# Intra- and inter-observer reliability of the facial expression in sows around farrowing

*Elena Navarro, Eva Mainau and Xavier Manteca*

*Universitat Autonoma de Barcelona, Campus de la UAB, Plaza Cívica, s/n, 08193, Bellaterra (Barcelona), Spain; elena.navarro@uab.cat*

Changes in facial expression have been shown to be a useful tool to assess the severity of pain in humans and animals. Although it is widely accepted that farrowing is likely to be painful, there are few studies on the assessment of pain caused by farrowing. The objective of this study was to evaluate the intra- and inter-observer reliability of the facial expression in sows around farrowing. The study was carried out on a commercial farm with a total of 21 Danbred sows. Seven nulliparous and 14 multiparous sows were selected based on their body size on the day of parturition. Sows' facial expression were recorded using a video camera per sow. Video recordings were obtained from the beginning of farrowing until the last piglet was born. On day 19 post-farrowing, sows were recorded again. A total of 263 images of facial expressions were obtained: 109 of them during active piglet delivery (considered to be associated with severe pain), 90 during the time intervals between the delivery of two consecutive piglets (considered to be associated with moderate pain) and 64 on day 19 post-farrowing (considered not to be associated with any significant degree of pain). Five facial expression items were identified by an observer who was blinded (Silver Observer, SO): Tension above eyes, Snout angle, Neck tension, Temporal tension and ears and Cheek tension. Each facial zone of all the images, was then ranked by the SO who was blinded to the period when each image was taken using a 4-point scale where = 0 meant 'no pain', 1 meant 'moderate pain', 2 meant 'severe pain' and 3 meant 'that the SO was not confident enough to score pain severity'. In a second phase of the study, eight animal welfare scientists with no previous experience on the study of facial expressions received a one-hour training session on how to assess sows' facial expressions and were then asked to rank a subset of 60 images using the same scale (from 0 to 3) used by the SO. To calculate intra- and inter-observer reliability, Simple Kappa Coefficient ($\kappa$) were calculated using the SAS software. Intra-observer reliability between the moment when the image was taken and the SO punctuation showed a very good agreement for Tension above eyes and Cheek tension zones ($\kappa=0.88$ and $0.90$, respectively), a good agreement for Snout angle and Neck tension zones ($\kappa=0.69$ and $0.64$, respectively) and a moderate agreement for Temporal tension and ears zone ($\kappa=0.56$). Inter-observer reliability (between naïve scientists and the SO) showed a good agreement for Tension above eyes ($\kappa=0.63$) and a moderate agreement ($\kappa$ values from $0.41$ to $0.49$) for the other items. In conclusion, intra- and inter- observer reliability ranged from moderate to very good for all facial expression zones and Tension above eyes showed the highest reliability.

## Sex differences in the behavioural response to tickling in juvenile Wistar rats

*Emma Tivey[1], Sarah Brown[1,2], Alistair Lawrence[2] and Simone Meddle[1]*
*[1]Roslin Institute- Royal (Dick) School of Veterinary studies, Easter Bush Campus, Edinburgh EH25 9RG, United Kingdom, [2]SRUC, Easter Bush Campus, Edinburgh EH25 9RG, United Kingdom; s1102478@sms.ed.ac.uk*

Good welfare is increasingly considered to include the presence of positive experiences. Tickling is a technique that can be used to mimic rough-and-tumble play in rats, particularly in juveniles. Tickling induces positive affective states as shown by 50 kHz ultrasonic vocalisations which are produced during other hedonic behaviours such as feeding. However, sex differences in the response to positive affect are not well understood. The present study investigated potential sex differences in the response to tickling in Wistar rats to test the hypothesis that there are sex differences in the behavioural responses to social play and the underlying neural circuitry regulating positive emotion. Wistar rats (n=32/sex) were pseudo-randomly assigned to a treatment group; tickled (n=16/sex) and untickled controls (n=16/sex). After 5 days of acclimation, rats were placed in an arena for 2 min/day for 10 days. Tickled animals received alternating 15 seconds of tickling and 15 seconds of release for the 2 minutes of testing. Control animals were placed in the arena, but the experimenter's hand was placed motionless on the side of the arena; the rat received no hand contact. The same experimenter was used for all rat interactions. Play behaviour was digitally recorded; the number of scampers (hopping and darting) and the number of hand approaches (forward motions towards the hand) were quantified. Ultrasonic vocalisations (USVs) were recorded by ultrasonic microphone and 50 kHz vocalisations produced during the 2 minute testing period were coded. Brain, faecal and blood samples were taken and are currently being analysed. Behavioural data was analysed using a two-way analysis of variance and Tukey pairwise comparison. Tickled animals, regardless of sex, scamper more than control animals ($F_{3,60}=22.15$, $P<0.001$). Female control rats approached the hand more than female tickled, male control or male tickled animals ($F_{3,60}=5.85$, $P=0.001$). Female tickled rats produced significantly more 50 kHz calls than male tickled and control animals ($F_{3,60}=20.85$, $P<0.001$). These data suggest that tickling was successful in eliciting a behavioural response in both sexes, and that there is a sex difference in USV production during tickling. Thus, the underlying neural and neuroendocrine regulation of this behavioural response to a positive affective state between males and females warrants further investigation.

## Heart rate variability patterns as indicators of stress and welfare in Leghorn hens: a comparative housing system study

*Ashley Bigge[1], Allison Reisbig[1], Sheila Purdum[1], Dola Pathak[2] and Kathryn Hanford[1]*
*[1]University of Nebraska-Lincoln, 1400 R Street, Lincoln, NE 68588, USA, [2]Michigan State University, 426 Auditorium Road, East Lansing, MI 48824, USA; ashley.bigge@huskers.unl.edu*

This study evaluated heart rate (HR) and heart rate variability (HRV) of laying hen pairs in different housing systems following a stressor event. It was hypothesized that aviary hens would have higher HRV than those in cages, and that relatedness would affect HRV patterns. Sixteen hens were used. Replicate 1 used 8 70wk old Hy-line Browns, while replicate 2 used 8 32wk old Bovans White Leghorns. Hens were reared in floor pens until maturity, then housed in conventional cages or aviaries (two pairs each with 2 aviaries and 2 cages per replicate). Cages were each stocked with 3 hens (0.069 m$^2$/hen), while aviaries contained up to 50 hens (0.18 m$^2$/hen). Wireless electrocardiogram (ECG) telemetry devices (M01) were implanted subcutaneously and data were collected using Ponemah 6.2 (Data Sciences International). Two event periods were measured 1 week apart: baseline and post feed failure (PFF). During baseline, hens were cared for normally. The PFF period had a chronic stressor applied to experimental units in a factorial design. This 23-hour feed failure was applied to 1 aviary and 1 cage per replicate, while the other aviary and cage acted as controls following the normal feeding schedule. An acute stressor simulating a swooping predator (a black bag tied to a broom handle swung quickly by the housing units) was applied in both periods to determine whether there was a difference in HRV during PFF compared to return to baseline. Hens were captured daily to turn on telemetry and returned to their housing. ECG was recorded continuously and a 30-minute segment of data, starting with the acute stressor, was chosen for analysis. A 2-second logging rate was used. Comparisons included relatedness (cagemates (CM), non-cagemates (NCM) in same or different housing type), period differences, and interactions. Data were analyzed using a non-linear model and recurrence plots were used to determine HRV patterns. Two values of interest from the plots are %DET (percentage of recurrent points forming diagonal lines due to a deterministic pattern) and LineMax (the length of the longest non-central diagonal line). A lower %DET or significant shortening of LineMax indicates a more chaotic condition of the data. %DET and LineMax were significantly higher during the acute stressor compared to return to baseline, %DET (P=0.0138) and LineMax (P=0.005). The acute stressor was successful in startling the hens. HR peaked in the first minute after the stressor was applied and hens returned to baseline within 2 minutes. When comparing all hens' responses as individuals there were no differences based on period, housing, or removal of feed (P>0.05). However, when comparing pair combinations of relatedness there was a significant difference in period, %DET (P=0.0406). %DET of NCM pairs increased from 94 to 96% between baseline and PFF, while CM remained more stable at 95% for both periods. Strain differences were not part of this analysis. More research needs to be done, but these results imply the presence of relationships among laying hens may be beneficial when responding to a stressor condition.

## A sampling strategy for the determination of infra-red temperature of relevant external body surfaces of dairy cows

*Jashim Uddin, David McNeill and Clive Phillips*
*The University of Queensland, Centre for Animal Welfare and Ethics, School of veterinary Science, Building Number 8134, Gatton 4343, Australia; j.uddin@uq.edu.au*

We investigated the minimum number of cows, thermography sessions, and thermograms within a session required to detect biologically important differences in the infrared temperature (IRT) of external body surface regions of interest. Previous research identified differences of at least 0.4 and 0.9 °C in the IRT of eyes and limbs, respectively, due to stress, so these were taken to be biologically meaningful. Thirty one cows were selected for this study from a subtropical and predominantly Holstein Friesian herd of 202 cows based on extreme left (relaxed cows, n=15) and right (stressed cows, n=16) laterality. Cows were managed with pasture mixed ration offered during the day under cover, gazing during the cooler hours of the day and through the night, and milked twice daily at morning and afternoon in a rapid exit herringbone parlour. Thermogram collection comprised 4 replicates of the head and 4 of the lower forelimbs individually on all cows at approximately the same time (15:00 to 18:00 h) of each day for 6 days whilst standing to be milked. From captured thermograms, minimum, average and maximum IRT were determined: within a box frame that incorporated the periorbital area of both eyes including the outer cantus of each and connected forehead area; right ear and left ear; and from an ellipsoid frames placed around the right eye, left eye, muzzle, coronary band of right and left forelimb. Maximum IRT of each of these regions was less varying than minimum or average, and the eyes and coronary band were more descriptive than ears and muzzle. Contralateral eyes and limbs were similar in their IRT (P>0.05). Consequently maximum IRT of both eyes (BEM = (Left eye + Right eye) / 2), and both limbs (BLM = (Left coronary band + Right coronary band) / 2) were selected for a power analysis. The effects of combinations of thermograms within a session and number of days (thermography sessions) on the precision of BEM and BLM were investigated initially by a restricted maximum likelihood mixed model (REML) in Minitab 18. Each REML identified variances for the factors cows, days, cows×days, cows×days×replicates, and total variances. From corresponding variances of each REML, margins of error (MoE) were calculated at P<0.05, and then least significant differences (LSD = MoE × √2). The LSD's indicated that at the same number of cows the LSD of BLM were approximately double that of BEM. Increasing the number of replicates beyond 2 did not appreciably improve the precision of measurement of BEM and BLM. However, increasing the number of days (thermography session) reduced the LSD of BEM and BLM. In conclusion, this study recommended the capture of 2 thermograms within a session using left or right or both eyes and forelimbs in quick succession. This should be over 2 thermography session on at least 15 cows or 3 thermography session on at least 11 cows to detect 0.4 and 0.9 °C biological changes in the IRT of eyes and limbs, respectively.

## Effects of injection of clove oil or isoeugenol on mechanical nociceptive threshold and behaviour during injection in calves

Anna Juffinger[1], Anna Stanitznig[2], Julia Schoiswohl[2], Reinhild Krametter-Frötscher[2], Thomas Wittek[2] and Susanne Waiblinger[1]

[1]Institute of Animal Welfare Science, University for Veterinary Medicine, Veterinärplatz 1, 1210 Vienna, Austria, [2]Clinic for ruminants, University for Veterinary Medicine, Veterinärplatz 1, 1210 Vienna, Austria; juffingera@staff.vetmeduni.ac.at

Disbudding, is one of the most frequently used interventions in European dairy cattle production. It negatively affects animal welfare due to associated pain and stress. Injecting clove oil (composed mainly of eugenol) or isoeugenol under the horn bud may be a more welfare-friendly alternative, due to cytotoxic and anaesthetic effects of eugenol. We investigated the effect of the injection of both substances under the horn buds on the behaviour of calves during the procedure and on the course of mechanical nociceptive threshold (MNT). Twenty-nine Simmental calves, 1-5 days old, were allocated randomly, but balanced for age and sex, to four treatments: injection (1.5 ml/side) of clove oil (Clove, n=7), isoeugenol (Iso, n=8), saline (Control, n=7) or hot iron disbudding with sedation (Xylazine 0.1 mg/kg i.m.) and local anaesthesia (Procaine hydrochloride 5 ml/side) (Burn, n=7). MNT was measured using von Frey Filaments (vFF, Aesthesio®, Ugobasile) lateral to each horn bud and a pressure algometer (PA; ProdPlus®) on four different locations around each horn bud. The measurements were performed before, 15 min, 6 h, 9 h, 24 h and 3, 7, 14 and 21 days after the treatment. Behaviour during the treatment (frequencies: movements of limbs, head or whole body, tail flicking, teeth grinding, vocalisations; 0-1: strong defensive movements) was observed directly by a person blinded to injection treatment. Frequencies of behaviours (in total and per minute) were analysed using Kruskal-Wallis-test and post-hoc pairwise comparison with Mann-Whitney-U tests. MNT of vFF was analysed with a linear mixed model with the fixed factors treatment, sex and time point, the two-way and three-way interactions. PA data are still under analysis. The duration of treatments differed (KW: P=0.001), being shortest for Control (mean±SD: 35.83±5.98 s), and longer for Iso (65.00±20.49 s), Clove (55.00±30.23 s) and Burn (125.71±74.21 s), which differed from each other, too. During the procedure, due to sedation, Burn calves showed less movements/min than the three injection treatments (KW: P=0.021, MWU: all P<0.05), which did not differ (MWU: all P>0.1). Regarding MNT, we found effects of treatment (P=0.003; $F_{3,14}$=7.426), time (P<0.0001; $F_{8,26}$=10.583), treatment×time (P=0.002; $F_{23,28}$=3.275), treatment×sex (P=0.002; $F_{3,14}$=8.067) and treatment×sex×time (P=0.012; $F_{23,28}$=2,455). Burn had highest sensitivity from time 6 h on until day 21 (overall mean±se 4.9±0.30), followed by Clove (5.6±0.21) and Iso (6.1±0.17), compared to Control (6.2±0.25). Sensitivity in Control increased only very slightly in some animals 15 min and 6 h after injection. Clove but not Iso calves showed the first drop in MNT only 6 h post injection. Thus, clove oil seems to exhibit a transient anaesthetic effect. Our preliminary data indicate some beneficial effects of clove oil and isoeugenol, as compared to hot iron disbudding with respect to MNT, but further parameters (behaviour, effectiveness) need to be taken into account for final evaluation.

## The development of stereotypic behaviours by two laboratory mouse strains housed in differentially enriched conditions

*Emily Finnegan[1], Anna Trevarthen[1], Michael Mendl[1], Elizabeth Paul[1], Agustina Resasco[2] and Carole Fureix[3]*
*[1]University of Bristol, Langford House, Bristol BS40 5DU, United Kingdom, [2]Laboratorio de Animales de Experimentación, Facultad de Cs Veterinarias, UNLP, Argentina, [3]School of Biological & Marine Science, Portland Square Drake Circus, Plymouth PL4 8AA, United Kingdom; emily.finnegan@bristol.ac.uk*

Stereotypic behaviours (SB) have previously been associated with sub-optimal housing conditions, often linked to a lack of stimuli within the captive environment. Conspecifics can vary in their SB response to such environments and differential behavioural patterns have been documented in various laboratory mouse strain, despite identical housing conditions. Prior studies indicate providing enrichment can reduce the extent to which SB is performed. We aimed to monitor the development of SB (patterned climbing on the cage lid, bar mouthing and route-tracing) by two commonly used laboratory mouse strains: C57BL/6J (C57) and DBA/2J (DBA) in differentially enriched housing conditions. Studies have suggested these strains differ behaviourally under certain stressors and have variable SB responses. Additionally, we monitored the effect of environmental adjustment on the performance of SB by individuals of each strain. 62 three-week-old females were mixed-strain pair housed and assigned to a treatment cage: Enriched (EE) (n=30) or Non-Enriched (NE) (n=32). Small NE cages contained basic bedding, a tunnel and cardboard to gnaw. Larger EE cages had various nesting and gnawing materials, hideaways and a running wheel. After baseline observations, the following treatment manipulation was implemented: Half of the EE (n=15) were moved to NE cages (EE-NE) and half remained in the original treatment (Long Term EE). Simultaneously, half of the NE (n=16) were moved to EE cages (NE-EE) and half remained (Long Term NE). We predicted there would be more SB performed by individuals subjected to an EE-NE treatment manipulation and more SB overall in NE cages. We hypothesised the DBA strain would exhibit more locomotor SB, consistent with previous findings. To test these hypotheses, the mice were observed within home cages for 12 days prior to treatment manipulation and 12 days after via scan sampling (24/animal/day). We excluded route tracing due to insufficient data. Preliminary results suggest prior to treatment manipulation, mice displayed significantly more bar mouthing ($t_{60}$=4.30, P<0.001) and patterned climbing ($t_{60}$=2.23, P=0.001) in NE than EE conditions. Following treatment manipulation, results show a significant effect of treatment on both patterned climbing ($F_{3,54}$=3.46, P=0.023) and bar mouthing ($F_{3,54}$=6.73, P=0.001). As predicted, SB increased in EE-NE whereas they decreased in NE-EE (patterned climbing: P=0.032; bar mouthing: P=0.003). Contrary to our hypothesis, initial analyses indicate C57s exhibited more patterned climbing than DBAs before treatment manipulation ($F_{1,54}$=3.37, P=0.072) but no apparent difference was observed for bar mouthing ($F_{1,54}$=0.51, P=0.478). Our results suggest enrichment can directly affect the development of SB in both strains and SB frequency adjusts accordingly to the EE and NE housing alterations. Environmental conditions may differentially influence oral and locomotor SB and warrants further investigation.

## Providing complexity: a way to enrich cages for group housed male mice without increasing aggression?

*Elin M. Weber[1], Sofia Elgåsen Tuolja[1], Charlotte Berg[1] and Joseph P. Garner[2]*
*[1]Swedish University of Agricultural Sciences, Department of Animal Environment and Health, P.O. Box 234, 532 23 Skara, Sweden, [2]Stanford University, Department of Comparative Medicine, Veterinary Service Center, 287 Campus Drive, Stanford, CA 94305-5410, USA; elin.weber@slu.se*

Group housing of mice used in research is mandated by law in many countries for both animal welfare reasons, and to ensure that experiments are performed on animals with a good general physical and mental health status. However, aggression is a major problem affecting nearly 15% of group housed male mice and causing pain, suffering and death. Environmental enrichment is often used to minimize negative stress and give animals more control. However, defensible enrichment items can trigger aggression in group housed male mice, thus increasing a welfare problem instead of alleviating it. One explanation could be cage size. Only a few enrichment items fit inside a conventional mouse cage, and shortage of highly valued resources, such as enrichment objects, can lead to competition. Instead of adding enrichment we therefore used an alternative cage design to increase cage complexity and provide visual barriers. To minimize the risk of territorial behaviour, escape routes were provided throughout the cage, and nest sites, food and water provided at two locations. The study was performed under a Stanford-approved IACUC protocol. We used forty eight C57BL/6 male mice arriving at the animal facility in the ages 6 or 9 weeks (relatedness unfortunately unknown). They were housed in groups of four in standard mouse cages (SC) for the first week, and then transferred to the complex cage (CC). Cages were video recorded, inspected at least once daily to detect signs of aggression, and all mice checked for wounds three times per week. Since aggression tend to increase after handling, aggressive behaviour was also observed from video recordings during 1 h after wound checks. Three groups had to be euthanized due to aggression, two in SC and one shortly after being moved to CC. Fighting was observed in six of the remaining groups, five from the 6 weeks group and one from the 9 weeks group. In total, 21 behaviour observations were made; nine in SC and 12 in CC. Two of the groups fought in both SC and CC, two fought only in SC, and two only in CC. In two cages, fighting was observed but no wounds detected during wound scoring. The amount of data was too small for statistical analysis but of the mice observed fighting, the number of fights were 1.6 fights/hour in SC and 0.6 fight/hour in CC. The mean duration spent fighting per hour was similar in both systems (29 sec in SC vs 35 sec in CC). Mice did not stop fighting completely in the complex cage, but compared to other studies reporting that aggression increase with enrichment, it did not seem to get worse. Mice in the complex cage were observed using shelves in the cage to move away from aggressive encounters, which might mimic a natural behaviour response when approach with an aggressor. However, housing male mice in groups is challenging, and more studies are needed to investigate if housing male mice in a complex cage might be an alternative to adding enriching items to decrease levels of aggression.

## Environmental controllability and predictability affect the behavioural adaptability of laying hen chicks in novel situations

*Lena Skånberg[1], Ruth C. Newberry[2], Inma Estevez[3], Nicolas Nazar[4] and Linda J. Keeling[1]*
*[1]Swedish University of Agricultural Sciences, Department of Animal Environment and Health, Box 7068, 750 07 Uppsala, Sweden, [2]Norwegian University of Life, Faculty of Biosciences, Department of Animal and Aquacultural Sciences, 1432 Ås, Norway, [3]Neiker-Tecnalia Basque Institute for Agricultural Research, Department of Animal Health, Vitoria-Gasteiz, Spain, [4]Instituto de Investigaciones Biologicas y Tecnologicas, Consejo Nacional de Investigaciones Científicas y Técnicas, Universidad Nacional de Córdoba, Instituto de Ciencia y Tecnología de los Alimentos, Facultad de Ciencias Exactas, Físicas y Naturales, Córdoba, Argentina; lena.skanberg@slu.se*

Animals can be stressed when aspects of their environment are changed or when they are transferred to a new environment. This is a particular problem in laying hens, possibly because under commercial conditions, they are often reared in relatively stable, unenriched environments. Low environmental variability may impair their capacity to learn new routines, use new equipment, and exploit new resources. In this study, we investigated the hypothesis that the levels of controllability (by varying the amount of choice) and predictability (by varying the degree of stability) of the early environment would affect the adaptive capacity of laying hen chicks. According to a 2×2 factorial design, chicks (Bovans Robust) were reared in one of four treatment combinations comprising no choice vs choice and a static vs dynamic environment, by manipulating the litter and perch types present. Treatments were balanced across 16 pens (1.2×2.4 m) with 20 chicks in each pen. Pens with 'choice' had four different types of litter and types of perches while 'no choice' pens had only one litter and perch type. The location or type of litter and perch was changed three times weekly in 'dynamic' pens but never changed in 'static' pens. After one month, chicks were exposed to various behavioural tests. Compared to chicks given no choice, chicks with a choice of perch and litter types took less time to solve a detour test and so reunite with their pen mates (mean±SE, 186±39 vs 298±39 s, P=0.02, LMM), and had a tendency for higher weight gain (P=0.09, Kruskal-Wallis rank sum test). Compared to chicks reared in a static environment, those reared in a dynamic environment had a shorter latency to their first movement when placed in novel new surroundings (9±5 vs 26±15 s, P=0.05, LMM), and moved more in an open arena (18±2 vs 14±2 lines crossed, P=0.05, GLMM). Our results suggest that a dynamic early environment that offers choices promotes behavioural plasticity in chicks, leading to greater adaptability in later challenging situations. We conclude that, at the levels investigated in this study, controllability and predictability in the early rearing environment had different but positive effects on the behavioural development of chicks, with implications for environmental enrichment strategies.

## Environmental enrichment, group size and confinement duration affect play behaviour in goats

*Regine Victoria Holt, Judit Vas and Ruth C. Newberry*
*Norwegian University of Life Sciences, Faculty of Biosciences, Department of Animal and Aquacultural Sciences, 1432 Ås, Norway; regine.victoria.holt@nmbu.no*

Given current interest in play behaviour as a potential indicator of positive welfare, it is important to understand how environmental factors affect play behaviour. We investigated three environmental factors hypothesised to stimulate the play behaviour of goats in an outdoor enclosure: (1) environmental enrichment (EE), due to an increase in opportunities to perform diverse behaviours, (2) increased group size (GS), due to greater social complexity and/or 'safety in numbers', and (3) increased duration of confinement indoors (CI), due to a rebound effect when released in a more spacious and complex environment. We also expected that play would be negatively associated with vigilance, based on hypothesised play suppression when alarmed. Twenty 6-8-month-old female dairy goats, housed together indoors in a slatted-floor pen, were observed in three experiments (Exp. 1-3, each 3 weeks long) whereby they were divided into temporary groups for 30-min observation sessions in a familiar 40 m$^2$ outdoor enclosure. In Exp. 1, varying groups of five goats were exposed to both a control condition (CC, no added enrichments) and an enriched condition (EC, addition of a climbing structure, suspended ball and bucket of small branches) in alternating observation sessions (6 sessions/goat/condition). In Exp. 2, we observed GS varying from 2-8 goats (3 sessions/GS, with the order of exposure to different GS balanced across individuals, all in EC). In Exp. 3, we observed six stable groups of 3-4 goats following CI for 2, 3, 4 and 5 days (in a balanced order across groups, all in EC). Locomotor play was defined by spontaneous, rapid, exaggerated movements in different planes. Social and object play were recognized by the occurrence of brief, easily interrupted, non-harmful attempts to physically manoeuver pen-mates or objects. Vigilance was characterised by standing motionless with head up, a fixed gaze and ears directed forward. Data were collected by instantaneous and 1-0 sampling and analysed using generalised linear mixed models with goat and group as random factors. Compared to CC, EC was associated with more locomotor (0.03±0.03 vs 2.4±0.5% of scans, P<0.001), social (0.9±0.3 vs 4.7±1.3%, P<0.001) and object (0.9±0.2 vs 7.6±1.3%, P<0.001) play, while vigilance was unaffected (39.5±3.8 vs 32.4±3.5%, P=0.124). In Exp. 2, object play (P<0.001), but not locomotor (P=0.630) or social (P=0.149) play , increased with increasing GS while vigilance declined (P<0.001). In Exp. 3, locomotor (P=0.002), social (P<0.001) and object (P=0.001) play declined with increased CI duration, accompanied by increased vigilance (P<0.001). Higher vigilance was associated with less play, though not in all contexts. Rather than showing a rebound in play following release outdoors, the Exp. 3 results suggest that goats became sensitized to the outdoor environment with increased time indoors. Overall, EE, GS and CI affected play, but not consistently across play types or always in the predicted direction. Our findings highlight the need to take the environmental context and play type into account when interpreting play frequencies.

## Assessing the behaviour of fast-growing broilers reared in pens with or without enrichment

Zhenzhen Liu[1], Stephanie Torrey[1], Ruth C. Newberry[2] and Tina Widowski[1]
[1]University of Guelph, Department of Animal Biosciences, 50 Stone Rd E, N1G 2W1 Guelph ON, Canada, [2]Norwegian University of Life Sciences, Department of Animal and Aquacultural Sciences, Faculty of Biosciences, 1432 Ås, Norway; zliu17@uoguelph.ca

Fast-growing broilers are at risk of developing welfare problems which may lead to negative affective states. Previous research suggests that providing broilers with suitable environmental enrichment could improve their welfare by reducing lameness and increasing mobility. However, the extent to which affective states of broilers can be improved by enrichment remains unclear. Play behaviour is often considered an indicator of welfare because its frequency is often reduced when animals are under biological challenge, and it is increasingly being used as an indicator of a positive affective state. Our objective was to assess the play behaviour of broilers reared in enriched (E) or non-enriched (NE) environments. We housed 456 Ross 708 broiler chickens in 12 floor pens (38 birds/pen; 19 females and 19 males) from 1 day until 43 days of age. Each pen (1.68×2.29 m²) was equipped with one feeder, nipple drinkers and wood shavings. E pens were enriched with a raised platform with a 25° ramp, a hanging scale, mineral pecking stone and suet feeder filled with wood shavings. Two tests performed at the pen level were intended to stimulate play behaviour: giving extra space and offering objects suitable for worm running. The 'extra space' tests conducted at day 8, 22 and 36 involved temporarily removing a feeder to create an area of unoccupied space. 'Worm running' tests conducted at day 10, 25 and 39 involved throwing a 'worm' made from twisted tissue paper into the pen. In both tests, video recordings were used to quantify behaviour for 5 minutes using continuous all-occurrence sampling. Generalized linear mixed models were used to evaluate effects of treatment, age and their interaction on play frequency/pen/5 min. During 'extra space' tests, total play behaviour (sum of run, wing-assisted run, wing-flap, play-fight) decreased as the chickens aged, but there was a treatment × age interaction (P<0.001). On day 8, the frequency of play behaviour was higher in NE (Least Squares Mean±SD: 3.47±0.25) than E pens (0.87±0.25; P<0.001) whereas the difference was non-significant on day 22 (NE: 2.04±0.25; E: 0.99±0.25) and day 36 (NE: 0.48±0.25; E: 0.51±0.25). During the 'worm running' tests, the total frequencies of worm chase (P<0.001), worm run (P=0.035) and worm exchange (P=0.034) were higher in NE (3.20±0.16, 1.02±0.05, and 0.40±0.04, respectively) than E pens (1.92±0.16, 0.85±0.53 and 0.28±0.04, respectively). Our findings indicate that NE birds played more than E birds during tests intended to stimulate play behaviour. The larger contrast between the NE environment before and after giving the test stimuli (i.e. opening up extra space and offering a 'worm') compared to that in the E pens may have led the NE birds to be more easily stimulated to play during the tests. According to this interpretation, the higher play behaviour in the NE birds reflected transiently higher responsiveness in the test context rather than reflecting an underlying state of greater positive welfare when kept in a non-enriched environment.

## Differences in laying behaviour associated with nest box design in commercial colony cage units

*Sarah Lambton[1], Federica Monte[2] and Christine Nicol[3]*
*[1]University of Bristol, Bristol Veterinary School, Langford House, Langford, North Somerset, BS40 5DU, United Kingdom, [2]ADAS, Spring Lodge, 172 Chester Road, Helsby, Cheshire, WA6 0AR, United Kingdom, [3]Royal Veterinary College, Hawkshead Lane, North Mimms, Herts, AL9 7TA, United Kingdom; sarah.lambton@bristol.ac.uk*

Pre-laying and nesting behaviours are considered behavioural needs for laying hens; furnished colony cages are intended to facilitate their performance. However, some eggs are laid outside nest boxes, raising questions over the extent to which hens' behavioural needs are met. Hens show substrate preferences and prefer seclusion during laying; this study investigated whether commercial conditions adequately satisfy these preferences. Forty-eight cages throughout a commercial laying house were each fitted with one of four commercially available nest box floor materials (hard plastic grid (PG); artificial turf (AT); textured rubber mat (RM); plastic coated wire (PW)) and one of two curtain lengths: long and short (LC and SC; 11 and 22 cm, respectively between the nest box floor and the bottom of the curtain). A 2×4 experimental design was used, thus six cages for each treatment combination. Visits took place at 23, 31, 40, 50, 59, 67 and 75 weeks of age. At each visit we recorded number of eggs laid inside and outside nest boxes, pre-laying behaviour, and plumage damage. Multilevel regression models, accounting for the hierarchy within the dataset (repeated measures within cage within visit), examined associations between the treatment, laying location, pre-laying behaviour and plumage score. Using egg counts from all visits (n=366) number of eggs laid outside nest boxes decreased with age ($\chi^2$=226.0, df=2, P<0.001), and was higher in SC cages ($\chi^2$=4.528, df=1, P=0.033). At 23 weeks of age (n=48) more eggs were laid outside the nest box in PW cages and fewest in PG cages ($\chi^2$=11.6, df=3, P=0.009). Analysis of pre-laying behaviour, based on 197 video observations found rate of entrances/exits of the nest box increased with age in all floor treatments, except PW, where rate began high relative to other treatments, and decreased with age ($\chi^2$=24.3, df 7, P=0.001). Rate of entrances/exits was lower in LC cages ($\chi^2$=5.83, df 1, P=0.016). Rate of sitting inside the nest box decreased with time since lights-on in PW and PG cages, and increased with time since lights-on in AT and RM cages ($\chi^2$=37.2, df 7, P<0.001). Plumage scores (n=1,980) increased (worsened) with age ($\chi^2$=223.6, df 3, P<0.001) and showed a tendency to vary with nest box floor type ($\chi^2$=7.43, df 3, P=0.059); more plumage damage was observed in PW cages, compared with PG and AT cages. Our data suggest hens are less likely to use nest boxes if they have short curtains or a PW floor, particularly early in the laying period. High rates of entrances/exits in PW and SC cages suggest hens perform more nest searching in those cages. Analysis of plumage damage suggested that birds in PW cages performed more injurious pecking, which could be suggestive of frustration or stress. These results raise questions about the ability of cages with PW nest box floors and/or short curtains to meet the behavioural needs of laying hens.

## Long-term effects of peat provision in broiler chicken flocks

*Judit Vas[1], Neila BenSassi[2], Guro Vasdal[3] and Ruth C. Newberry[1]*
[1]*Norwegian University of Life Sciences, Faculty of Biosciences, Department of Animal and Aquacultural Sciences, P.O. Box 5003, 1432 Ås, Norway,* [2]*Neiker-Tecnalia, Campus Agroalimentario de Arkaute, Apto 46, 01080 Vitoria-Gasteiz, Spain,* [3]*Norwegian Meat and Poultry Research Centre, P.O. Box 396, 0513 Oslo, Norway; judit_banfine.vas@nmbu.no*

Under commercial conditions, environmental enrichments may be provided infrequently and at limited locations in chicken houses, raising questions about the extent to which they are enriching. We hypothesized that broilers in flocks given limited access to peat would nevertheless exhibit more intense use of peat at a later age compared to broilers in flocks never previously given access to peat. We observed 27 commercial broiler flocks (Ross 308), 9 control flocks with no previous experience with peat and 18 flocks given access to peat starting from 7 days of age in limited locations and often not available all the time. During a visit to each flock at approximately 28 days, we placed 10 l of fresh peat in a central location of the house (peat condition) and pretended to do so at another comparable location (sham condition). We observed behaviour in the peat-covered patch (or equivalent sized patch in sham condition) for 20 min. Latencies of the first 5 birds to perform ground pecking, ground scratching, vertical wing shaking, standing and lying in each patch were recorded and averaged per flock and condition. We also recorded the number of birds lying, and total number present, in each patch based on instantaneous scan sampling at 1-min intervals. Data were analysed in generalised linear mixed models incorporating previous flock experience of peat (vs no previous experience), current peat exposure (vs sham) and their interaction, with flock as a random factor. Due to scarcity of ground scratching and vertical wing shakes in sham patches, only the previous experience effect was examined (in peat patches only) for these variables. We found that flocks with previous experience of peat had a tendency for a shorter latency to ground peck (70±28 vs 141±45 s, P=0.055), shorter latencies for ground scratching (308±79 vs 813±123 s, P<0.001) and vertical wing shaking (730±89 vs 1,088±68 s, P=0.010), similar latencies to stand and lie (P>0.05), a similar number of birds present (P>0.05) and a similar proportion of birds lying (P>0.05), when compared with flocks without previous experience of peat. In peat (vs sham) patches, there was a shorter latency to ground peck (93±24 vs 197±36 s, P=0.012), a longer latency to stand (180±56 vs 37±8 s, P=0.001) and lie (250±49 vs 97±22 s, P=0.001), a lower number of birds present (8.4±0.2 vs 9.1±0.3, P<0.001) especially in flocks without peat experience (P<0.001), and a lower proportion of birds lying (0.49±0.02 vs 0.84±0.01, P<0.001). Thus, birds in the peat experienced flocks were quicker to exploit the fresh resource by performing natural behavioural elements of foraging and dust-bathing whereas flocks without previous peat experience were more hesitant to interact with the peat. We conclude that sparse and ephemeral provision of peat as an enrichment material was sufficient to contribute to long-term memory of the resource.

## Which rooting materials make a weaner most happy?

*Marko Ocepek, Ruth C. Newberry and Inger Lise Andersen*
*Norwegian University of Life Sciences, Faculty of Biosciences, P.O. Box 5003, 1432 Ås, Norway;*
*marko.ocepek@nmbu.no*

Although European pig producers are legally required to provide rooting material as enrichment for pigs, this requirement has been primarily oriented towards reducing negative aspects of welfare. Information is lacking on the impact of different types of rooting materials on the promotion of positive affective states. We hypothesised that intermittent daily access to rooting materials, especially different types provided in combination, would stimulate positive affect in weaned pigs. We offered pig litters 10 l of rooting material (silage, straw or peat, or a combination of all three) twice daily, in comparison to no added rooting materials (control condition, wood shavings litters present in all pens). Behaviours considered indicative of positive (exploration, play, tail curled, wagging) and negative (aggression, ear or tail biting, tail down) affective states in this context were quantified from video recordings by 1-0 sampling. Over five weeks, 10 litters of weaned pigs (TN70 sows × Duroc boar) were assigned to one condition weekly (order balanced across litters), and behaviour was assessed during the 30 min before and after delivery of rooting materials on Days 1 and 4 each week. The effect of condition (control, silage, straw, peat or combo) on positive (mean ± SE: exploration, 43.1±0.9% of scans; play, 1.8±0.1%; tail curled, 49.2±0.8%; wagging, 15.2±0.5%), and negative (aggression, 1.1±0.1%; ear or tail biting, 0.9±0.1%; tail down, 3.8±0.3%) behavioural expressions was analysed using a generalised linear mixed model with binominal distribution and pig nested within pen as a random effect. Time period (before vs after), time of day (morning vs afternoon), day of week (1 vs 4), sex (male vs female), condition by time period and condition by time of day were included in the model as fixed effects and litter size (n), starting bodyweight (kg) and week (1-5) were continuous covariates. The peat and combo conditions resulted in higher levels of exploration, play, tail curled and wagging (P<0.001) and lower levels of aggression (peat, 0.3±0.1; combo, 0.4±0.1; vs control, 1.9±0.3%; P<0.001), ear or tail biting (peat, 0.8±0.2; combo, 0.4±0.1; vs control, 1.8±0.3%; P<0.001) and tail down (P<0.001) compared to control, with the silage and straw being intermediate. Pigs showed more exploration, tail curled (P<0.001) and wagging (P<0.001), and less aggression and ear or tail biting (P<0.001), after than before material provision whereas an increase in levels of play and tail wagging after material provision occurred only in the peat and combo conditions (P<0.001). More exploration, play, tail curled and wagging (P<0.001), and less aggression (P=0.009), occurred in the afternoon than the morning. More exploration, play, tail curled and wagging also occurred on Day 1 than 4 (P<0.001) and, with increasing age, exploration, play and tail wagging declined (P<0.001). Sex, litter size and bodyweight were not consistently associated with either positive or negative behavioural expression (P>0.05). Our results suggest that peat, or peat in combination with straw and silage, were the most effective rooting materials for inducing positive, and reducing negative, affective states in weaned pigs.

# Nest-building material affects pre-partum sow behaviour and piglet survival in crate-free farrowing pens

*Ellen Marie Rosvold[1,2], Ruth C. Newberry[1] and Inger Lise Andersen[1]*
*[1]Norwegian University of Life Sciences, Faculty of Biosciences, Department of Animal and Aquacultural Sciences, P.O. Box 5003, 1432 Ås, Norway, [2]Nord University, Faculty of Biosciences and Aquaculture, P.O. Box 2501, 7729 Steinkjer, Norway; ellen.m.rosvold@nord.no*

The objective of this study was to evaluate the effects of providing different nest-building materials to loose-housed sows on nest-building behaviour and activity budget before farrowing, and on farrowing duration and piglet mortality. Sows (Norsvin Landrace × Swedish Yorkshire) were loose-housed in individual farrowing pens with wood-shavings bedding. They were provided with peat (n=18), long-stemmed straw (n=17) or the control condition (n=18; no additional substrate), from two days before expected farrowing until farrowing. From video recordings of each sow, behaviour in the last 12 h before onset of farrowing was instantaneously scan sampled at 5-min intervals and farrowing duration registered from first to last piglet born. From birth to weaning, all dead piglets were subjected to post mortem examination. Based on generalized linear mixed models, sows provided with straw or peat spent a greater proportion of the observations nest building compared to control sows (straw: 16.5±1.4, peat: 14.5±1.3, control: 12.3±1.1% of scans, P<0.001). The sows with straw exhibited the highest number of nest-building elements (straw: 3.9±0.2, peat: 2.9±0.2, control: 2.7±0.2, P<0.001), rested more (straw: 62.1±2.5, peat: 57.8±2.4, control: 60.8±2.6% of scans, P<0.001), and had fewer observations with stereotypies (straw: 0.8±0.2, peat: 2.1±0.4, control: 2.2±0.4% of scans, P<0.001) than sows in the other two groups. They also had the shortest farrowing duration closely followed by sows with peat, whereas control sows had the longest (straw: 295.8±41.1, peat: 322.7±46.7, control: 438.1±82.6 min, P<0.001). The percentage of stillborn piglets was lowest among sows in the straw group compared to the other two (straw: 2.8±1.0, peat: 6.0±1.8, control: 8.1±1.9% of total born, P<0.001). Our results demonstrate that provision of straw or peat in comparison to the control stimulated more nest-building behaviour, and eased the birth process resulting in shorter farrowing duration. Although peat gave better results than the control condition, our results suggest that straw was the most appropriate nest-building material as it elicited a wider range of nest-building elements, increased resting and reduced stereotypies in the last 12 h pre partum, and resulted in the lowest percentage of stillborn piglets.

## Brush use by dairy heifers

*Jennifer M.C. Van Os[1,2], Savannah A. Goldstein[2], Daniel M. Weary[2] and Marina A.G. Von Keyserlingk[2]*
*[1]University of Wisconsin, Madison, WI, USA, [2]University of British Columbia, Vancouver, BC, Canada; jvanos@wisc.edu*

For cattle, grooming is an important behavior that can be directed toward objects in their surroundings. The use of man-made brushes by confined dairy cattle has been studied in adult cows, but little research has focused on younger animals. Our objective was to describe the usage patterns of brushes by naïve dairy heifers. Groups of 4 heifers (n=11 sequential groups, 144±7.4 d old) were introduced to the experimental pen; visual contact with conspecifics in an adjacent pen was blocked with plywood. Each group was provided 4 deck-scrub brushes (25-cm long) with either stiff or extra-stiff bristles, mounted on the walls either horizontally or vertically (1 of each bristle stiffness × brush orientation combination in a 2×2 factorial design, balanced by location). Continuous video recordings (24 h/d) were scored and averaged for 2 focal heifers/group on d 1, 2, and 6 of exposure for their use of the brushes for oral manipulation (brush contact with the mouth, tongue, or nose) or grooming (rubbing the head, neck, or rest of the body with a brush); these measures were summed for total brush use. Linear mixed models were used to evaluate patterns of brush use across days (1, 2, 6) and preferences (based on total use of each brush) for bristle stiffness, brush orientation, and their interaction, as well as for brush location (close to the feed bunk, the adjacent pen, and their interaction). Latency to use any brush after entering the pen was 3.4±4.9 min (mean±SD; range: 8 sec to 17 min 45 sec among individuals). Heifers used the brushes for oral manipulation (39.7±17.5% of brush use, mean±SD) and grooming (60.3±17.5%), predominantly of their heads (89.9±5.4% of grooming). Total brush use (summed for all 4 brushes/group) was highest on d 1 [45.9 min; 95% confidence interval (CI): 33.2-63.3 min, back transformed from natural log values], and decreased on d 2 (25.0 min, CI: 18.4-34.0 min; day effect: $P=0.004$; pairwise: $P=0.010$), remaining similar on d 6 (21.0 min, CI: 15.4-28.5 min; pairwise: $P=0.41$). This pattern across days was mainly driven by decreases in grooming (day effect: $P<0.001$) but not in oral manipulation of the brushes ($P=0.13$). On d 1, the latency with which heifers first used each brush indicated a preference for those mounted closer to either the feed bunk ($P=0.027$) or the adjacent pen ($P=0.004$; no interaction, $P=0.85$). On d 2 and 6, after heifers were no longer naïve, they spent more time using brushes mounted near the adjacent pen of heifers ($P=0.010$), but they showed no other preferences for proximity to the feed bunk ($P=0.45$, no interaction). Heifers also showed no preferences for bristle stiffness or brush-mounting orientation ($P≥0.13$, no interactions), whether for oral manipulation or for grooming on d 2 and 6. Despite having no previous experience, dairy heifers began using the brushes immediately. Overall brush use decreased after the first day of exposure, but did not further decrease 4 d later. Heifers spent more time using brushes mounted near an adjacent group of conspecifics, but showed no preference for brush attributes or mounting orientation. These findings suggest that young dairy heifers will readily use small, stationary brushes for both grooming and oral manipulation.

**Impact of environmental enrichment on circadian patterns of feedlot steer behavior**

*Rachel Park and Courtney Daigle*
*Texas A&M University, Animal Science, 474 Olsen Blvd, College Station, TX 77845, USA; rachelpark@tamu.edu*

Environmental enrichment (EE) has the potential to increase the environmental complexity of the feedlots and to enhance cattle welfare by providing mental and physical stimulation. The objectives of the present study were to: (1) identify circadian patterns of feedlot steer behaviour; and (2) evaluate the impact of EE on these circadian patterns. Predominantly British and British continental crossbred steers (n=54) were shipped to Texas A&M AgriLife Feedlot in Bushland, Texas, blocked by weight and assigned to pens. Pens were randomly assigned to one of two treatments: (1) No enrichment (CON; n=3 pens at 9 steers/pen); and (2) BRUSH (Cattle brush; n=3 pens at 9 steers/pen). Sample size was deemed sufficient through power analysis of previous feedlot cattle behavioral research. Video recordings were decoded using continuous observation to determine the hourly frequency of headbutting, mounting, kicking, bar licking, tongue rolling, allogrooming and brush usage from 08.00 to 17.30 on d -2, -1, 0, 1, 2, 4, 8, 16, 32 and 64 relative to brush implementation. All ten research days were combined for analysis. Impact of time (hour), treatment and their interaction on cattle behavior was evaluated using a General Linear Mixed Model (PROC MIXED) in SAS. A treatment by time interaction was observed for mounting (P=0.002) and allogrooming (P=0.01) frequency. Cattle in BRUSH pens performed fewer mounts (0.01±0.001) than CON cattle (0.04±0.001) at 16.00, however, cattle in CON pens performed fewer mounts (0.001±0.001) than BRUSH cattle (0.01±0.001) at 17.00. Allogrooming increased throughout the day with BRUSH cattle performing significantly fewer allogrooming bouts (0.15±0.004; 0.11±0.004) compared to CON cattle (0.33±0.004; 0.33±0.004) at 14.00 and 15.00. Cattle performed the most (P<0.001) headbutts at 09.00 (2.38±0.01), 10.00 (3.11±0.01), 16.00 (4.17±0.01) and 17.00 (3.05±0.01). Throughout the entire day, BRUSH cattle engaged in fewer bar licking bouts (0.02±0.01) compared to CON cattle (0.06±0.001; P=0.02). Time of day impacted bar licking (P<0.001) and tongue rolling (P<0.001) with the lowest frequency occurring at 08.00 (0.01±0.001; 0.11±0.004) and 17.00 (0.01±0.001; 0.14±0.004). Cattle interacted with the brush most frequently at 13.00 (0.20±0.002), 14.00 (0.31±0.002), 15.00 (0.23±0.002) and 16.00 (0.36±0.002; P<0.001). Knowledge of behavioral patterns is critical to advancing feedlot cattle welfare research and provides information to husbandry technicians regarding cattle behavioral expectations.

## Brush use and displacement behaviors at a brush in Angus crossbred feedlot cattle

*Xandra Christine Meneses, Rachel Park and Courtney Daigle*
*Texas A&M University, Department of Animal Science, 2471 TAMU, College Station, TX 77843-2471, USA; xandm8@tamu.edu*

Environmental enrichment (EE) provides mental and physical stimulation to animals housed in captivity. Brushes as EE present cattle the opportunity to perform diverse grooming behaviors. Previous research demonstrated that feedlot cattle engage in fewer stereotypic and aggressive behaviors when supplied a brush. However, competition to access and use a novel object may unintentionally compromise animal welfare. If EE is to be adopted at the commercial scale, there is a need to understand the level and intensity of resource competition. The objective of this study was to evaluate patterns of displacement behaviors at a brush (BD) performed by feedlot cattle. Twenty-seven yearling British and British-Continental crossbred steers were sorted by body weight into a light ($283.95\pm13.75$ kg) and heavy block ($320.69\pm12.97$ kg). Steers were then allocated to one of three pens (n=3 pens with 9 steers/pen). Each pen was $25.5\times7$ m ($19.83$ m$^2$ per steer) and included earthen flooring, a partial roof covering ($5\times7$ m; 5 m$^2$ per steer), one mounted L-shaped cattle brush, one automatic water trough, and nine individual Calan head gate feeders. Video recordings were decoded from 08.00 to 17.30 for displacement frequency, as well as brush use frequency and duration on d 0, 1, 2, 4, 8, 16, 32 and 64 relative to brush implementation. For each displacement event, the actor and reactor were recorded. On average, $90.75\pm1.74\%$ of cattle used the brush (d 0, d 64 = 96%; d 1, d 2, d 4 = 93%; d 8, d 16, d 32 = 85%). Brush usage bouts averaged $4.15\pm0.27$ counts/day (range: 0 – 22). Brush usage duration averaged $464.31\pm38.27$ seconds/day (range: 0 – 3,297.28). Impact of research day on cattle behavior was evaluated using PROC GLIMMIX in SAS. The frequency of BD initiated ($P<0.0001$) and BD received ($P<0.0001$) decreased over time. On d 0, each steer initiated $3\pm0.70$ displacements and received $3\pm0.44$ displacements; however, by d 64, steers initiated $0.6\pm0.20$ displacements and received $0.6\pm0.20$ displacements. The frequency of BD initiated was positively associated with BD received (PROC CORR; $r=0.49$, $P<0.0001$), brush usage duration ($r=0.15$, $P=0.03$) and brush bouts ($r=0.26$, $P=0.0001$). Cattle maintained sustained interest in cattle brushes as EE, while competition and displacements decreased over time. The data suggests that cattle brushes provided at this density do not create competitive environments, and all animals can access the EE as needed.

## Behavior of recently weaned organic Holstein calves exposed to a mechanical calf brush

*Ana Velasquez-Munoz[1], Diego Manriquez[1], Sushil Paudyal[1], Gilberto Solano[1], Hyungchul Han[2], Robert Callan[3], Juan Velez[4] and Pablo Pinedo[1]*
*[1]Colorado State University, Animal Sciences, Fort Collins, CO 80523, USA, [2]Polytechnic University Pomona, Pomona, Pomona, CA 91768, USA, [3]Colorado State University, Department of Clinical Sciences, Fort Collins, CO 80523, USA, [4]Aurora Organic Dairy, Platteville, CO 80651, USA; pinedop@colostate.edu*

Calf stress at weaning and during transition to group pens represents a concern in dairy farms. Favoring natural behaviors, such as grooming, may help on reducing this challenge. Our objective was to describe the use and the effects of a mechanical grooming brush on the behavior of recently weaned calves (94±7 d), after transferring from individual to group housing. Two treatment groups (non-exposed control [CON, n=81]; automated brush [AB, n=81]) were compared enrolling organic Holstein heifers that were monitored for 3 weeks after transferring. Four cohorts, comprising one CON and one AB group (19-20 calves/pen) were enrolled sequentially. At enrollment, each calf was affixed with a previously validated 3-D accelerometer sensor attached to the ear. Continuous measurements (min/h) were generated for the following behaviors: not-active, eating, and ruminating time. In addition, interactions (contact) of calves with the brush were assessed by video recording. Each calf was weighted using a mobile crate scale at enrolment and at the end of the observation period. Calf health was evaluated daily by trained farm personnel. Behavioral data were summarized as daily averages (min/h for 24 h periods) for the whole study period. Data were examined using repeated measures analysis, with day or hour as the time unit. Overall, 97% of the calves used the brush, having a first interaction within 60 h with a mean (SE) of 7 (±9.6) h after being transferred to collective pens. In this first interaction, 57% of the calves spent ≥1 minute in contact with the brush. The first area of contact was the nose and head, being followed by other body areas such as neck, thorax and the abdomen. The use of the grooming brush persisted during day and night. The repeated measures analyses indicated a significant effect for the interaction day by treatment group for the not-active and eating time. Control calves spent more time not-active in comparison with AB calves (22.8±0.82 vs 21.7±0.82 min/h; P=0.014) and less time eating (6.43±0.40 vs 7.01±0.40 min/h; P=0.012). Calf average daily weight gain were similar for both groups (CON=0.75±0.03 vs AB=0.77±0.03 kg/d; P=0.69). A tendency was determined (P=0.06) for greater odds of sicknesses (OR=2.17, 95% CI=0.95-5.18) in CON calves compared with AB calves. No differences by group were found for the time to the first disease (9 d and 6 d for CON vs AB; P=0.12). We conclude that recently weaned calves consistently used the available automated brush and this interaction reduced inactive time, while increasing eating time.

## Behavioural differences in a Novel Object Test between male and female turkeys – a pilot study

*Jenny Stracke, Katja Kulke and Nicole Kemper*
*University of Veterinary Medicine Hannover, Foundation, Institute for Animal Hygiene, Animal Welfare and Farm Animal Behaviour, Bischofsholer Damm 15 (Building 116), 30173 Hannover, Germany; jenny.stracke@tiho-hannover.de*

Feather pecking and cannibalism are considered to be serious animal welfare issues in turkeys. Providing novel objects as enrichment material can alleviate these behaviours, novelty seems to be decisive for a positive effect. The reaction to a novel object is generally considered to be an indicator for fear and is traditionally measured by the Novel Object Test (NOT). However, there is a strong relation between fear and exploration– both behavioural systems share stimuli (novelty) and responses (orientation). Furthermore, behaviour in the NOT can also be interpreted with regard to emotional reactivity or personality. The presented study is based on a NOT in turkeys proposed by Erasmus & Swanson. The aim was to apply their test design to on-farm conditions and analyse differences in the behaviour between male and female animals. The study was performed on a farm, keeping both, male and female turkeys simultaneously in two stables. Birds were kept separately at all ages. Data was obtained from three batches. The flock sizes and densities varied between batches (~3.000 animals in total; female ~5.5/m$^2$; male~3.0/m$^2$ at the date of stalling-in), management was determined by the farmers practice. The first NOT was performed in the 7$^{th}$ week of life (LW) and then repeated in a 14-day rhythm (9$^{th}$ LW / 11.LW / 13.LW / 15.LW). At each test date one of four different novel objects (see reference for a detailed description) was presented at six different positions (P1-P6) per barn, for ten minutes each. Objects were presented alternately, balanced across barns and batches. The latency to the first approach, the latency to the first peck, the pecking frequency and the number of animals approaching the object were recorded by direct observation. Data was analysed using the GLIMMIX-procedure in SAS, analysing each parameter separately. The sex (male/female), the novel object (1-4) and their interactions were included as fixed effects. The different positions (P1-P6) nested in LW, nested in the batch were included as random factor. There was a significant difference in the behaviour of male and female animals (all F>15.3, all P<0.001), with females showing a shorter latency to the first approach and the first peck, a higher frequency of pecking and more animals approaching the novel object. In addition, a significant effect of the novel object could be found for the latency to the first peck (all F>8.7, all P<0.001) as well as an interaction effect of object and gender in all parameters except the latency to the first approach (all F>4.1, all P<0.01). With regard to management procedures concerning feather pecking and cannibalism, the results may contribute to a better understanding of the respective needs of both, male and female turkeys. Furthermore, sex differences in the NOT might hint to differences in the underlying motivation of those behavioural disorders, requiring further validation of the NOT in this species.

## The influence of enrichment on leg parameters in a conventional strain of broiler chicken

*Midian Nascimento Dos Santos, Daniel Rothschild, Tina Widowski and Stephanie Torrey*
*University of Guelph, Animal Biosciences, 50 Stone Road E, Guelph, N1G 2W1, Canada;*
*mnascime@uoguelph.ca*

The association among high body weight, low activity, and leg disorders causing lameness has been well studied in commercial broiler chickens raised in standard environments. In addition to the economic implications, lameness is a significant welfare concern as it is associated with pain. Therefore, several strategies have been proposed to decrease the incidence of leg disorders in commercial broiler chickens, including the addition of enrichment. Enrichments may increase activity level and locomotion in chickens, especially in earlier stages of growth. However, the impacts of enrichment on leg disorders (as evaluated by morphology and bone quality) is unclear. The objective of this study was to evaluate the impact of an enriched environment on tibial dimensions, ash content, bone breaking strength and tibia dyschondroplasia in a commercial strain of broiler chickens raised indoors. The enriched material consisted of an elevated platform (30 cm above the litter, attached to a 25° ramp), a mineral pecking stone, shredded nylon rope, and automated hanging scale. A total of 456 male Ross 708 birds were equally distributed in 12 pens (38 birds per pen, $30\,kg/m^2$), divided into 2 treatments: enriched environment (n=6) and control without enrichment (n=6). On day 43, 4 birds per pen were selected based on body weight (one low, two average, and one high) The birds were euthanized by cervical dislocation and both left and right tibia were removed and dimensions (length and diameter) were obtained. The left tibia was used to estimate dry matter content and bone ash concentration (%) as an indicator of bone mineralization. The right tibia was used to evaluate tibia breaking strength using Instron testing machine, followed by the assessment of tibial dyschondroplasia. Data were analyzed as a randomized complete block design with treatment (enriched and non-enriched) as main factors and final body weight as a covariate using Proc Glimmix in SAS. The enrichment materials provided did not alter tibia length (P=0.38), tibia diameter (P=0.84), tibia dry weight (P=0.71), tibia ash content (P=0.57) or tibia breaking strength (P=0.29). None of the tibiae from either treatment exhibited tibia dyschondroplasia. Similar tibial morphometric and bone quality indicators have been reported in previous studies, indicating that the values found in our study were within normal range for commercial broiler chickens. Walking ability and leg strength were assessed as part of another study with these birds, in which no difference was found between the treatments, supporting the findings of our study. We suggest that the low levels of activity in modern strains of broiler chickens reduces the ability of the chosen enrichments to improve leg heath. Furthermore, additional materials should be studied as enrichment for broiler chickens raised indoors to verify their ability to improve in locomotor activity and leg health.

## The effect of enrichment on organ growth, cardiac myopathies, and bursal atrophy in a conventional strain of broiler chicken

*Daniel Rothschild, Midian Nascimento Dos Santos, Tina Widowski, Stephanie Torrey and Zhenzhen Liu*

*University of Guelph, Animal Biosciences, 50 Stone Road East, N1G 2W1, Guelph, Canada; drothsch@uoguelph.ca*

One by-product of fast growth in conventional broiler chickens is a mismatch in nutrient allocation, resulting in reduced organ growth compared to body weight, which may lead to poor health and reduced activity which are major welfare concerns. Although the connection between exercise and health in avian species has not been well studied, the link has been made in various other farmed animal species. The objective of this study was to determine if broiler cardiovascular and immune health, as well as organ growth, can be improved with the provision of enrichments that offer different opportunities for exercise. Animal use was approved by the University of Guelph Animal Care Committee in accordance with the Canadian Council on Animal Care. There were 12 pens in total, with half including enrichments (E; n=6), and half with no enrichment as a control (NE; n=6). A total of 456 Ross 708 male broilers were used and randomly allocated into each pen (38 birds/pen; 30 kg/m$^2$) at 1 d of age. Enriched pens had a 25° ramp to a raised platform (30 cm above litter), and a hanging round scale. Both objects encouraged birds to walk and climb. Additionally, a suet feeder with wood shavings was used to encourage oral and locomotor behaviour. On D43, 4 birds per pen were selected based on body weight (1 heavy, 2 average, 1 light), and euthanized via cervical dislocation. Carcasses were dissected and the heart, lungs, liver, kidneys, and bursa of Fabricius were weighed. Ventricles were separated, and weighed to determine the right ventricle to total ventricle (RV:TV) weight ratio. Data were analyzed as a complete randomized block design with treatment (E vs NE) as a fixed effect using Proc Glimmix in SAS 9.4 with final BW as a covariate. BW did not differ between treatments. There was a trend (P=0.051) for NE birds (0.38%; SEM±0.008) to have heavier heart weights, as a percentage of BW, than E (0.35%; SEM±0.009). No other variables were influenced by enrichment. Of all the sampled birds, 40.6% had abnormal RV:TV ratios (0.14<healthy<0.24), suggesting subclinical heart disease, and 14.1% had some degree of bursal atrophy (healthy>0.11). In conclusion, the enrichment used in this trial had minimal to no effect on organ growth, cardiac myopathies, and bursal atrophy. A large proportion of the broilers sampled from both treatments had signs of impaired cardiovascular and immune functioning, which is a concern in conventional broiler production due to genetic selection for muscle growth.

## Effects of enrichment objects on piglet growth and behaviour

*Hayley Bowling[1], Cyril Roy[1,2], Jennifer Brown[1] and Yolande M. Seddon[2]*
*[1]Prairie Swine Centre, Ethology, Box 21057, 2105 8th Street East, Saskatoon, SK, S7H 5N9, Canada, [2]Western College of Veterinary Medicine, Large Animal Clinical Sciences, 52 Campus Drive, Saskatoon, SK, S7N 5B4, Canada; jennifer.brown@usask.ca*

The objective of this study was to evaluate the effects of enrichment objects on piglet growth and behaviours associated with enrichment use, aggression and exploration. Thirty litters (median 13 pigs/litter, range 11-15) were selected over three replicates at 4 days after farrowing and assigned to three treatments (10 litters/treatment): enrichment provided in nursery (EN, 4 to 8 weeks of age), enrichment provided in farrowing and nursery (EFN, 1 to 8 weeks of age), and no enrichment provided (Control). Enrichment consisted of a series of objects (cotton rope, commercially produced objects: 'PorkyPlay' and 'Bite-Rite', rubber mat, hay cubes and PVC pipe); three or four enrichments were provided at once and the set of objects was rotated twice weekly. At weaning, pigs were mixed within treatment into 30 pens of 10-11 pigs/pen (318 total). Pigs were individually weighed at litter selection, before weaning and at nursery exit. Enrichment interaction was recorded on all pigs for 1 h on days 8 and 10 after weaning and transcribed using 3-min scan sampling. The behaviour of six focal pigs/pen (3 small and 3 large per litter) was assessed using one-hour videos recorded at weaning and 21 d post-weaning and transcribed using 5-min scan sampling. Skin lesions were scored before weaning and at 24 h and 4 weeks post weaning (score 0-4). Data were analyzed using Mixed, Glimmix and Logistic regression models in SAS 9.1. Enrichment treatments did not affect average daily gain during pre-weaning or nursery periods (P>0.10). Cotton rope was the most frequently contacted enrichment (LS Mean ± SEM; 20.88±1.44 contacts/hr), followed by the 'PorkyPlay', 'Bite-Rite', rubber mat, hay cubes and PVC pipe (P<0.05). At weaning, EFN pigs showed reduced pen-mate manipulation compared to EN and Control treatments (Odds ratio and CI: EFN 0.30 [0.10-0.86] vs Control; P=0.004). EN and EFN piglets showed increased frequency of pen exploration at d 21 post-weaning compared to Control pigs (Odds ratio and CI: EFN 1.97 [1.06-3.63]; EN 2.20 [1.26-3.84] vs Control; P=0.012). Pigs given enrichment in nursery (EN) had fewer skin lesions on the head and shoulders at 4 weeks post weaning compared to EFN and Control (LS Means: 0.32, 0.53 and 0.65±0.14, respectively; P=0.004). In conclusion, provision of enrichment to piglets before weaning and in the nursery period increased exploratory behaviours and reduced pen mate manipulation. Enrichment in the nursery period only had the greatest impact on lesion scores indicating that timing of enrichment can influence aggression. Hanging enrichment objects may be viable alternatives to substrate enrichments for commercial farms with slatted floor systems.

## A combination of rooting stimuli reduces fear of novelty and enhances collaboration in groups of weaned pigs

*Benedicte M. Woldsnes, Marko Ocepek, Ruth C. Newberry and Inger Lise Andersen*
*Norwegian University of Life Sciences, Faculty of Biosciences, Department of Animal and Aquacultural Sciences, P.O. Box 5003, 1432 Ås, Norway; benedicte.marie.woldsnes@nmbu.no*

Materials provided as environmental enrichment for pigs may vary in their effectiveness to promote confidence and cooperativeness. We hypothesized that pigs stimulated with rooting material, and especially a combination of rooting materials, twice daily from 5-10 wk of age, would have less fear of unfamiliar people and a novel object, and collaborate more with penmates in a cooperative exploration task at 10 wk of age. We predicted that latencies to contact the test stimuli would be lowest in pigs regularly given a combination of peat, straw and roughage pellets (combo treatment), intermediate in pigs regularly exposed to a comparable volume of only one of these types of materials, and highest in control pigs (wood shavings, given to all groups). We also predicted that pigs in the combo treatment would show the highest, and control pigs the lowest, stimulus interaction times and levels of cooperative exploration (i.e. two pigs working together to open a trapdoor and access material in the same container). Pigs were housed in groups of 4 (n=6 groups/rooting material treatment) and tested in their home pens, and behavioural data were extracted from video recordings. In successive tests conducted over 2 days, we examined pig latency to approach and time spent interacting with 2 unfamiliar people (1-min exposure/person) and a novel object (10-min exposure to a birch log). We also observed responses to containers of familiar material behind trapdoors (5-min exposure to 2 containers/ group; latency to approach, time nosing container, latency to lift trapdoor, time interacting with material, latency to work together, frequency of working together). Treatment effects were investigated in generalised linear mixed models. As predicted, we found that latency to contact an unfamiliar human was shortest in the combo treatment (25.8±3.0 vs control: 33.5±2.3 s, P<0.001). Time spent interacting with a human was highest in the pellets and combo groups (P<0.001). Pigs in the peat and combo groups had shorter latencies to contact a novel object than pigs in the other treatments (e.g. peat: 10.7±3.2, combo: 13.7±3.4, control: 40.2±20.2 s, P<0.001), and control pigs spent the least time interacting with the novel object (P<0.001). In the cooperative exploration task, the peat and the combo groups had the shortest latencies to contact the container (P<0.001), and those in the combo treatment were the quickest to collaboratively gain access to the material inside the container (i.e. by lifting the trapdoor enabling one or more to get their nose inside the container; e.g. combo: 196.5±17.8, control: 242.9±15.9 s, P<0.001). In the combo treatment, pigs collaborated more frequently than pigs in the other treatments (P=0.001), and spent the most time interacting with the material in the container (combo: 16.2±2.9, control: 0.4±04% of test time, P<0.001). We conclude that, relative to having wood shavings litter only, giving pigs access to a combination of different rooting stimuli supplied freshly twice a day for 5 wk was effective in reducing fear of humans and novel objects, and enhancing collaboration with pen mates.

## Individual variation in enrichment use in finishing pigs and its relationship with damaging behaviour

*Jen-Yun Chou[1,2,3], Rick B. D'Eath[2], Dale A. Sandercock[2] and Keelin O'Driscoll[3]*
*[1]University of Edinburgh, Royal (Dick) School of Veterinary Studies, Easter Bush, EH25 9RG, United Kingdom, [2]SRUC, Animal & Veterinary Sciences Research Group, Roslin Institute Building, Easter Bush, EH25 9RG, United Kingdom, [3]Teagasc, Pig Development Department, Moorepark, Co. Cork, P61 C996, Ireland; jenyun.chou@ed.ac.uk*

Enrichment studies often use group-level treatment comparisons whereas variation between individual pigs is less explored. As differences in individual personality and temperament can have an impact on behaviour, this study investigated enrichment use at an individual level and explored its relationship with other behaviours and lesions. Finishing pigs (11 weeks old) were housed in 40 groups of seven pigs (wk0, n=280). The main aim of the experiment was to compare use of various commercially relevant enrichment materials on fully slatted floor in tail docked pigs, and due to a higher risk of tail biting, pigs were docked under veterinary advice. One of four different enrichment items (one Spruce/Larch/Beech wood post or rubber floor toy) was randomly assigned to each group. The behaviour of each pig (individually marked) was observed continuously from video recordings taken on six different occasions (during wk2-3, 4-5 & 7-8; 1 h/occasion). Individual tail and ear lesion scores were recorded every 2 weeks. Three focal pigs/group were classified as 'Bold', 'Neutral' or 'Shy' in wk2, based on the latency to approach the experimenter during the first saliva sampling, then sampled again every 2 weeks. Data were analysed using linear mixed models (fixed effects: rep, age, treatment, sex and type where applicable; random effect: block) and Spearman's correlation (SAS 9.4). Time spent using enrichment was higher with Spruce (104.29±14.37 s/h/pig) and Rubber toy (100.73±14.13 s/h/pig) than with Larch (48.66±14.37 s/h/pig) and Beech (32.36±14.13 s/h/pig, P<0.01). Pigs were ranked from 1 (highest) to 7 (lowest) based on total time using enrichment within group. Ranks were not correlated between sessions in Spruce groups (i.e. high users were not always consistently high users), indicating Spruce was used by pigs more equally. There was no correlation between enrichment use and tail biting or ear biting, and neither was there a relationship between enrichment use and lesion scores (P>0.05). No sex difference was found in enrichment use (P>0.05). Salivary cortisol tended to be higher in 'Shy' than 'Bold' pigs (0.146±0.009 vs 0.123±0.009 µg/dl, P=0.05), but enrichment use, tail biting and lesion scores did not differ between pig types (P>0.05). By investigating individual variation rather than just group means of behaviour, a more complete picture was obtained of how individual pigs interacted with the enrichment. Although no relationship was found between an individual's enrichment use and damaging behaviours, the overall higher use of Spruce did not appear to be attributable to just the high user pigs within a group.

# Economic impact of an on-farm innovation to enhance novelty of enrichment materials for fattening pigs

*Emma Fàbrega[1], Antonio Velarde[1], Joaquim Pallisera[1], Xènia Moles[1], Aranzazu Varvaró[1], Cecilia Pedernera[1], Aida Xercavins[1], Pau Batchelli[1] and Agata Malak-Rawlikowska[2]*
*[1]IRTA, Animal Welfare Program, Veïnat Sies, 17121, Spain, [2]Warsaw University of Life Sciences, Faculty of Economic Sciences, Nowoursynowska 166, 02-787, Poland; emma.fabrega@irta.cat*

Provision of enrichment material is a key strategy to allow pigs to perform exploratory behaviour, and is mandatory according to EU legislation. However, farmers often claim that enriching pens can be costly, with no return on productivity. The aim of this case study was to evaluate the economic impact of an innovation designed by a farmer to increase sense of novelty of enrichment for fattening pigs. The innovation comprised an automatic rotation system to move different objects (ropes, straw container, wood, a bite-rite) between consecutive pens of pigs on a daily basis. It was implemented in a farm of 660 pigs (12 pigs/pen), with fully-slatted floor, producing the local breed 'Porc Ral Avinyó' (PRA). Data regularly collected in the PRA company management system were used to compare 3 production batches of this farm (1.837 slaughtered pigs) to the average of 60 production batches from 20 other PRA farms (APRA) of the same company (61,650 pigs). APRA farms used the same genetics, feed composition, stocking density and fully slatted floors as the case study farm, but with other point-source enrichments not changed between pens. An interview of farmers to collect additional information on labour and veterinary costs was undertaken. Calculations were performed using the INTERPIG economic model calibrated to average PRA farm data. Costs and benefits of this system were applied considering the assumed changes in technical performance parameters of the case study farm (with all other parameters unchanged). The benefits found for this farm compared to the APRA farms were a lower mortality rate (1 vs 2.2%); a 8% higher finishing live weight gain and a better feed conversion ratio (3.13 vs 3.36). Veterinary costs were reduced by 5%. In contrast, labour costs were increased due to the 1.5 h additional labour input per day. Overall, the variable production costs in this farm compared to APRA farms were lower by 5% per kg of slaughter weight, mainly due to lower mortality, vet costs and better weight gain, but enlarged by costs of enrichment materials and objects. The fixed costs were 23% higher per kg of slaughter weight, due to labour and depreciation costs of equipment. As a result, the total costs of the farm were a 1.3% lower as per hot slaughter weight compared to the APRA farms. Therefore, although enrichment novelty may not be the single reason for this increased efficiency, these case study farm results indicate that, up to a certain extent, a return on productivity may be expected from appropriate enrichment strategies. They also support the role of bottom-up approaches as a tool to bring innovation to commercial production farms.

## Nest-building behaviour in crated sows provided with a jute-bag – an exploratory case study

*Martin Fuchs, Anke Gutmann, Christoph Winckler, Christine Leeb and Christina Pfeiffer*
*University of Natural Resources and Life Sciences, Division of Livestock Sciences, Department of Sustainable Agricultural Systems, Gregor-Mendel-Straße 33, 1180 Wien, Austria; martin.fuchs@students.boku.ac.at*

Sows crated in farrowing pens are often unable to perform adequate nest-building behaviour as their freedom to move is restricted and usually no suitable material is provided for management reasons. Jute-bags have been proposed as enrichment for sows in the pre-farrowing phase as they are manipulable, deformable, movable, safe for sow and piglets, and compatible with slatted floors. We were interested how crated sows make use of a jute-bag during the pre-farrowing phase and how pen-directed and bag-directed nest-building related behaviours were connected. Therefore, 23 sows at two farms were provided with jute-bags attached to the farrowing crates and observed by continuous video-recording from 12 h ante partum until farrowing. We assessed durations of the behavioural categories 'rooting', 'pawing', and 'biting', directed either to the objects 'pen' or 'bag', and analysed them separately for the two farms at different aggregation levels using general linear mixed models with 'behaviour' and/ or 'object' and 'breed (Large White and Large White × Landrace)' as fixed factors and 'sow' as random factor. Additionally, we calculated Spearman correlations at the different aggregation levels. All results are given in minutes as arithmetic mean ± SD (min-max). Total time spent rooting, pawing or biting in farm 1 and 2 amounted to 88±37 (32-164) and 73±72 (5-178), respectively (ANOVA F=0.43, P=0.520). At both farms, sows showed more rooting than biting and pawing (farm 1: 58±21 (25-102), 24±17 (4-65), 5±6 (0-21), F=84.5, P<0.001; farm 2: 46±42 (3-106), 24±30 (1-77), 3±4 (0-9), F=5.3, P=0.027). In total, more time was spent with the bag than with the pen, but not at a statistically significant level (farm 1: 51±33 (4-116) vs 37±22 (4-82), F=2.1, P=0.158; farm 2: 42±50 (1-143) vs 31±35 (4-106), F=0.2, P=0.673). There was one significant interaction effect (farm 1: biting-pen 6.4±5.5 (0-19) vs biting-bag 18±15 (0-46), F=7.5, P=0.008), and no statistically significant effects of breed were found. Generally, rooting and biting were highly correlated (r=0.772, P<0.001), however behaviours summed up to bag- and pen-directed behaviours were not correlated. At behaviour × object-level, pawing and rooting the pen (r=0.491, P=0.017) as well as pawing and biting the bag (r=0.424, P=0.044) were moderately, and rooting and biting the bag highly correlated (r=0.931, P<0.001). Our results show that all sows directed nest-building related activities to both the bag and the pen. The large variation in duration and styles of these activities was due to individual differences rather than breed or farm. Observed behaviours were partly (and meaningfully, e.g. rooting-biting the bag) correlated within objects, but always independent from each other between objects, indicating an additive quality of jute bags in the nest-building phase.

## Effects of environmental enrichments on reproductive performance and behavior pattern in gestating sows

*Hyun-jung Jung, Yong-dae Jeong, Du-wan Kim, Ye-jin Min and Young-Hwa Kim*
*National institute of animal science, Swine science division, 114, sinbang 1-gil, cheonan-si, 31000, Korea, South; hyjjung@korea.kr*

Many countries have growing interest into animal welfare by the consumers. Environmental enrichments (EE) are useful tool improving the welfare. Natural enrichment as rice straw (NE) is readily available because of low price and accessibility in aspect to farmers, whereas artificial enrichment with plastic material (AE) does not. Furthermore, research related to the EE is relatively limited in sows than growing or finishing pigs. Thus, the EE in this study was chosen to rice straw block as the NE and commercial product as the AE for investigating reproductive performance and behavior pattern in gestating sows. A total of thirty pregnant sows (Landrace) were randomly allotted into three treatment groups as control without enrichment, AE and NE. The enrichments (AE and NE) were situated respectively in the middle of experimental pens (1,160×600 cm). Trial period was approximately 12 weeks. Body conditions, body weight and backfat thickness, of the sows were measured at initial or terminal days of the trial, and cortisol is analyzed at the end date. Reproductive performance as litter size and piglet's mortality was checked at next days after parturition. Behavior properties were recorded for 24 h on d 91 gestation, and then analyzed using Vegas pro (ver. 13.0) program. All data were statistically analyzed by using ANOVA in performances and by using Kruskal-Wallis test in behavior properties. The backfat thickness was reduced (15.73±0.66 vs 16.56±0.42 mm; $P=0.01$) in AE than control, but terminal body weight, total litter size and alive piglets did not differ among treatment groups. Born dead piglets showed decreased tendency (1.00±0.38 and 0.63±0.32 vs 1.50±0.89 heads, respectively; $P=0.09$) in AE and NE than control. Similarly, farrowing mortality of piglets was slightly lower (6.86±2.72 and 3.40±1.69 vs 8.68±4.91%, respectively; $P=0.25$) in AE and NE than control. In the behavior patterns, feed intake was higher ($P<0.05$) in control than AE and NE. The playing or digging were only showed ($P<0.05$) in AE and NE, respectively, but rubbing only observed ($P<0.05$) in the control. Our study indicates that EE supplied to the sows should be not influenced on the performance and may be resulted to improve farm animal's welfare by rich or welfare-friendly behavior patterns.

## Preference for and behavioral response to environmental enrichment in a small population of sexually-mature, commercial boars

*Lara Sirovica, Maggie Creamer and Kristina Horback*
*UC Davis, Animal Science, 1 Shields Ave, Davis, CA 95616, USA; mlcreamer@ucdavis.edu*

In most commercial swine facilities, sexually-mature boars are housed individually to prevent potential injuries from mounting or aggressive behavior. While commercial boars do engage in limited social interaction during the day, the lack of social pen mates greatly reduces the environmental complexity for these animals. To date, there is no research specific to the applicability of environmental enrichment for commercial boars, nor on commercial boar welfare in general. Although provision of straw has been documented to be the most successful enrichment item for sows and piglets, based on engagement and reduction of stereotypic behaviors, straw is very undesirable for production systems using manure-flush systems. Based on previous research which identifies preferred enrichment objects for weaning piglets and gestating sows, this study examined whether individually-housed, sexually-mature boars (age range = 12-17 months) preferred to interact with hanging cotton rope or hanging rubber chew sticks. Each boar (n=8) was individually housed in a pen (4.9×4.9 m) with solid concrete flooring and a gutter in the back of the pen for the manure flush system. Each boar was filmed for one hour between 09.00-11.00 and 14.00-16.00 for four days with no enrichment in order to record a baseline activity budget. During the enrichment treatment, two 30 cm strands of twisted cotton rope and two 30 cm rubber chew sticks (BiteRite™) were suspended on opposite sides of the pen at the same height from 08.00-18.00. Each boar was also filmed for one hour between 09.00-11.00 and 14.00-16.00 for four days with enrichment present. Seven observers (Cohen's kappa>0.80) conducted continuous observations of the video files using The Observer v. 11 (Noldus Information Technology).Duration spent in each body posture (sitting, standing, laying) and all-occurrence of behavioral events (such as, interact with enrichment, sham chew, and oral manipulation of pen) were recorded. The boars interacted with the hanging cotton rope enrichment for significantly more of the observation period (13.73±7.04%), as compared to the hanging rubber chew sticks (7.54±7.04%) when given access to both items at the same time (Wilcoxon signed rank tests, Z=-2.40, P<0.05). There was a significant negative correlation ($r_s$=-0.83, P<0.05) between the amount of time boars spent interacting with rope enrichment and the amount of time they spent performing stereotypic pen manipulation (bar bite, floor lick). These results suggest that cotton rope may be an effective enrichment choice for producers looking to enhance the environmental complexity for individually-housed boars.

## Endotoxin challenge and environmental enrichment changes the behaviour of crated boars

*Thiago Bernardino, Leandro Sabei and Adroaldo Zanella*

*University of São Paulo, Department of Preventive Veterinary Medicine and Animal Health, Avenida Duque de Caxias Norte 225, 13635900, Brazil; thiagobernardino@usp.br*

In commercial farms, breeding boars are often exposed to stressful situations, such as individual housing, inadequate ambient temperature, food restriction, lack of social interaction, diseases, among others. The consequences of these situations for the welfare of breeding animals and possible changes in semen quality are unknown. Epigenetic changes in semen, as acetylation of histones, DNA methylation and the population of small non-condign RNAs, are important areas of study because they are the possible mechanism for the heredity and behavioural and physiological modulation already reported in the offspring. Moreover, these changes can modulate the resilience of the offspring and, consequently, their welfare. Most studies on foetal programming have been conducted in females, because they play an extremely important role in modulating the offspring adaptation mechanisms. Males, in mammalian species, have more limited interaction with their offspring than females, so much of the transmitted changes come through their germ cells. The aim of this study (São Paulo Research Foundation FAPESP/CAPES grant 2017/05604-2), reviewed and approved by the Ethics Committee on the Use of Animals (CEUA 3612010616), was to evaluate the impact of environmental enrichment in boars subjected to a disease challenge with lipopolysaccharide (LPS). We studied 26 boars, 13 Large Whites and 13 Landrace. All of them were housed in crates, in a commercial farm. We observed the behaviour 1 hour in the early morning (7:30 at 8:30), late morning (10:30 at 11:30) and late afternoon (18:00 at 19:00). We observed for 13 weeks, during two consecutive minutes, three times each hour. The endotoxin challenge was performed after two basal behavioural measures. All boars were inoculated in lateral ear vein. Half of the animals were inoculated with LPS (2µg/kg) and half with sterile saline solution. We measure the rectal temperature hourly for 6 hours, and then every two hour for 18 hours. Half of the individuals were brushed daily, for 2 minutes, starting the day after the LPS challenge. According to this arrangement, we had four treatments: LPS with brushing; LPS without brushing; saline with brushing; and saline without brushing. All the data were tested for residual normality, with Shapiro-Wilk test. For the variables that had residual normality, we used ANOVA and for non-parametric variables we used Kruskal Wallis test. The rectal temperature was higher for the boars inoculated with LPS until three hours after the inoculation ($P<0.05$). After the treatments, the animals of the group brushing (both challenge with LPS or saline) spent more time lying laterally ($P<0.05$). The mean value for brushed animals was 45.99 seconds and 32.14 for control animals. We did not find any differences in the other observed behaviours. Based in this research, brushing can modulate the behaviour of boars housed in crates, a well know stressful housing. The next steps are to study different house conditions.

## Experience of ramps at rear is beneficial for commercial laying hens

*Kate Norman[1], Claire Weeks[1] and Christine Nicol[2]*
*[1]University of Bristol, Animal Behaviour and Welfare, Dolberry Building, Langford House, BS40 5DU, United Kingdom, [2]Royal Veterinary College, Pathobiology and Population Sciences, Hawkshead Lane, North Mymms, Hatfield, Hertfordshire, AL9 7TA, United Kingdom; kate.norman@bristol.ac.uk*

In commercial systems laying hens must move between levels to reach resources. Recent studies have shown the importance of early life experiences on the development of behaviours. However, more research is needed to look at whether providing ramps during rear can improve the ability of birds to transition between levels at lay. This study examined 12 organic commercial flocks (2,000 birds/flock) on a multi-age site, studied from 1 to 17 weeks of age. Two flocks were reared at a time separately in two rearing sheds, then moved to two laying sheds of identical facility design. At 4 days of age, both rearing sheds were provided with 6 identical elevated structures (ES). Each ES comprised 9 perches (length 302 cm, width 3.5 cm), with three perches (25 cm apart) at three heights (43, 73 and 103 cm). Each ES had two additional plastic slats (width 60 cm, length 115 cm) on top of perches. In one rearing shed three ES were additionally fitted with plastic ramps (width 60 cm, length 74 cm, angle 35.5°) leading up to the low perch and the remaining three ES had ramps (width 60 cm, length 115 cm, angle 40°) leading up to the middle perch. Data were analysed using T-tests or Mann-Whitney U tests if normal distributions could not be achieved. Scan samples were taken to count the number of chicks on the ES at rear. At 1 and 3 weeks of age the chicks with access to ramps were recorded using the ES significantly more than chicks without ramps. By 16 weeks of age at the end of rear, there were no differences between the groups. At lay, after a week of acclimatisation, focal behaviours were recorded at 17 weeks of age looking at movements up and down the level between the litter and raised slatted area, which was fitted with intermittent ramps. The non-ramp reared groups showed significantly more behaviours indicative of hesitancy when transitioning down a ramp for percentage of crouching (P=0.002), pacing (P=0.038) and wing flapping (P=0.03). There was no difference between treatments in the number of birds transitioning, but overall more transitions upwards were recorded in the ramp area (18.54±11.30) compared to areas with no ramps (3.71±2.78; P=0.001). There was also a tenfold greater use of ramps for downward transitions at an average of 23.83±10.80 vs no ramps at 2.38±2.30 (P=0.001). Rearing with access to ramps appears to increase the use of elevated structures at 1 and 3 weeks of age. When transferred to the laying shed, flocks reared without ramps showed more behaviours indicative of hesitancy when transitioning such as crouching, pacing and wing flapping. Generally, a higher number of birds transitioned using ramps, suggesting a preference to use ramps to move between levels in a commercial system. Rearing experience with ramps reduced hesitancy type behaviours when transitioning. This may be important to encourage movement on arrival in the laying shed for early access to important resources.

## Do laying hens have a preference for 'jungle' light?

*Anette Wichman[1], Rosan Degroot[1], Olle Håstad[2], Helena Wall[3] and Diana Rubene[4]*
*[1]SLU, Dep. of Animal Environment and Health, P.O Box 7068, 750 07 Uppsala, Sweden, [2]SLU, Dep. of Anatomy, Physiology and Biochemistry, P.O Box 7011, 750 07 Uppsala, Sweden, [3]SLU, Dep. of Animal Nutrition and Management, P.O Box 7024, 750 07 Uppsala, Sweden, [4]SLU, Dep. of Ecology, P.O Box 7044, 75007 Uppsala, Sweden; anette.wichman@slu.se*

Vision is possibly the laying hens´ most important sense and the light environment influences their behaviour and welfare. The ancestor of the laying hen, the Red Jungle fowl, originates from forested habitats in SE Asia, where the light is filtered through the green canopy. We tested the hypothesis that modern laying hens might still prefer and have a higher welfare if housed in their ancestral light habitat, i.e. more green light. Three different light environments, control (C, 425-700 nm, warm white), daylight (D, 330-700 nm, more blue light) and Jungle (J, 360-650 nm, more green light) were used. The D and J light were created with LEDs with a spectral composition similar to light measured in the natural habitat of jungle fowl as perceived by the chicken eye and contained UV, and the C light was a standard LED lamp without UV. 192 Bovans (white) chicks were housed in the designated light treatments from week 5 to 27 with 4 groups/light treatment and 16 birds/group. The birds were kept in floor pens (3.6×3.6 m) with perches, litter and ad lib food and water. Behavioural observations in the home pens were done from week 8 to 21. The behaviour performed by each bird in the group (e.g. feeding, perching, standing and resting) was recorded using scan sampling in combination with continuous observations assessing the frequency of behaviours such as aggressions, severe and gentle feather pecking. During week 16 to 24 the birds´ preference for the different light spectra was tested in three separate pens. Each pen was divided into two similar sections where each section was illuminated with one of the different light treatments. The birds could move freely between the two sections and be tested for their preference for each light combination (C + J, C + D and J + D). All groups of birds were tested three times, once for each light combination and recordings of the number of birds and their behavior in each section were carried out repeatedly over three days. Plumage condition was assessed at 25 weeks and data on egg production were collected. We found no differences in behaviour performed in the home pens between light treatments. Instead, the patterns in bird behavior were to a large extent determined by age and e.g. resting decreased over time (Glimmix; F=13.9, P=0.000). In the preference test there was a tendency for a difference between light treatments (Glimmix; F=2.77, P=0.072). Pairwise analyses revealed that birds spent more time in the J compared with the C light (55±1.8% vs 45±1.8%; F=13.32, P=0.0014). The best feather score was found in D groups (GLM; F=4.47, P=0.044) and these birds also tended to have a lower egg production at start of lay at 18 weeks (GLM; F=3.40, P=0.079). Altogether, some subtle effects of the light treatments were found indicating benefits both with the J and D light. This would rather support that the modern laying hen benefits from light spectra similar to natural light that contain UV, perhaps not specifically green 'jungle' light, but also natural daylight under open sky.

**Welfare of FUNAAB Alpha chickens as affected by environmental enrichment**

*Oluwaseun Iyasere, Damilola Oyetunji, Taiwo Adigun, Kehinde Mathew and James Daramola*
*Federal University of Agriculture, Abeokuta, Animal Physiology, Federal University of Agriculture,*
*P.M.B 2240, Alabata, Ogun state, 110001, Nigeria; iyasereos@funaab.edu.ng*

FUNAAB Alpha is the first Nigerian improved dual purpose (meat and egg) indigenous chicken. This study is the first to investigate the effect of environmental enrichment on the egg quality and behaviour of FUNAAB Alpha chickens. Sixty four (64) chickens at 22 weeks of age were randomly allocated into two treatments (control vs enriched) with 2 cockerels and 6 hens in a deep litter pen (2×1.5 m). Birds in adjacent pen were prevented from seeing each other by using trampoline to cover the sides of the pen halfway. The enriched treatment had a perch, three curtained nest boxes, sand bath containing sand mixed with 100 g wood ash and provided with 100 g fresh *Talinum triangulare* foliage in a tray daily for 5 weeks and provided with feed and water while the control pen had only three open nest boxes and provided with feed and water. Each treatment was replicated 4 times. The behaviours of the birds were monitored through videos (nesting, preening, foraging and dust bathing) were recorded (three times per week between 9:00 and 11:00 am, and data on the number of birds performing each of the 4 behaviours were extracted during playback of the video at 5 min intervals. Egg qualities (egg weight, length, breadth, albumen height, yolk weight, yolk colour and yolk height, Haugh unit) were analysed weekly. Behavioural data were analysed using Mann-Whitney U test while those on egg quality were analysed using independent T- test of SPSS (version 23).The percentage of birds nesting, preening and foraging were similar in both treatments. However, the percentage of birds dust-bathing was greater in the enriched than control pens. Eggs laid by chicken kept in the enriched pen had deeper ($P<0.05$) yolk colour but the egg were had reduced egg weight, egg length and egg breath than the eggs from the control pens. Other egg quality parameters such as albumen height, yolk weight, yolk height and Haugh unit were not significantly different between treatment. In conclusion, FUNAAB Alpha chickens housed in environmental enrichment pens displayed more of dust bathing behaviour and laid eggs of deeper yolk colour.

## Role of range use in infections with parasites in laying hens

*Monique Bestman[1], Thea Van Niekerk[2], Elske N. De Haas[3], Valentina Ferrante[4] and Stefan Gunnarsson[5]*

[1]*Louis Bolk Institute, Kosterijland 3-5, 3981 AJ Bunnik, the Netherlands,* [2]*Wageningen Livestock Research, De Elst 1, 6708 WD Wageningen, the Netherlands,* [3]*Institute for Agricultural and Fisheries Research, Scheldeweg 68, 9090 Melle, Belgium,* [4]*University of Milan, Via Celoria, 10, 20133 Milano, Italy,* [5]*Swedish University of Agricultural Sciences, P.O. Box 234, 532 23 Skara, Sweden; m.bestman@louisbolk.nl*

In organic layer farms a free-range area is provided for animal welfare reasons. Both higher and lower worm burden (*Ascaridia (Asc), Heterakis (Het) and Capillaria (Cap)) are described* in hens housed in free range systems compared to other systems. Parasite infections can reduce health, welfare and productivity. We investigated the role of the range area in helminth infections: (1) Is infection of manure different for samples being collected in the free-range or inside the house, assuming to distinguish 'outdoor hens' from 'indoor hens'? (2) Is there an association between the proportion of hens using the range area and parasite eggs in soil and manure? (3) Is there an association between parasite eggs in manure, health and production parameters? Forty one flocks are being visited once when hens >45 weeks old and >3 weeks after a deworming. Together with farmers the proportion of hens using the free-range was estimated assuming optimal conditions (%HensOut), as well as health status (score on scale 1 (=bad) to 10 (=perfect)). Lay % at 60 weeks and mortality % till 60 weeks were collected too. Six soil samples per farm were taken at 5, 20 and 50 m from the pop-holes. Seventy individual manure droppings, pooled into 7 samples were collected inside and 70, pooled into 7, outside. On the free-range, manure samples were collected >50 m from the pop-holes, assumed to originate from 'outdoor hens'. Manure samples inside were taken from the inner part of the barn, away from the pop-holes, assumed to originate from 'indoor hens'. All soil and manure samples, 20 per farm, were analysed for parasite eggs per gram (EPG; McMaster method). This abstract contains preliminary results from 14 farms. From the soil samples (n=84) 7% was infected with *Asc*, 5% with *Het* and 0% with *Cap*. From the manure samples collected outside (n=98), 76% was infected with *Asc and* 26% with *Het*. From the manure samples collected inside (n=98), this was respectively 68 and 14%. There was no difference in number of positive manure samples between outside and inside, neither for *Asc*, nor for *Het*. A negative correlation between %HensOut and soil samples infected with *Asc* was found (-0.57; P=0.034). A tentative explanation may be that the hens' behaviour changes the soil surface into an environment detrimental to parasite egg survival. No correlation was found between %HensOut and soil samples infected with *Het*. Furthermore, no correlation was found between %HensOut and manure samples infected with *Asc,* nor with *Het*. No correlation was found between manure samples positive for *Asc* and health status, lay% 60 weeks or mortality till 60 weeks. Furthermore, no correlation was found between manure samples positive for *Het* and health status, lay% 60 weeks or mortality % till 60 weeks. These preliminary results indicate that range use may not be a risk factor for parasite infections in laying hens.

## Are breeding rabbits motivated for bigger cages?

*Arantxa Villagra[1], Martinez-Paredes Eugenio[2], Martinez-Talavan Amparo[2], Estelles Fernando[2] and Cervera Concha[2]*
*[1]Centro de Tecnologia Animal-Instituto Valenciano de Investigaciones Agrarias CITA-IVIA, Poligono La esperanza 100, 12400 Segorbe, Castellon, Spain, [2]Instituto de Ciencia y Tecnologia Animal- Universitat Politecnica de Valencia ICTA-UPV, Camino de Vera s/n, 46022 Valencia, Spain; villagra_ara@gva.es*

Choice and motivation tests have been used for the establishment of animals' preferences, and they are considered as a tool for the evaluation of animal welfare. The improvement of housing to enhance animal welfare is a hot topic in rabbit production. The aim of this study was to assess the possible preference of rabbit does for three different cages. A motivation test was performed in 40 breeding rabbit does in their third parturition (beginning 2 days post-insemination). Rabbit does were originally housed in four types of individual cages (10 rabbit does each): · CC: Control group (50×70×38 cm, 3,500 cm$^2$) and a 22×35×37 cm exterior nest (770 cm$^2$) · LC: Longer cage, 40×98×38 cm cage (3,920 cm$^2$) and interior nest (40×25×47, 1000 cm$^2$) · WC: Wider and higher cage, 50×85×50 cm (4,250 cm$^2$) and exterior nest (40×25×37, 1000 cm$^2$) · PC: Platform cage, 40×98×60 cm cage provided with a platform (30×26 cm, 1,040 cm$^2$) and interior nest of 40×22×39 cm (total surface of 4,960 cm$^2$) Measures are given in widthxlengthxheight. Motivation was assessed in a pen in which ballasted push-doors gave access to three different cages which corresponded to the dimensions of the original cages: LC, WC and PC (CC was considered only as rabbits with no previous contact with the experimental cages and exterior nests' areas were not considered). During three weeks, all animals were subjected to a training process to push the doors of the motivation pen (giving access to the studied cages). Then, rabbit does were tested once a day during 3 days (in the morning, with the lights off and temperature of 21.3±1 °C) in the motivation cage. During the test, three increasing resistances were used: 5, 10 and 15% of rabbit does' average weight. Categorical variables were compared using Chi-squared test and quantitative variables were subjected to a GLM using Statgraphics Centurion®. Differences between test days were found (P<0.01), so results are presented for each day. Rabbits made fewer attempts to open LC and PC (P<0.01) than WC (57.50% of the total attempts) in day 1, and day 3 (P<0.0001, 67.27%), although these significant differences were not found in day 2 (P=0.0771). However, although 85.71% of the total rabbit does finally opened heavier doors (day 3) to access to the WC, chi-squared was not statistically significant (P=0.4511). No trend was observed during day 1 and 2. The time that the animals spent inside each cage was not statistically significant (P=0.9273): 105.99±18.12 sec for LC, 108.6±20.15 sec for PC and 110.043±21.20 sec for WC. In addition, the previous experience on the home cages did not affect the studied variables (P=0.397 and P=0.5728). These results might indicate a preference of the rabbit does for the wider and higher cage (WC), although several aspects should be improved in this motivation cage, as the high level of attempts to reach a cage combined with the low number of animals which finally opened the doors, may imply that ballast was too heavy for the tested animals.

## Effect of environmental enrichment rotation on juvenile farmed mink behaviours

*Isabelle Renault and Miriam Gordon*
*Dalhousie University, Animal Science and Aquaculture, 58 Sipu Awti, B2N 5E3, Canada;*
*miriam.gordon@dal.ca*

Various benefits are observed when paired-juvenile mink are provided enrichments, including decreased display of abnormal behaviours and reduced fearful states. However, mink tend to lose interest in enrichments that are static in their environment. This study investigated the effect of a weekly enrichment rotation for four weeks in juvenile mink (*Neovison vison*) compared to a control where the same enrichment (loose pvc ring) was provided for four consecutive weeks. For the treatment, three enrichments were rotated: loose pvc ring (15.2 cm diameter, 2.5 cm width), a suspended pvc ring looped through a 10.1 cm length cable (swing), and a golf ball (4.3 cm diameter). The ring was presented the first and last week to examine the effect of rotation on habituation. A crossover design was used: all mink experienced treatment and control. Weaned male-female siblings were pair-housed (n=13) in cages (76.2×30.48×38.1 cm) containing a nest box and a wire shelf. Mink were fed a typical meat-based diet twice daily and *ad libitum* water through a drinking nipple. Instantaneous sampling of behaviours occurred every 30 s for 2 h pre and post pm feeding from continuous camera (Speco Technologies® VL62W) recordings. ANOVA, using repeated measures, was performed on the behaviour proportions obtained. Individual mink behaviours were recorded and nested within cage in the model. Males and females behaved differently but didn't affect enrichment interaction. Mink temperament was evaluated using the stick test, but no effect was observed. Compared to the control, rotation increased the proportions of enrichment interaction (1.3±0.2% vs 1.0±0.2%; P<0.0001) and social interaction (0.6±0.1% vs 0.5±0.1%; P=0.02). Rotation halted habituation, as there was no difference in enrichment interaction between the initial and final presentations of the ring (P=0.06), but it did not reduce the proportions of any abnormal repetitive behaviours. Mink spent more time interacting with the swing (1.9±0.3%) compared to the ring (1.3±0.3%) and golf ball (0.2±0.5%; P<0.0001). Mink were most inactive with the swing (58.2±3.9%) compared to the ring (42.5±3.9%) and the ball (26.3±6.0%; P<0.0001). Pacing was most often observed prior to feeding (P=0.002). Mink spent more time in their nest box when presented with the golf ball (64.9±5.9%) compared to the ring (42.5±3.9%) or swing (58.2±3.9%; P<0.0001). Overall, mink behaved differently to a rotation of enrichments and to different enrichments. Evaluation of behaviours through the entire day, a larger N size, a longer enrichment rotation, and in adult mink may provide more information regarding enrichment effectiveness.

**Physiological response of pet dogs to different methods of provision of classical music**

*Megumi Fukuzawa[1], Takuya Takahashi[1], Takayasu Kato[2], Tomio Sako[2], Takahiro Okamoto[2], Ryo Futashima[2] and Osamu Kai[1]*
[1]*Nihon University, College of Bioresource Sciences, Fujisawa, 252-0880, Japan,* [2]*NOK Corporation, Fujisawa, 251-0042, Japan; fukuzawa.megumi@nihon-u.ac.jp*

There are reports on environmental enrichment for dogs that are being managed in limited spaces, such as shelters, using auditory stimulation. Although there are multiple ways of providing music, there has hitherto been no research on the effect of the method of provision of the same music using different sound sources. We examined the effects of different methods of provision of classical music on the dogs' ECG waveform recordings using rubber electrodes for detecting biomedical signals (ver. 4; NOK Corporation, Kanagawa, Japan). Six healthy pet dogs (11-51 months, 21.7-25.4 kg) were used in the study. No dogs showed any fear response to sound or to pre-treatments such as shaving. Each dog was then held in a supine position with no anesthesia, and three rubber electrodes (positive, negative, and ground electrode) for recording biomedical signals (connected to a BITalino biomedical recording system; Lisbon, Portugal) were attached to the dog's chest (right and left intercostal and costal-cartilage junction). Lead II was used to derive a reference point for the unipolar chest leads. ECG waveforms were recorded continuously using Bitadroid (version 1.0). After confirmation of stable waveform recording, each dog was exposed, for three successive days, to one of three types of trials: 'Speaker' (playing classical music recorded on a CD), 'Piano' (played by a familiar person), or 'No sound' (no sound playing). Each played back volume was same dB level, and the dog could not visually confirm the sound source area. The interval between the provision of these different types of trials was set at 4 days. Each trial, lasting 30 minutes, comprised 10 minutes each of no sound ('before exposure'), sound provision as a recording or live performance ('during exposure') and no sound ('after exposure'). The waveform was continuously recorded during each trial. The average heart rate (bpm) in each 10-minute period varied according to the conditions: under 'Speaker' conditions, it was significantly lower 'during exposure (68.22±28.95)' than 'before exposure (70.51±28.66)' or 'after exposure (69.94±30.00)' but under 'Piano' conditions, it was significantly higher 'during exposure (72.02±28.21)' than 'before exposure (69.45±26.99)' or 'after exposure (69.30±27.77)' (Tukey, $P<0.05$, respectively). The effect of the condition in each 10-minute period was also different: 'Speaker' gave a lower bpm than the other two conditions 'during exposure,' and during 'no sound' after exposure, it was higher than under the other two conditions. A significant difference was also observed in the R-R interval for all conditions. Similarly to previous reports, no aversive response of dogs to wearing rubber electrodes was observed, and stable measurement was possible for over 30 minutes while they were attached. The dogs were aware of the differences in method of provision of music, because their physiological responses to sound differed accordingly. It appears that the use of recorded classical music is more likely to have the desired stress-reducing effect than playing the piano live.

## Effect of different forms of presentation of an interactive dog toy on behaviour of shelter dogs

*Bernadette Altrichter[1], Sandra Lehenbauer[1,2], Susanne Waiblinger[1], Claudia Schmied-Wagner[2] and Christine Arhant[1]*

*[1]Institute of Animal Welfare Science, Vetmeduni Vienna, Veterinärplatz 1, 1210 Wien, Austria, [2]Specialist Unit for Animal Husbandry and Animal Welfare, Veterinärplatz 1, 1210 Wien, Austria; christine.arhant@vetmeduni.ac.at*

Interactive dog toys might counteract boredom and promote interactions with humans. However, a hasty introduction may lead to low success and cause frustration. Therefore, we tested whether a hasty ('complete') presentation of an interactive dog toy with four boxes with different opening mechanisms (Poker Box 1 Strategy Game; ©Trixie) compared to a gradual ('stepwise') presentation increases stress-related behaviour and reduces success during the task. In this study, 28 dogs were tested in two different conditions in a counterbalanced order. Both conditions consisted of three trials interspersed with two breaks of 5 minutes. In the stepwise condition (S), the dog was introduced to only two out of the four different boxes in the first two trials (T1 & T2). Each of the two boxes was filled with one treat six times per trial. In the third trial (T3) all four boxes were used simultaneously and filled three times in a row. The complete condition (C) consisted of three trials using the same setup as T3 in the stepwise condition. In both conditions, 12 treats per trial were used. Additionally in T1 and T2 of condition S the dog was allowed to observe two additional demonstrations of the opening mechanism of the boxes. Dogs in the order S/C were tested in condition S on the first day and in condition C two days later (recommended order), for dogs in the order C/S, the sequence was reversed ('hasty' introduction). The study was discussed and approved by the institutional ethics and animal welfare committee in accordance with GSP guidelines and national legislation. During the experiment, video recordings of dog behaviour were conducted. Continuous sampling was used to record success in reaching the treat and frequencies and durations of dog behaviour (Inter-rater reliability: mean: $r_s$=0.99, P<0. 01; min: $r_s$=0.85, P<0.01). Principal component analysis was used to reduce the number of dog behaviour variables (resulting in five components). Linear mixed effects analyses (lme; R-package 'nlme', Pinheiro *et al.* were calculated for success and behaviour component scores in T3 with order (S/C, C/S), condition S, C) and order×condition as fixed factors, and dog as random effect. On the second day, dogs tested in the order S/C had a significantly higher rate of success (condition C: mean ± SD=5.4±3.5 success/min) compared to dogs in the order C/S (condition S: 3.3±2.7; results lme: order: P<0.001; condition: P<0.001; order×condition: P<0.001). Regarding behaviour, dogs had higher factor scores for arousal (C/S: mean ± SD=0.6±1.6, S/C=- 0.3±0.5; lme order: P<0.001) and displacement behaviours (C/S=0.2±1.1, S/C=- 0.2±0.4; lme order: P<0.001) in the order C/S compared to the order S/C on both days. Our results support a gradual introduction of this interactive dog toy to avoid negative effects such as frustration caused by low success.

## Evaluation of environmental enrichment influence on animal welfare in captive *Panthera onca* through non-invasive methods

*Lucia Fernández[1], Juan Manuel Lomillos[2] and Marta Elena Alonso[1]*
*[1]University of Leon, Animal Production Department, Veterinary Faculty of Leon, Campus de Vegazana, 24071 Leon, Spain, [2]Cardenal Herrera University, Animal Production and Health, Public Health and Food Science and Technology Department, C/ Tirant lo Blanc 7, 46115 Alfara del Patriarca, Valencia, Spain; marta.alonso@unileon.es*

This is the first step carried out in Spain with the aim to determine the influence of an Environmental Enrichment Program (EEP) in the welfare of *Panthera onca* keep under captivity conditions, through the study of ethological parameters and the variations in the Faecal Cortisol Metabolites (FCM) levels. Observational study was conducted on three individuals (two males and one female) from Safari Madrid (Spain), with 24-hour video surveillance during a total of 68 days, in order to assess any differences demonstrated by each individual when EEP was applied versus the lack of it. Additionally, faecal samples were collected during the last 15 days of each part of the study (with and without EEP). These samples were stored at -20 °C and then analyzed with the LKCO1 Immulite® Cortisol Kit, using Radioimmunoassay (RIA) in the 'Laboratorio de Técnicas Instrumentales' at the University of León. Likewise, a Polymerase Chain Reaction (PCR) test was conducted for the individual identification of the stool samples from the female and her cub, both living in the same area. EEP included environmental, food, sensorial and motor. Statistical analysis were carried out with paired samples T-test and correlation matrix using SPPSS® with $P<0.05$ as significant level. 405 hours of videotaped behaviour were visualized and results were presented with Excel 2007 Dynamic Tables graphics. Only data obtained from the male were used in the present results because the differentiation between female and cub were not possible with 100% security using the PCR test due to absence or very little presence of DNA in some fecal samples. RIA technique was validated in this work for FCM quantification using cortisol LKCO1 kit by Immulite in *P. onca*. Data samples obtained were within the control curve of the above mentioned kit, been reliable the quantification. Male daily FCM values were significant different between the two study periods, $T_{(13-1)}=2.212$ $P<0.045$, with higher mean values during the EEP period (73,54 ng/g) compared with the non EEP period (19.85 ng/g), but no correlations were found between the behaviour categories and FCM. During the EEP cortisol values had higher variations than during non EEP period (Std. Dev. 7.04 vs 1.61, respectively) and we could infer that elements used as environmental enrichment activated the Hypothalamus-Pituitary-Adrenal pathway. Male activity interaction with EE elements could elevate the FCM without been under a negative stress situation. Behaviour corroborated this found because during the EEP period mean daily stereotypic movement (repeated walking and tail self-licking) duration were significantly $T_{(13-1)}=6.197$ $P<0.0001$ lower than in non EEP period (29 vs 124 minutes, respectively). As conclusion, and taking into account that this is a preliminary study, we consider that EEP are beneficial for the studied male *P. onca* welfare based on the reduction in stereotypic behaviours.

## Genetics and epigenetics of domesticated behaviour – what we can learn from chickens and dogs

*Per Jensen*

*Linköping University, IFM Biology, 58183 Linköpiung, Sweden; per.jensen@liu.se*

Darwin realized how powerful an example of evolution we have in domestication. He therefore allocated not only the first part of 'The Origin of Species' to this, but returned later to complete an entire volume on the subject. With this in mind, it is somewhat puzzling that domestication was then almost exclusively left for archeologists and historians to pursue scientifically for almost 150 years. But biologists are now reclaiming the territory and the knowledge of the behavioural consequences of domestication as well as the mechanisms underlying these are growing. In the presentation I will focus on the oldest domesticated species, the dog, and a somewhat younger member of the group, the chicken. Each of them has a unique history and relationship to humans while at the same time they share many of the general traits referred to as the domestication syndrome. It is particularly interesting that domestication in both species has modified social behaviour and stress susceptibility. Unravelling the mechanisms underlying these changes may therefore provide new insights into important fundamentals of animal welfare in addition to serving as an elegant model of evolution. In the talk I will provide examples from various approaches. Using cross-breeding between ancestral Red Junglefowl and a modern domesticated White Leghorn egg layer, we have been able to dissect the genetic architecture underlying many behavioural differences, and by experimental 're-domestication' of Red Junglefowl we have unraveled central mechanisms in the emergence of the domestication syndrome. Studies of social behaviour in dogs have allowed genome-wide mapping of genes involved in dog-human interactions, and shown how domestication has 'hi-jacked' genetic mechanisms common to both species. Both in chickens and dogs we have found evidence for the immense importance of epigenetic mechanisms, primarily DNA-methylation, on domestication. Since DNA-methylation is to a large extent affected by environmental challenges, this offers a previously overlooked route for experiences and stress to modify the genome of an individual. Over the last decades it has become clear that some transient epigenetic modifications can be transferred to the next generation, and our results indicate that methylated genetic loci may sometimes be prone to mutations. Hence, in contrast to established theory, it appears that the environment may sometimes shape the genome, perhaps in an adaptive way. The data presented demonstrate the importance of ethology for the understanding of domestication. Furthermore, they show how fundamental and applied research can act in fruitful coalescence.

## Science and sentience: assessing the quality of animal lives

*Francoise Wemelsfelder*
*SRUC, Animal and Veterinary Sciences, Roslin Institute Building, Easter Bush Campus, Roslin EH25 9RG, United Kingdom; francoise.wemelsfelder@sruc.ac.uk*

This paper will review recent research on qualitative behaviour assessment (QBA), asking how human and non-human animals might interact and understand each other as whole sentient beings. The conference theme '*Animal lives worth living*' suggests a view of animals as sentient beings, implying that life has value *for animals* – in other words, that animals, like humans, have a perspective, a point of view of their own, where things acquire a personal, 'felt' quality and are enjoyed or disliked. Such 'feeling' does not have to be regarded, as animal scientists often do, as essentially private; rather it indicates the possibility of a framework that complements and enriches mechanistic ones. Sentience, simply put, is a notion implying that animals are not just 'things', and so accommodating sentience is to ask how we might assess animals 'not as things'. It was this question that drove the development of qualitative behaviour assessment (QBA), a method designed to assess the expressive qualities of an animal's demeanour (e.g. as relaxed, fearful, agitated or content). Many years of research by research groups, scientists, and students across the world have helped establish that such an approach can be reliable and valid, and has significant potential to support the assessment of animal welfare states. Unease persists, however, about the status of the descriptors on which QBA relies. Does describing a pig as relaxed and happy really speak to us of how the animal feels, or is this a delusion, imposing anthropomorphic language on mute patterns of physical movement shown by the pig? Such dilemmas can bring to light points of friction that arise when science and sentience meet. Recent studies indicate that although QBA can indeed provide valid information about animals, in uncontrolled on-farm environments agreement between observers on what animals feel is considerably harder to reach, and various kinds of observer bias may come into play. This may reinforce scientists' unease that QBA is essentially a form of 'subjective', not 'objective' assessment; however we should consider such apparent contrasts with care. QBA's sensitivity to dynamic human-animal-environment relationships and their context gives it immediate meaning, and is basically an interpretative strength. Assessing meaning is not in itself anthropomorphic; it raises the possibility of engaging with an animal's world. Yet when anthropocentric bias threatens this effort, what we need is *more*, not less focus on 'meaning for the animal', through observation, training, and informed debate. The insight this creates may be different – less certain and fixable – compared to knowledge produced by mechanistic measurement, but it is no less real and useful. It may particularly benefit stockpeople, keepers and caretakers who work with animals every day, and so can practice and apply their QBA skills to improve the quality of animals' lives. Projects are currently underway exploring such practical implementation of QBA, working together with dairy farmers, zoo keepers, and donkey specialists. That applied ethology increasingly accepts such 'grounded perspectives' (a social science term) as part of science, is a sign that animal and human perspectives are beginning to meet.

## Parturition in two sow genotypes housed under free-range conditions

*Sarah-Lina A. Schild[1,2], Lena Rangstrup-Christensen[1], Cecilie K. Thorsen[1], Marianne K. Bonde[3] and Lene J. Pedersen[1]*
*[1]Aarhus University, Department of Animal Science, Blichers Allé 20, 8830 Tjele, Denmark, [2]Swedish University of Agricultural Sciences, Department of Biosystems and Technology, Sundsvägen 16, 23053 Alnarp, Sweden, [3]Center of Development for Outdoor Livestock Production, Marsvej 43, 8960 Randers, Denmark; sarah.lina.schild@slu.se*

High prolificacy has affected sow parturition and resulted in prolonged parturitions and a consequently increased risk of dead piglets. In several countries sows used in outdoor production are the same hyper prolific genotypes as those used in conventional, indoor systems. Indoors, birth assistance is used during long parturitions to avoid negative impacts on piglet survival. However, in outdoor systems where sows give birth in huts, use of systematic birth assistance is difficult. Thus, the primary aim of the current study was to investigate the course of parturition and maternal protectiveness in two sow genotypes. Topigs Norsvin (TN, Landrace × Z-line) sows were, amongst other, bred for the sow being able to nurse her own litter whereas Danbred (DB, Landrace × Yorkshire) sows, amongst other, were bred for number of live piglets at day five *postpartum*. Giving the different breeding goals, DB sows were expected to have longer parturitions than TN sows, due to larger litter sizes. Furthermore, as the landrace part of the TN sows was selected under loose-housed conditions, TN sows were expected to be more protective of their piglets. The study comprised 87 parturitions from 48 sows, followed in their first two parities. The sows were individually housed on pasture in the farrowing field and gave birth in huts. Parturitions were video filmed and sow behaviour was recorded during the first 3 hours of parturition. Maternal protectiveness was recorded by direct observation of sow behaviour on day 1 and 3 *postpartum* during piglet handling. Parturition duration and behaviour during parturition were analysed using negative binomial general linear mixed modelling (GLMM), behaviour during piglet processing by GLMM and piglet mortality by binomial GLMM. Parturition duration (median 4.3 h) was not influenced by genotype. During the first 3 h of parturition, TN sows had more posture changes ($X^2$(1df)=21.5, P<0.001) and were lying more in sternal position ($X^2$(1df)=29.9, P<0.001) (raw mean±SD, 12±10 posture changes and 11±12 min sternal) compared to DB sows (5±6 posture changes and 2±4 min sternal). Additionally, in first parity TN sows spent less time lying in lateral position (34±18 min) compared to DB sows (51±12 min; $X^2$(1df)=5.7, P=0.017). There was no significant effect of sow genotype on agitation score, but TN sows remained closer to their piglets during piglet handling than did DB sows ($F_{1,47}$=14.72, P=0.0004). The median number of total born piglets was 17.0. Longer parturitions increased stillbirth risk (odd ratio 1.05 95% CI [1.00;1.11]; $X^2$(2df)=8.7, P=0.003) and tended to increase early postnatal mortality (odd ratio 1.29 95% CI [1.05;1.57]; $F_{1,35}$=4.1, P=0.051). Furthermore, postnatal mortality was not affected by posture changes during parturition despite crushing being the main cause of early mortality. In conclusion, sow genotype did not affect the duration of parturition, whereas behaviour during parturition depended on both sow genotype and parity.

## Growing slowly with more space: effects on 'positive behaviours' in broiler chickens

*Annie Rayner[1,2], Ruth C. Newberry[3], Judit Vas[3], Andy Butterworth[1] and Siobhan Mullan[1]*
*[1]University of Bristol, Bristol Veterinary School, Division of Food Animal Science, Langford BS40 5DU, United Kingdom, [2]FAI Farms, Northfield Farm, Wytham OX2 8QJ, United Kingdom, [3]Norwegian University of Life Sciences, Faculty of Biosciences, Dept. of Animal & Aquacultural Sciences, 1432 Ås, Norway; annie.rayner@faifarms.com*

Broiler welfare has increasingly been under the scrutiny of the media and non-governmental organisations such as the European 'Better Chicken Campaign'. Much of the focus has surrounded welfare issues associated with breeding for rapid growth. However, rearing conditions, including stocking density, have also raised concerns. This study explored the welfare of broilers in 4 treatments representing 4 commercial systems with varying combinations of stocking density and breed. The treatments were: (1) Breed A, 30 kg/m$^2$; (2) Breed B, 30 kg/m$^2$; (3) Breed B, 34 kg/m$^2$; (4) Breed C, 34 kg/m$^2$. Breeds A and B were 'slower-growing' breeds and Breed C was a typical 'fast-growing' breed. Treatments were pseudo-randomly allocated to 4 houses (similar in layout, size and orientation) over 4 crops on one farm. The day before processing, 100 randomly-selected birds/flock were gait scored using a 6-point system (0=no detectable abnormality; 5=incapable of sustained walking). We also compared the levels at which behaviours considered indicative of positive welfare (i.e. appearing playful, exploratory, comfort-related or safety-related in the context observed) were expressed in the different flocks. We sampled behaviour using 'behaviour transect' methodology, whereby successive 15-s scans were made of 57 'disturbed' (D) patches and 54 'undisturbed' (U) patches per house. A D patch was a defined area within a transect through which the assessor had slowly walked 75 s previously, while a U patch was a defined area located two transects across from, and ahead of, the assessor's current location. The assessor recorded numbers of birds/patch/scan performing the following mutually-exclusive behaviours: running, jumping, wing-flapping, play-fighting, ground-scratching, dustbathing and perching, from which the overall percentage of birds engaged in 'positive behaviour'/patch/scan was calculated. Behavioural transect sampling was undertaken at 14 and 28 days old and again 2 days before processing. Clear differences in the behaviour and health measures were detected between treatments. There was a difference in the distribution of gait scores between treatments (Kruskal Wallis, $X^2$ (3)=368.7, P<0.001; mean scores: 1.09, 1.42, 1.38, 2.01 in treatments 1-4, respectively) and post-hoc pairwise comparisons revealed significant differences between treatments except between treatments 2 and 3. Thus, at least for slower growing breed B, stocking density did not influence the gait scores detected. The percentage of birds engaged in 'positive behaviour' was lowest in U patches in treatment 4. Across all ages, 1.37, 1.62, 1.44 and 0.68% of observed birds (treatments 1-4, respectively) displayed 'positive behaviour' in the U patches, and 6.17, 5.02, 4.07 and 1.47% in the D patches. The results suggest that better welfare outcomes, as demonstrated through 'positive behaviours' and walking ability, can be achieved in systems that utilise a combination of slower growing breeds and lower stocking densities.

## Range use and plumage condition of two laying hen hybrids in organic egg production

*Fernanda M. Tahamtani[1], Atefeh Berenjian[2], Stefan Gunnarsson[3] and Anja Birch Riber[1]*
*[1]Aarhus University, Department of Animal Science, Blichers Alle 20, Tjele 8830, Denmark,*
*[2]University of Tehran, Department of Animal and Poultry Science, Pakdasht, Tehran, Iran,*
*[3]Swedish University of Agricultural Sciences, Department of Animal Environment and Health,*
*53223 Skara, Sweden; fernandatahamtani@anis.au.dk*

The FreeBirds project aims to generate more insight into the relationship between organic chickens' free-range use and bird health, welfare and performance, as well as soil nutrient load. One completed study of the project compared the range use, welfare and performance of Dekalb White (DW) and Bovans' Brown (BB) laying hens on an experimental setting simulating organic on-farm conditions. Six hundred non-beak-trimmed hens of each hybrid (n=1,200) were housed in groups of 100, from 17 to 38 weeks of age, in 12 pens according to EU requirements for organic egg production. Indoor pens measured 4.5×4.5 m. Each pen provided 12 nest boxes and 18 cm/hen of elevated perch length. Wood shavings was provided as litter and the indoor net stocking density was 5.1 hens/m$^2$. The outdoor range for each pen measured 4.5×90 m, providing an outdoor density of 4 m$^2$/hen. The ground on the outdoor range had grass covering but no trees; shelter was available in the form of four 10-meter long tarps running along the length of the range at 10-meter intervals. Feed and water were provided *ad-lib* indoor and maize silage and/or carrots was provided as roughage under the first shelter of the outdoor range three times a week. Live observations of range use were carried out weekly, recording the number of hens outdoors and their distance from the pophole. Furthermore, assessment of plumage condition, as part of a welfare assessment protocol, was performed by four observers on all hens at 17 and 38 weeks of age. Preliminary statistical analysis were carried out in the software SAS using mixed models. The models included the fixed factors hybrid and age and the random factors pen and observer. Tukey test was used for *post hoc* testing. The range use in week 38 showed that while the total number of hens observed outside did not vary with hybrid, DW used shelters more than BB (mean ± SD: 15.1±12.1 vs 10.8±7.5 hens; P=0.002). Furthermore, BB ranged further away from the house than DW (mean ± SD: 4.47±5.34 BB hens observed >60 m away from the pophole vs 1.23±2.61 DW hens observed >60 m away from pophole; P<0.0001). Regarding plumage condition, there was an interaction of hybrid and age: while both hybrids had lower scores, (i.e. better plumage condition) at 17 weeks compared to 38 weeks of age (P<0.001), BB had better plumage condition than DW at each time point (P<0.001; lsmeans plumage score ± SE: BB17=0.98±0.29; DW17=1.37±0.29; BB38=4.2±0.29; DW38=5.8±0.29). These preliminary results indicate that DW are less explorative/more fearful compared to BB and highlights the need for breed-specific management such as greater shelter provision for DW. This, combined with the better plumage condition of BB, suggests that BB is a breed more robust to the environment of organic production compared to DW hens.

## Genetic parameters for maternal traits in sows recorded by farmers

*Christoph Winckler, Birgit Fuerst-Waltl, Christine Leeb, Alfons Willam and Christina Pfeiffer*
*Division of Livestock Science, Gregor-Mendel Straße 33, 1180 Vienna, Austria;*
*christoph.winckler@boku.ac.at*

In livestock breeding programs, behavioural traits are rarely included. Breeding goals usually focus on production traits. Nevertheless it has to be assumed that animals expressing normal behaviour, such as exploration, may also show better performance. Apart from the effect of improved animal welfare on biological performance and health, behavioural traits are also subject of public interest. In the case of pigs, free farrowing systems will be mandatory in Austria from 2033 on. The new farrowing systems require a revision of the current breeding goals with a special emphasis on maternal behaviour. Therefore a working group, consisting of scientists, members of breeding organisations and breeders were set up to define new maternal traits for on-farm recording. The aim of the study was to analyse the genetic background of five maternal behavioural traits (nestbuilding (NB), farrowing behaviour (FB), aggressiveness toward piglets (AP), aggressiveness towards stockman (AS), overall impression of good maternal abilities (MA)) and fertility traits (number of total born piglets and number of weaned piglets, respectively) in the Large White and Landrace population. Sows were provided with jute bags 24 h before expected farrowing date. Manipulation of the bags was scored based on their signs of usage after completed farrowing. Scores ranged from 1 (no use) to four (extreme use). Pre-FB could be scored as calm or agitated behaviour and were just recorded when farmers could observe it. AP and AS as well as MA (including lying, resting and nursing behaviour) were recorded as binary traits (yes or no). From June 2017 – July 2018 all traits were recorded by farmers after intensive training. In total, 2,900 records of 1,671 Large White and Landrace sows from 22 farms were used to fit a uni- or bivariate linear animal model using an average information REML algorithm. Fixed effects comprised farm, year, season, breed and farrowing class (parity and age of sow in month). The additive genetic effect of the sow and the permanent environmental effect of the sow were fitted as random effects. The pedigree included 9,772 pigs. Heritabilities ranged from almost zero for NB to 0.18 for FB. Genetic correlations between maternal traits ranged from -0.43 between NB and AP and 0.71 between NB and number of weaned piglets. For example, sows with MA weaned more piglets ($r_g$=0.67), whereas sows which were more aggressive towards piglets weaned fewer ($r_g$=-0.16). In general genetic correlations between traits were meaningful, indicating a genetic background between maternal traits as well as fertility traits. For the implementation of a routine genetic evaluation a maternal trait which can be easily recorded is needed. However, it still has to be discussed which trait is the most suitable one and correlated to other important traits e.g. piglet mortality.

## Space requirements for fast and slow growing chickens on elevated grids

*Julia Malchow, Jutta Berk and Lars Schrader*
*Friedrich-Loeffler-Institut, Institute of Animal Welfare and Animal Husbandry, Dörnbergstraße 25/27, 29223 Celle, Germany; julia.malchow@fli.de*

Young chickens show a high motivation for perching at young age. However, broiler housings rarely offer elevated structures. To estimate appropriate space requirements for elevated grids we used chickens differing in growth performance. A total of 173 Ross 308 (meat strain, fast-growing; rearing period: 5 wks) and 168 Lohmann Dual (dual-purpose strain, slow-growing; rearing period: 10 wks) male chickens were kept in groups of 56 to 58 birds in 3 pens (each 2×3 m) per strain. Pens were equipped with elevated grids (h×l×w: 50×300×60 cm, grid area: 18,000 cm$^2$) which could be reached by a ramp (l×w: 120×60 cm; angle: 24.6°). Single-phase feed and water were given *ad libitum*. From day 3 of life, pens were artificially lighted (20 lx) from 04:00am to 08:00pm. First sub-study: Elevated grids were videoed by an infrared camera from 04:00am to 09:00pm and number of chickens on grids were counted for 2 days each week in 15 min time sampling (light period: 65 time points, dark period: 4 time points). Second sub-study: At each week of age, for planimetric measurements photos were taken from recordings of a second camera installed at a right angle above grids during light period. From these, the body area of 100 single resting chickens were completely surrounded by line using the software 'AxioVision' (Carl Zeiss Microscopy, Thornwood, United States). The created pixel area by line were converted into the covered area in cm$^2$. Photos were randomly selected but equally distributed across strains and pens. Effects of time of day, age, and strain were tested on the use of elevated grids. Effects of age and strain were analyzed on area covered by body. Both analyses were done with GLM and pen number was included as random factor. First sub-study: Chickens of both strains increasingly used the elevated grids with increasing age ($P<0.0001$) both during light and dark period ($P<0.0001$). Second sub-study: Area covered by the chickens increased with age ($P<0.0001$) and interaction age×strain was significant as well ($P<0.0001$). At 10 weeks of age, the mean area covered by a Dual chicken was 558.2±75.4 cm$^2$ (body weight at slaughter: 2,333.5±222.3 g) and for Ross chicken at 5 weeks of age 403.2±65.1 cm$^2$ (body weight at slaughter: 1,770.7±298.9 g). During the entire fattening period, we observed a maximum of 42 Dual and 26 Ross chickens on the elevated grids. Compared to the available grid area of 18,000 cm$^2$ a higher number of Dual chickens were observed on the elevated grids than expected based on area covered by single birds. Thus, they often rested in close contact to each other on the elevated grids. In contrast, Ross chickens did not fully occupy the area of elevated grids. Based on the mean number of birds for Ross chicken the space on elevated grids should be sufficient for about 15% of birds. However, the proportion of chickens using elevated grids may alter in relation to group size and accessibility of these structures.

## A comparison of tonic immobility reactions between domesticated *Columba livia* and *Columba guinea*

Abdelkareem Ahmed

*University of Nyala, 583, Sudan; kareemo151@gmail.com*

The rock doves (*Columba livia domestica*) were domesticated from the wild several thousand years ago as food, pets as well as carrier pigeons particularly in first and Second World War as a result of their capabilities to find their habitat from long distances. However, the speckled pigeons (*Columba guinea)* or African rock pigeon are a resident breeding bird in much of Africa south of the Sahara countries and is still wild seen around human habitation and cultivation where food is available. No information about fearful behavior in these strains in Sudan. We hypothesized that the fearfulness behavior may be differ between *C. livia* (domesticated) and *C. guinea* (wild). In the present study, we investigated the differences in tonic immobility (TI) response between the *C. livia* and *C. guinea* form the same genus to determine which strain is more fearful. Five months old 20 mature birds 10 each breed were obtained from the Veterinary Research Laboratories (Nyala, Sudan), included male=3 and female=3 for each breeds were used to measure TI reaction as a response to physical restraint. The birds were allowed free access diet and water for adaptation. After 7 days of adaptation, we followed the consistent and reliable method to induce tonic immobility using the cradle as the most effective apparatus to do this experiment. After 7 days of TI test, blood samples were collected and total erythrocytes and leukocytes parameters were measured using the dilution flask method and the counting of corpuscles using a Bürker chamber, while plasma Corticosterone (CORT) were determined using enzyme immunoassay (EIA). SPSS software used to analyze the data. The experiment procedures were approved by the Animal Ethics Committee of University of Nyala. The results showed that both *C. livia* and *C. guinea* males displayed decreased TI responses compared speckled females (P<0.05). However, *C. livia* females showed increased TI responses associated with elevated plasma corticosterone concentration when compared to C. *guinea* female (P<0.05). TI in both *C. livia* and *C. guinea* females was longer than males (P<0.05). High haemoglobin concentrations together with high red blood cell counts were observed in C. *guinea* compared to *C. livia*. Likewise, higher white blood cell count was found in C. *guinea* vs *C. livia*. Moreover, heterophil/lymphocyte ratio was high in *C. guinea* compared to *C. livia*. We conclude that fearfulness behavior differences may be due to the strain differences.

## Awareness and use of canine quality of life assessment tools in UK veterinary practice

*Claire Roberts[1], Joanna Murrell[2], Emily Blackwell[1] and Siobhan Mullan[1]*
*[1]University of Bristol, Bristol Veterinary School, Langford House, Langford, BS40 5DU, United Kingdom, [2]Highcroft Veterinary Group, 615 Wells Rd, Bristol, BS14 9BE, United Kingdom; claire.e.roberts@bristol.ac.uk*

A number of formal canine quality of life (QoL) assessment tools exist in the literature, and their use in veterinary practice has been recommended by several veterinary associations. An online survey was used to investigate the awareness, current use and potential use of QoL assessment tools amongst veterinary surgeons and nurses in the UK. Potential barriers to the use of these tools in veterinary practice were also identified. An anonymous 24-question survey was available online between February and May 2018. Links to the survey were shared in the veterinary literature, social media and through regional sections of the British Veterinary Association. Respondents were self-selected practicing veterinary surgeons and veterinary nurses. Questions investigating the demographics and background of participants were included, along with questions relating to the awareness and use of QoL assessment tools in dogs. Potential barriers to using QoL assessment tools in practice were listed with four available options for the respondent to select: 'not a barrier', 'somewhat of a barrier', 'a moderate barrier' and 'a strong barrier'. Chi-square tests were used to analyse whether respondent age, role (veterinary surgeon or nurse) and possession of additional qualifications influenced the awareness, willingness to use and potential use of canine QoL assessment tools. Ninety veterinary surgeons and twenty veterinary nurses responded to the survey. Thirty-two respondents (29.1%) were aware of the existence of formal QoL assessment tools for dogs, with a further fifteen (13.6%) 'not sure'. Of the three QoL assessment tools listed in the survey, current use was reported as less than 4%. Most veterinary professionals reported that they would surely (18.2%), probably (37.3%) or maybe (35.5%) want to use a QoL assessment tool in some aspect of their practice. The strongest barrier to the use of formal QoL assessment tools was reported to be lack of time; 40% of respondents reported time needed to complete the tool as a 'strong barrier' to their use in practice. Time to discuss and record results were also 'strong barriers' for 38.9 and 29.1% of respondents respectively. Other 'strong barriers' included a perceived resistance from owners (20.4%) and having insufficient information on QoL assessment tools (19.3%). No statistically significant influences of respondent characteristics on the awareness or use of QoL assessment tools were found (all P<0.05). Although the current use and awareness of canine QoL assessment tools in UK veterinary practice is low, veterinary professionals appear willing to use the tools within their daily practice. This discrepancy implies that QoL assessment tools are not well disseminated to veterinary surgeons and nurses in practice. The main barrier to their use is the lack of available time, followed by a perceived resistance from owners.

## Owner caregiving style and the behaviour of dogs when alone

*Luciana Santos De Assis[1], Oliver Burman[1], Thomas Pike[1], Raquel Matos[2], Barbara Georgetti[3] and Daniel S. Mills[1]*

[1]*University of Lincoln, School of Life Sciences, Joseph Banks Laboratories, LN6 7DL, Lincoln, United Kingdom,* [2]*University Lusofona of Humanities and Technologies, Faculty of Veterinary Medicine, Campo Grande 376, 1749-024 Lisboa, Portugal,* [3]*Federal University of Minas Gerais, School of Veterinary, Avenida Antônio Carlos 6627, 31270-901 Belo Horizonte, Brazil; lassis@lincoln.ac.uk*

There is evidence that dogs develop an attachment bond with their owners that, according to the human literature, is characterised by two behavioural systems directionally defined as Attachment (from infant to parent) and Caregiving (from parent to infant). The two systems are the product of differing, but compatible emotional bonds directed towards personal survival (attachment) and maximising genetic investment (caregiving), with the infant's attachment behaviour being a response to the parent's caregiving style. A large body of research focused on infant development, describes four different styles of attachment and caregiving: Secure, Avoidant, Ambivalent and Disorganised. Hence, given the significance of caregiving styles on the form of attachment bond that develops, it seems reasonable to suppose that owner behaviour may play an important role in the attachment bond formed by their dogs and so how it copes with being separated or left alone. Therefore, this study aimed to define the different profiles of dogs' reactions when left alone and evaluate their relationship with owner caregiving style. Two online surveys were carried out to develop valid instruments to assess both behaviour and caregiving styles: (1) The behaviour of dogs showing one of the following signs: destruction, elimination and vocalization, when left alone (161 items, n=5,122) allowed the definition of dog profiles using 54 items following principal component analysis (PCA), with hierarchical cluster analysis used to group subjects; and (2) an assessment of caregiving style for owners in line with the maternal care literature (40 items, n=973), which resulted in 15 reliable items that differentiated the four types of caregiving mentioned above. The relationship between caregiving styles and behavioural groupings could then be compared in a final test population (n=211) using binomial Generalised linear models. Four different profiles were identified: A (high destruction and high vocalization during pre-departure and when alone); B (high vocalization during pre-departure and when alone, and high vocalization when unfamiliar people/dog approach), C (High vocalization when unfamiliar people/dog approach); and D (low rates of all behaviours). Owners' caregiving styles were significantly associated with groups B, C and D (P<0.05), e.g. Disorganised style for group B (estimate: 9.24; SE: 2.99; Z-value: 3.09; P=0.01). Hence, owners' caregiving style and, consequently, the style of attachment in the dog seems to be an important feature in how it reacts when left alone. These findings emphasise the importance of consistency and safety reassurance provided by humans in the development of a healthy owner-dog relationship.

# What can short-term fostering do for the welfare of dogs in animal shelters?

*Lisa M. Gunter[1], Erica N. Feuerbacher[2], Rachel J. Gilchrist[1] and Clive D.L. Wynne[1]*
*[1]Arizona State University, Department of Psychology, 950 S McAllister Avenue, Tempe, Arizona, TX 85287, USA, [2]Virginia Polytechnic Institute and State University (Virginia Tech), Department of Animal and Poultry Sciences, 175 W Campus Drive, Blacksburg, VA 24061, USA; lgunter@asu.edu*

One of the greatest stressors for dogs living in animal shelters is social isolation. Many studies have demonstrated that human interaction reduces cortisol in shelter dogs, with the possibility that longer periods of interaction may yield greater effects. More recently, animal shelters are utilizing short-term fostering programs to provide relief from the perceived stresses of kennel life; however the effects of these programs are not well understood. This study assessed the impacts of one- and two-night fostering programs on the urinary cortisol levels, resting pulse rates, longest bout of uninterrupted rest, and proportion of time spent resting of dogs awaiting adoption. Five animal shelters from across the United States participated in the study: Best Friends Animal Sanctuary (BFAS), Arizona Humane Society (AHS), Humane Society of Western Montana (HSWM), Lifeline Animal Project (LAP), and the SPCA of Texas (SPCATX). During the study, dogs' urine was collected in the morning before, during, and after fostering stays for cortisol: creatinine analysis. Non-invasive health monitors were worn by the dogs, which collected heart rates and activity levels. In total, 207 dogs participated in the study, and 1,076 cortisol values were used in our analysis. Mean cortisol values from before, during, and after the sleepovers were calculated and analyzed using a linear mixed model. Significant differences were found among these phases, $F(2, 209.04)=31.71$, $P<0.001$. Dogs had significantly lowered mean cortisol values during the sleepover as compared to before the sleepover in the shelter ($P<0.001$) and after ($P<0.001$), with no difference in the before and after comparison ($P=0.799$). Dogs' longest bouts of rest before, during, and after their sleepovers were analyzed with a linear mixed model. A main effect of phase was found, $F(2, 193.51)=16.99$, $P<0.001$, demonstrating dogs' resting bouts differed during the study. Post-hoc comparisons revealed that dogs had their longest bouts of uninterrupted rest during sleepovers ($P<0.001$). Resting bouts in the shelter after sleepovers were longer than before they left ($P=0.012$) but shorter than during sleepovers ($P=0.010$). Lastly, in-shelter cortisol values differed at our five shelters ($F(4, 196.41)=7.14$, $P<0.001$). Dogs at BFAS were found to have significantly lower cortisol than dogs at AHS ($P<0.001$), HSWM ($P=0.003$), and DeKalb ($P=0.004$) with dogs at SPCATX having lower cortisol values than dogs at AHS ($P=0.002$). These differences were, in some cases, greater than the impact of the fostering intervention itself. Considering the diversity of facilities that participated in this study, it is possible that as yet unstudied, shelter-specific, environmental factors could be contributing to the overall welfare of shelter dogs. Thus while a reprieve from the shelter is impactful for dogs awaiting adoption, mitigating the stressors present in kennel conditions should also be addressed to improve the lives of shelter dogs.

## Farmer perceptions of pig aggression compared to animal-based measures of fight outcome

*Rachel Peden[1], Irene Camerlink[2], Laura Boyle[3], Faical Akaichi[1] and Simon Turner[1]*
*[1]Scotland's Rural College, West Mains Rd., Edinburgh, EH9 3JG, United Kingdom, [2]University of Veterinary Medicine, Institute of Animal Welfare Science, Veterinärplatz 1, 1210 Vienna, Austria, Austria, [3]Teagasc, Pig Development Department, Animal & Grassland Research and Innovation Centre, Moorepark, Fermoy, Ireland; rachel.peden@sruc.ac.uk*

Most farmers do not perceive pig aggression to be a problem that needs to be addressed, despite the fact that stress and injuries are common and a number of management strategies have been identified. This study investigated whether routine exposure to pig aggression disrupts farmer responses to aggression. Pig farmers (n=90) were shown video clips of pigs immediately after a dyadic contest (clips 1-3) and during a contest (clips 4&5). Video footage was obtained from a separate research project whereby 168 growing pigs were video recorded in aggressive encounters. For each pig measures of skin lesions (injury) and blood lactate (physical fatigue) were taken pre- and post-fight to indicate relative change as a result of aggression. A stepwise selection process was used to choose each video clip. Based on the severity of skin lesions and blood lactate we identified from the entire dataset the encounters whereby both pigs obtained low (lower quartile, clip 1), medium (interquartile range, clips 2, 4&5) or high (upper quartile, clip 3) severity measures. Participants viewed all clips, and the order was randomised at each recruitment session. Following each clip, farmers were asked to place a downward line through 100 mm visual analogue scales at a point they felt best represented: (1) how much of a negative emotional reaction they had; (2) how exhausting they believe the fight was; and (3) how severe they believed the fight was. Their judgements were compared to the objective measures of severity, and against control groups with similar pig experience (10 pig veterinarians) and without experience (26 agricultural students; 24 animal science students) of working with pigs. Farmers did not show desensitisation to aggression; they expressed greater emotional response scores (mean=45.6, SE=1.5) and judged fight exhaustion (mean=55.7, SE=1.5) to be higher than agriculture students (emotion: mean=38.3, SE=2.3; exhaustion: mean=49, SE=2.4) (P<0.05). However, all groups underestimated the severity of aggression when they saw pigs after the contest had ended as compared to witnessing a fight in progress. All participants scored emotional response, exhaustion and severity as greater for both of the 'during fight' clips than for the 'post-fight' clips (P<0.05), even when post-fight outcomes were severe as indicated by objective measures. Farmers a unlikely to witness fights as frequently as they actually occur. They are expected to witness the injuries from such interactions during their regular inspections; however, results suggest that they are unlikely to fully recognise these outcomes. We recommend that farmers are made aware of how to accurately determine the physical impact of aggression by scoring skin lesions on affected animals. Furthermore, researchers should calculate the economic and welfare impact of aggression as indicated by the lesions. All human and animal experimentation conforms with the ISAE Ethical Guidelines.

## Consistency of blue fox behaviour in three temperament tests

*Eeva Ojala[1], Hannu Korhonen[2] and Jaakko Mononen[3]*
*[1]Kannus Research Farm Luova Ltd., Turkistie 6, 69100 Kannus, Finland, [2]Natural Resources Institute Finland (Luke), Production Systems, Teknologiakatu 7, 67100 Kokkola, Finland, [3]Natural Resources Institute Finland (Luke), Production Systems, Halolantie 31 A, 71750 Maaninka, Finland; eeva.ojala@luovaoy.fi*

Temperament tests are important tools in animal welfare research, as well as in on-farm welfare assessment protocols. An essential feature of a valid temperament test is that individual animals show consistent test-retest behaviour. We studied consistency of blue fox behaviour in three temperament tests. The Stick Test (ST) is included in the current WelFur on-farm welfare assessment protocol for foxes, whereas the inclusion of the other two assessments, the Titbit Test (TBT) and subjective evaluation of human-animal relationship (SEH), is being considered. The ST is supposed to measure exploratory behaviour. In the test, the assessor inserts 20-30 cm of a 150 cm stick into a fox´s cage, and records whether the fox touches the stick or not within 10 seconds. The TBT and SEH are supposed to measure the human-animal relationship (HAR). In the TBT, the assessor records whether a fox takes a titbit (a dog biscuit) from the assessor's hand within 30 seconds. In the SEH, the response of a fox to an assessor approaching its cage is assessed subjectively on a six-point scale from 0 (approaches) to 5 (withdraws). The experimental animals were juvenile (J, 46-49) and adult (A, 91-144) breeding female candidates on the experimental farm of Luova Ltd. The foxes were housed singly in cages measuring 1.2 $m^2$ (J, November), 2.4 $m^2$ (J, December) or 0.8 $m^2$ (A, November-December). The number of tested animals decreased in the course of the study due to the selection of the final breeding animal stock. Assessor A1 carried out the three temperament tests twice in late November with 5-7 days between the repeats, and assessor A2 once in late December. Test-retest consistencies were analysed for the ST, TBT and SEH between the first and second tests by A1 and between the first tests by A1 and the only tests by A2 with Cohen's ordinary (ST and TBT) or weighted (SEH) kappa. Also the percentages of agreement (PA) are reported. The kappa for the A1-A1 comparison of the ST was 0.484 (PA=75.5%). Forty-nine foxes explored the stick in both tests, 93 foxes explored the stick in neither of the two tests, and 46 foxes had inconsistent responses. The respective numbers for the A1-A2 comparison were 40, 65 and 32, and the kappa was 0.522 (PA=89.1%). The kappa for the A1-A1 comparison of the TBT was 0.735 (PA=83.6%), the numbers of foxes being: 45 took the titbit in both tests, 126 took the titbit in neither of the two tests, and 21 had inconsistent results in the two tests. The respective numbers for the A1-A2 comparison were 35, 82 and 23, yielding the kappa 0.631 (PA=76.6%). The kappas for the SEH were: A1-A1 comparison 0.404 (n=189, PA=59.4%) and A1-A2 comparison 0.318 (n=137, PA=35.3%). The TBT has more potential as a new HAR test in the WelFur protocol than the SEH. The consistency of the ST was rather low, indicating that probably another exploration test should be considered in the WelFur fox protocol. Longer-term behavioural consistency of the foxes in these three tests remains to be confirmed.

## Owner perspectives of cat handling techniques used in veterinary clinics

*Madalyn Abreu[1], Carly Moody[2] and Lee Niel[1]*
*[1]University of Guelph, Department of Population Medicine, 50 Stone Rd E, Guelph, ON N1G 2W1, Canada, [2]Charles River Laboratories, Global Animal Welfare and Training, 22022 Trans-Canada Highway, Senneville, QC H9X 3R3, Canada; carly.moody@crl.com*

Recent literature suggest that cats show signs of fear and stress during handling, which is a cat welfare concern. Many cat handling guidelines exist with the aim to reduce negative experiences during handling at the veterinary clinic, however no research has examined cat owner perceptions of how their cats are handled. Owner satisfaction is important for maintaining a good veterinarian-client relationship, satisfaction with veterinary care received, and owner willingness to provide their cat with routine veterinary visits to ensure cat health and welfare. The current study used a cross-sectional survey to examine: (1) cat owner agreement with the use of 12 common handling methods used on calm, fearful, and aggressive cats during routine examinations and procedures, and (2) describe factors influencing agreement or disagreement with various restraint methods. The questionnaire consisted of questions pertaining to cat ownership, veterinary care, agreement with handling techniques, assessment of cat-owner attachment, and participant demographics. Cat-owner attachment was measured using the Lexington Attachment to Pets Scale. The questionnaire was open to all geographic locations and was distributed using social media and snowball sampling to recruit current cat owners that were 18 years of age or older. Only full responses (n=1,754) were included in analyses using SAS software (v 9.3). Descriptive statistics were used to assess agreement frequencies for the 12 handling techniques, and mixed logistic regression models examined associations between participant demographics (including cat-owner attachment), and agreement with use of full-body with scruff and minimal restraint, on fearful cats. Overall, the results show that participants strongly disagreed with the use of a scruffing tool and a cat muzzle, and strongly agreed with the use of minimal restraint and loose towels, when handling cats of all demeanor types (calm, fearful, and aggressive). Cat owners with a high cat-owner attachment score were at a greater odds (OR: 1.04 [95% CI: 1.02, 1.07], P=0.002) of disagreeing with the use of full-body with scruff restraint on a fearful cat. There was high agreement with the use of minimal restraint on fearful cats, but this decreased with a lower cat-owner attachment and a greater number of cats owned ($F_{1,735}$=7.92, P=0.005). The results suggest that use of minimal restraint and loose towels are preferred by cat owners, as well as reduction of scruffing tools and cat muzzles for cats of all demeanor types. It is important for veterinarians to understand owner perceptions about cat handling methods used, and factors influencing perceptions, such as the level of cat-owner attachment. Client perceptions should be used to help inform best practices, in addition to scientific-based recommendations to improve cat handling and restraint methods.

## A survey of visitors' views on free-roaming cats living in the tourist town of Onomichi, Japan

*Aira Seo[1], Stephany Nozomi Ota[2], Yuki Koba[3] and Hajime Tanida[1]*
*[1]Hiroshima University, Graduate School of Integrated Sciences for Life, 1-4-4, Kagamiyama, Higashi-Hiroshima, Hiroshima, 7398528, Japan, [2]Cummings School of Veterinary Medicine at Tufts University, 200 Westboro Rd., North Grafton, MA 01536, USA, [3]Teikyo University of Science, Faculty of Education & Human Sciences, 2-2-1, Senjusakuragi, Adachi-ku, Tokyo, 1200045, Japan; airaseosan@hiroshima-u.ac.jp*

In recent years, the old town of Onomichi has attracted attention from tourists as 'the town of cats.' Approximately 200 free-roaming cats live in the area, mostly feral, and many visitors feed them. However, many residents are annoyed by the cats, because their feces litter the paths and gardens. Additionally, from an animal welfare point of view, the cats' health is very poor. In spite of the problems created by the cats, research on the impacts of the cats of Onomichi has not been conducted. This study therefore aimed to analyze visitors' awareness of free-roaming cats in the town to find viable solutions for the related problems. We conducted interviews with 111 domestic and 45 foreign adult visitors on the streets of the town in the summer of 2017. Each anonymous interview consisted of an introduction explaining the purpose of the study, an inquiry into the interviewee's demographic characteristics, and questions gauging attitudes about free-roaming cats. The data were analyzed by Chi-square test. There were 55 male interviewees and 101 females. Significantly more Japanese visitors (76.8%) than foreign visitors (51.1%) were aware of the cats before arriving in Onomichi ($P<0.01$: Chi-square analysis). They were aware of the cats from Internet and TV programs (48.5%), guide books and magazines (28.7%), and through word of mouth (12.9%). When asked what cats need in order to be healthy, interviewees answered food or water (44.4%), medical treatment (11.2%), to be spayed/neutered (7.1%), and other (37.2%). Significantly fewer Japanese visitors (42.5%) than foreign visitors (64.3%) supported controlling the cat population ($P<0.05$); of such controls, spaying and neutering was suggested significantly more by foreign visitors (72.7%) than by Japanese visitors (48.5%) ($P<0.01$), and significantly more male visitors (62.0%) than female visitors (41.8%) supported controlling the cat population ($P<0.03$). Additionally, significantly more Japanese visitors (89.9%) than foreign visitors (62.2%) indicated that they did not think the cats impact the wildlife population ($P<0.01$). Finally, the majority (58.3%) of visitors indicated that they had no concerns about the cats' impact on human health. All this shows that views toward the free-roaming cats differed significantly between Japanese and foreign visitors. Japanese visitors were more inclined to think that free-roaming cats do not have much impact on society, and most did not feel it was necessary to control the cats. Additionally, about half of the interviewees were not concerned about zoonotic diseases being spread by cats, even though cat feces can contain bacteria, viruses, and parasites that infect humans. The results suggest that an effective educational campaign directed at both Japanese and foreign tourists is necessary to solve the problems related to the free-roaming cats in the old town of Onomichi.

## Risk factors for aggression in adult cats that were fostered through a shelter program as kittens

*Kristina O'Hanley, Lee Niel and David L. Pearl*
*University of Guelph, Population Medicine, 50 Stone Road East, N1G 2W1, Canada;*
*kohanley@uoguelph.ca*

Aggressive behaviour in cats is a threat to human and animal safety and can also impact cat welfare if breakdown of the human-animal bond leads to neglect, relinquishment or euthanasia. The influence of early and ongoing management factors on cat aggression was examined for cats aged 1 to 6 years that were adopted following shelter fostering as kittens (n=262). Early management details were extracted from shelter records, and adoptive owners completed an online survey concerning the frequency and severity of fear and aggression, owner demographics, the home environment, and training methods. Factor analysis on fear and aggression questions identified five outcome variables for further analysis using logistic regression. The odds of aggression towards the owner were significantly greater in female cats (OR=1.754, P=0.049), and lower in households with three or more cats (OR=0.192, P≤0.001), or when owners reported using positive reinforcement (OR=0.280, P=0.002). For novel people, objects, and situations, the odds of aggression were significantly greater when owners reported using various forms of positive punishment (i.e. verbal correction, holding the cat), and female cats raised without mothers had lower odds of aggression compared to male cats raised with mothers (OR=0.066, P=0.008). Furthermore, the odds of severe aggression towards people were significantly greater when the owner reported using various forms of positive punishment (i.e. making a loud noise, verbal correction, holding the cat) and lower when there were three or more cats in the household (OR=0.307, P=0.016). The odds of aggression towards other cats were increased in female cats (OR=2.377, P=0.004), in cats with equal indoor/outdoor access compared those without outdoor access (OR=4.885, P=0.017), and in cats that were owner surrendered as compared to stray kittens (OR=2.209, P=0.031).Lastly, the odds of severe aggression towards other animals were significantly increased when the owner indicated there were other pets, of any type, living in the household (OR=2.655, P=0.002) and when owners reported using positive punishment (i.e. verbal correction) (OR=3.517, P=0.006),and lower when provided with training enrichment (OR=0.280, P=0.034). Increased animal-directed aggression with other pets in the household likely corresponds with increased opportunity, although owners reporting 'no opportunity to observe' resulted in removal from the analysis. Surprisingly, we did not find any associations between aggression and management factors related to early social exposure that have been anecdotally suggested to influence aggression (e.g. bottle-reared, singleton, early rehoming age), possibly via increased fearfulness. Thus, while mother-rearing was associated with aggression during exposure to novelty, other factors associated with the various forms of aggression were related to the adoptive home environment and training methods. These results highlight several potential areas for future research, and for owner education to reduce cat aggression, particularly for kittens acquired through shelter-run fostering programs.

## Bit vs bitless bridle – what does the horse say?

*Inger Lise Andersen and Nina Kalis*
*Norwegian University of Life Sciences, Department of Animal and Aquacultural Sciences, Faculty of Biosciences, P.O. Box 5003, 1432 Ås, Norway; inger-lise.andersen@nmbu.no*

Horses show a range of fine graded behaviour and social signals that we can use to infer affective states during training and riding. The objective of the present experiment was to assess behavioural expressions (of ears, eyes, mouth, tail, and head/neck position), gait quality and symmetry, and protest behaviours (bucking, vertical head tossing, downward tail lashing) when horses were ridden in a standard snaffle bridle, bitless side pull bridle or simple halter. Fourteen horses were subjected to a 20-minute riding programme comprising walking, trotting, cantering and gait transition phases, 3 times weekly over a period of three weeks, with one week per bridle type. The same rider rode all horses, and the order of applying the bridle types was balanced across horses. Behaviours were scored from video by one observer who was blind to the bridle types, and scoring reliability was ensured by observer training for scoring consistency with another experienced observer. Behavioural expressions during walk, trot, canter and trot-canter transitions were scored from 1 (most tense) to 5 (most relaxed), and the protest events were counted. Gait quality (evenness of cadence/rhythm) and vertical symmetry were assessed using a motion sensor ('Equisense' program in a mobile app) attached to the girth. Data were analysed in a generalized linear mixed model incorporating effects of bridle type, phase in the riding programme and their interaction, as well as day nested within bridle type, with horse as a random factor. The snaffle bridle resulted in the most tense mouth expression (i.e. lowest score) compared to the bitless bridle and the halter ($P<0.001$). Overall, the bitless bridle resulted in the highest sum of behavioural scores (implying the most positive affective state; $P=0.034$) and the lowest number of tail lashes ($P=0.002$) whereas the halter resulted in the highest total number of protests ($P<0.001$). There were significant interactions between bridle type and phase of the riding programme, with more tail lashing and total protest events occurring with a snaffle bridle or halter during trotting than in the other bridle-gait combinations. When using a bitless bridle, the frequencies of protest behaviours were generally low and similar across gaits. There was no significant effect of bridle type on vertical symmetry or gait quality reading. However, the lower was the head and neck position of the horse during riding (i.e. higher the score), the higher was the reading for gait quality ($Rho=0.6$; $P<0.001$). Fewer total protests were also correlated with higher gait quality readings ($Rho=0.5$; $P<0.001$). There was a high individual variation between horses in the number of protests shown during riding We conclude that, in general, the bitless bridle gave the most positive results of the three bridle types assessed, whereas results with a snaffle bridle or halter were more variable depending on the horse's willingness to collaborate with the rider (as indicated by lack of protests).

## Pet ownership among people with substance use disorder: implications for health and use of treatment services

*Andi Kerr-Little[1,2], Ruth C. Newberry[3], Anne Line Bretteville-Jensen[4], Linn Gjersing[4], Ingeborg Bolstad[2], Jenny Skumsnes Moe[1,2], Stian Biong[5] and Jørgen G Bramness[1,2]*
[1]*Arctic University of Norway, Institute of Clinical Medicine, 9019 Tromsø, Norway,* [2]*Norwegian National Advisory Unit on Concurrent Substance Abuse & Mental Health Disorders, Løvstadvegen 7, 2312 Ottestad, Norway,* [3]*Norwegian University of Life Sciences, Faculty of Biosciences, Dept. of Animal & Aquacultural Sciences, 1432 Ås, Norway,* [4]*Norwegian Institute of Public Health, P.O. Box 222 Skøyen, 0213 Oslo, Norway,* [5]*University of South East Norway, Faculty of Health Sciences, Grønland 58, 3045 Drammen, Norway; andikerrlittle@gmail.com*

Information about the impact of pet ownership on the lives of people with substance use disorder (SUD) is lacking. SUD is frequently associated with other mental health problems, difficulties in forming and/or maintaining supportive networks, and high treatment drop-out risk. Pet ownership may help by reducing anxiety and providing emotional support, companionship, and physical health benefits, but at the same time imposes demands for animal care. Our aim was to investigate how owning a pet relates to substance use patterns and use of SUD treatment services. Data were taken from two cross sectional questionnaire studies among people with SUD not in treatment (study 1) and in treatment (study 2), conducted from 2018 to 2019 in Norway. Study 1 data were gathered outside the clean needle centre in Oslo (n=214) and study 2 was undertaken at 3 SUD treatment centres in South-East Norway (n=90). Among respondents in Study 1, dog owners were less likely to inject heroin than non-dog owners (84 vs 58%, P=0.026) and pet owners were more likely to take heroin by other means (38 vs 57%, P=0.041, chi-sq. tests). In general, the pet owners were less likely to have sought medication-assisted treatment than the non-pet owners (26 vs 49%, P=0.023, chi-sq.-test). Study 2 showed that, compared to non-pet owners, pet owners tended to have a higher level of mental distress at treatment start as measured by the Hopkins Symptoms Checklist-10 (on a scale ranging from 1: no symptoms to 4: high) (mean±SD: 2.29±0.86 vs 1.95±0.59, p=0.065, Student's T-test), and were more likely to drop out of treatment within 6 weeks (5 vs 17%, P=0.010, chi-sq.-test). Our results suggest that pet owners with SUD may have a less destructive pattern of heroin use than non-pet owners. However, utilisation of and retention in treatment services may be an area where difficulties are experienced. Pet owners may maintain themselves better at street level but, by the time they seek treatment, they may have poorer mental health and a higher drop-out risk. These findings suggest that owning a pet could be perceived as both a salutogenic factor and an obstacle in the treatment process. While provisions are made for child care during treatment, provisions for pet care have not previously been considered in this context. We conclude that greater attention to pet ownership is warranted when planning SUD treatment strategies.

## Sleeping through anything: the effects of predictability of disruption on laboratory mouse sleep, affect, and physiology

*Amy Robinson-Junker[1,2], Bruce O'Hara[3], Abigail Cox[1] and Brianna Gaskill[4]*
*[1]Purdue University, Comparative Pathobiology, 625 Harrison Street, Lafayette, IN 47907, USA, [2]University of California, San Francisco, Laboratory Animal Resource Center, 513 Parnassus Ave, San Francisco, CA 94143, USA, [3]University of Kentucky, Biology, 101 T.H. Morgan Building, Lexington, KY 40506, USA, [4]Purdue University, Animal Science, 270 S. Russell St, Lafayette, IN 47907, USA; amy.robinson.junker@gmail.com*

Many aspects of the laboratory environment are not tailored to the needs of rodents, which may cause them stress. These stressors can cause ulcers, prolonged pituitary-adrenal activation, and anhedonia. Similarly, pain has been demonstrated to slow wound healing, and mice experiencing pain exhibit altered behavior. However it is unknown how the timing of husbandry, which occurs when the mice would normally be inactive, and lack of analgesia, specifically in a punch biopsy procedure, affects animal physiology, behavior, and welfare. We were interested in these effects particularly as they relate to sleep fragmentation. We hypothesized that sleep fragmentation, induced by unpredictable husbandry and lack of pain management, would slow wound healing. This experimental protocol was approved by the Purdue Animal Care and Use Committee. Two main treatments were tested in a factorial design in C57BL/6 mice of both sexes (32 mice total): (1) analgesia (carprofen and saline); and (2) sleep disruptions (random and predictable). Mice were randomly assigned to analgesia and disruption groups, and singly housed in a non-invasive sleep monitoring apparatus on arrival (Day -4). Single housing was necessary due to the nature of the apparatus. Disruption treatments were applied from Day -3 to 2. All mice received a punch biopsy surgery (Day 0) with topical lidocaine gel and their analgesic treatment prior to recovery, with further analgesia treatment on Days 1 and 2. Nesting behavior was assessed daily and a sugar cereal consumption test, as a measure of anhedonia, was conducted on Days -1 to 2. On Day 3, mice were euthanized with inhaled $CO_2$ and wound tissue and adrenal glands were collected for sectioning and histopathology. We found that the disruption predictability had no effect on mouse sleep (GLM, $F_{(1,21)}$=0.0009, P=0.976), wound healing rate (GLM, $F_{(1,24.9)}$=0.0182, P=0.894), or adrenal cortex:medulla ratio, a measure of chronic stimulation of the hypothalamus-pituitary-adrenal axis (GLM, $F_{(1,24.87)}$=0.0932, P=0.344). It's possible that the disruption period, the disruptions themselves, or both, were not long enough or severe enough to induce chronic stress. However, male mice who received analgesia slept more than their female counterparts (GLM, $F_{(1,21)}$=6.38, P=0.02); this may be related to sex differences in pain perception. Additionally, mice slept less on Day 1 than Day 2 (GLM, $F_{(3,184)}$=25.99, P<0.0001), possibly reflecting a response to restraint for analgesia or saline injection. Overall, it does not appear that the predictability of disturbance affects sleep fragmentation or stress responses in C57BL/6 mice, indicating that their husbandry activities do not need to occur at set, predictable times to improve welfare.

## Human beliefs and animal welfare: a cross-sectional survey on rat tickling in the laboratory

*Megan R. LaFollette[1], Sylvie Cloutier[2], Colleen Brady[3], Marguerite O'Haire[4] and Brianna Gaskill[1]*
*[1]College of Agriculture, Purdue University, Department of Animal Sciences, 270 S. Russell St., West Lafayette, IN 47907, USA, [2]Canadian Council on Animal Care, 800-190 O'Connor, Ottawa, Ontario K2P 2R3, Canada, [3]College of Agriculture, Purdue University, Department of Agriculture Sciences Education & Communication, 915 W State Street, West Lafayette, IN 47907, USA, [4]College of Veterinary Medicine, Purdue University, Department of Comparative Pathobiology, 725 Harrison Street, West Lafayette, IN 47907, USA; lafollet@purdue.edu*

Optimization of laboratory rat welfare is influenced by the behaviors of laboratory animal personnel. Rat tickling, which mimics aspects of rat rough-and-tumble play, is a promising enrichment technique that can contribute to improving welfare, but may be difficult to implement. The theory of planned behavior can be used to study implementation by measuring intentions and beliefs, including behavioral attitudes (whether rat tickling is good or bad), subjective norms (whether there is social/professional pressure to provide rat tickling), and control beliefs (whether they feel in control of providing rat tickling). Therefore, the objective of this study was identify current prevalence and predictors of rat tickling. Laboratory animal personnel were recruited from widespread online promotion. A total of 794 personnel (M=40±11 years, 80% Caucasian, 80% female) completed at least 50% of an online survey and met inclusion criteria of currently working with laboratory rats in the USA or Canada. The survey included questions about demographics, enrichment practices and beliefs, attitudes towards rats, general positive behaviors, and beliefs about rat tickling. Qualitative data were coded using thematic analysis. Quantitative data were analyzed using general linear models. Laboratory personnel reported low levels of rat tickling implementation, with 89% of participants reporting using it never or rarely. Laboratory personnel report 2 key benefits (handling: 61%, welfare: 55%) and 3 key barriers (time: 59%, personnel: 22%, and research: 22%) to rat tickling using qualitative analysis. Current and planned rat tickling were positively associated with more positive beliefs (social/professional pressure P<0.0001, and control of providing tickling P<0.0001) and familiarity with tickling (P<0.0001). Future rat tickling was also positively associated with more positive attitudes about rat tickling (P<0.0001) and a desire to implement more enrichment (P<0.01). Current rat tickling was also positively associated with more positive general behaviors (e.g. talking to laboratory animals, P<0.0001). Our findings show that implementation of rat tickling is currently low. Furthermore rat tickling is positively associated with personnel beliefs, familiarity, general attitudes, and a desire for more enrichment. That is, personnel were more likely to provide rat tickling if they were more familiar with it, thought providing it was good, under their control, and subject to social/professional pressure, as well as if they wanted to provide more enrichment. There is potential to increase rat tickling and thereby improve rat welfare by increasing familiarity with the procedure through training, decreasing the time required, and changing personnel beliefs.

## A comparison of reactions to different stroking styles during gentle human-cattle interactions

*Annika Lange[1], Vera Wisenöcker[1], Andreas Futschik[2], Susanne Waiblinger[1] and Stephanie Lürzel[1]*
*[1]Institute of Animal Welfare Science, Department for Farm Animals and Veterinary Public Health, University of Veterinary Medicine, Vienna, Veterinärplatz 1, 1210 Vienna, Austria, [2]Department of Applied Statistics, JK University Linz, Altenberger Str. 69, 4040 Linz, Austria; annika.lange@vetmeduni.ac.at*

The quality of the animal-human relationship (AHR) can be improved by gentle interactions such as stroking. Previous studies showed that stroking was most effective and elicited most positive reactions when cows were being stroked at the ventral neck compared with stroking of the withers or chest region. Nevertheless intra-specific allogrooming – social licking – in cattle includes different body regions probably guided at least partly by the receiver. Thus we compared dairy heifers' (n=28; *Bos taurus*, Austrian Simmental) reactions to different stroking treatments, with the experimenter either responding to perceived momentary preferences of the heifer by stroking in a *reactive* way or exclusively stroking the *ventral neck*. Each test comprised three phases: pre-stroking (baseline, PRE), stroking (STROKE), and post-stroking (POST). In the *reactive* condition the experimenter stroked the whole head/neck region and responded to the presumed momentary preference of the heifer. Preference was determined by the animal leaning towards the experimenter, presenting a body part, (partly) closing the eyes or stretching the neck. All tests were video recorded and analysed for behaviours associated with different affective states, such as neck stretching (main variable) and establishing physical contact with the experimenter. All tests were performed while heifers were lying. Each animal was tested six times with alternating conditions to increase robustness of the data. The results were averaged instead of doing a repeated measurement analysis because of the non-normal responses. Nonparametric test statistics were computed and the false discovery rate was controlled by the Benjamini-Hochberg method resulting in a different, adjusted significance threshold for each test. In both conditions, neck stretching increased from PRE to STROKE (median, min-max: duration of neck stretching in s: *reactive* PRE 0.61, 0-52.25, STROKE 29.86, 0.61-58.09, Wilcoxon test P<0.001; *ventral neck* PRE 0.81, 0-6.15, STROKE 20.67, 0-93.15, P<0.001), indicating a positive perception of the interactions. When comparing the change from PRE to STROKE between conditions, we found a trend towards a stronger increase of duration of contact during *reactive* stroking (duration delta of contact in s: *reactive* 16.51, -1.99-63.43, *ventral neck* 7.58, -4.51-37.99; V=310, P=0.014), leading to the tentative conclusion that *reactive* stroking may be perceived as more positive than stroking exclusively the *ventral neck*; however the increase in other behaviours did not differ between the two conditions. We found no significant differences between PRE and POST. The absence of more significant results is likely due to only small differences in the perception of the two stroking treatments. We conclude that both ways of stroking can elicit positive emotions in cattle and increase the animals' wellbeing.

## Improving the cow-human relationship – influence of restraint during gentle interactions

*Stephanie Lürzel[1], Annika Lange[1], Anja Heinke[1], Kerstin Barth[2], Andreas Futschik[3] and Susanne Waiblinger[1]*
*[1]Institute of Animal Welfare Science, Vetmeduni Vienna, Veterinärplatz 1, 1210 Vienna, Austria, [2]Johann Heinrich von Thünen Institute, Institute of Organic Farming, Trenthorst 32, 23847 Westerau, Germany, [3]JK University Linz, Department of Applied Statistics, Altenberger Str. 69, 4040 Linz, Austria; stephanie.luerzel@vetmeduni.ac.at*

Gentle tactile and vocal interactions can improve animals' relationship towards humans up to a good relationship where animals are confident in humans, who may even provide social support. However, in fearful animals, close human presence and touch elicit stress, as a certain level of confidence in humans is necessary to allow a positive perception of tactile interactions. We tested whether imposing gentle interactions on restrained animals (where interactions are probably perceived as aversive in the beginning) or offering gentle interactions to free-moving animals (where it may take longer to establish physical contact) leads to a more efficient improvement of the animal-human relationship. We tested 36 Holstein-Friesian cows from two herds on a research farm with an avoidance distance (AD) of at least 0.3 m, balanced for herd and AD, among other factors. The restrained group (LOCK) was stroked and talked to in a gentle voice for 3 min/d while they were restrained in the feeding rack, the free-moving group (FREE) was approached, talked to and, if possible, stroked while they were free in the loose-housing barn, and the control group (CON) did not experience any additional tactile contact. All cows were fed some concentrate during the first five days of treatment to make it easier to approach the FREE animals; the CON group also received the additional feed to make sure that any differences between the groups were due to the gentle interactions and not due to feeding. We conducted AD tests in the barn, with a familiar (= stroker of the same herd, FAM) and an unfamiliar, blinded person (= stroker of the other herd, UNF), and approach tests towards FAM. In the AD test, the person approached the cow and estimated the distance between the cow's muzzle and the person's hand in steps of 10 cm at the moment of avoidance and recorded how long the cow could be touched. After recording the baseline AD (AD1), the animals were treated for 10 d out of 14 d and then retested; this was repeated twice (AD4 after 6 weeks). The animals had a median AD1 of 1.6 m (min 0.3 m–max 2.5 m) with FAM and of 1.4 m (0-2.5 m) with UNF. With FAM, AD decreased in CON, LOCK and FREE (Wilcoxon test, $P<0.01$), resulting in a significant difference between groups in AD4 (Kruskal-Wallis test, $P=0.02$, FREE median 0.0 m, LOCK 0.2 m, CON 0.7 m). AD decreased in LOCK and FREE ($P\leq0.03$) and tended to decrease in CON ($P=0.06$) with UNF, but the groups did not differ significantly after the treatment period, although the general pattern was similar to the results with FAM. The treatment improved the animal-human relationship most effectively in the FREE cows, possibly due to the perception of the situation as controllable. The treatment was less effective in LOCK cows, which might have perceived it as negative during the first days. The decrease in CON might be due to habituation to the person, the positive association with feed in the first 5 days and possibly social learning.

## Is cows' qualitatively assessed behaviour towards humans related to their general stress level?

*Silvia Ivemeyer[1], Asja Ebinghaus[1], Christel Simantke[1], Rupert Palme[2] and Ute Knierim[1]*
*[1]University of Kassel, Farm Animal Behaviour and Husbandry Section, Nordbahnhofstr. 1a, 37213 Witzenhausen, Germany, [2]University of Veterinary Medicine, Department of Biomedical Sciences, Veterinärplatz 1, 1210 Vienna, Austria; ivemeyer@uni-kassel.de*

The stress level is considered as one important aspect of dairy cow welfare. Beside aspects of housing, management and social herd stressors, stress might also be related to the human-animal relationship (HAR). An established and non-invasive physiological method to assess medium-term stress is the measurement of fecal cortisol metabolites (FCM), reflecting the adrenocortical activity over several hours with a delay of 8-10 hours due to gut passage time. For the assessment of the HAR, behavioural tests recording the cows´ responses towards humans can be used. Beside quantitative methods, e.g. recording the cows´ avoidance distances towards humans, also the qualitative behaviour assessment (QBA) assessing the cows´ body language have been shown to be reliably applicable. Thereby, QBA might reflect the cows´ responses in a more differentiated way than quantitative measures. Using data from 316 dairy cows on 25 German organic dairy farms regarding QBA during a standardised tactile human-animal interaction and FCM recorded on the same day, but with time differences varying between 0-10 h (during winter 2015/16 and 2016/17), we asked whether the cows' qualitatively assessed behaviour towards humans is related to their general medium-term stress level. For QBA we used a fixed list of 20 descriptors, which had specifically been developed for this purpose. A principal component analysis (PCA) resulted in two components: PC1 explained 67% of variance and appeared to reflect 'positive' (pos) and 'negative' (neg) valence (characteristic descriptors pos: e.g. trustful, relaxed; neg: e.g. fearful, distressed). PC2 explained 7% of variance and appeared to reflect the level of activation ('high' activation: e.g. contact-seeking, aggressive; 'low' activation: e.g. patient, insecure). Dividing the sample into four groups (pos_low, pos_high, neg_low, neg_high) using the medians of PC1 and PC2 as cut-points, we compared FCM levels (11,17 dioxoandrostanes, enzyme immunoassay method) by Kruskal-Wallis and post-hoc Wilcoxon rank-sum tests. Groups of cows classified by QBA differed regarding FCM ($P=0.010$, Kruskal-Wallis test). Cows reacting pos_high during the human-animal interaction (n=65), pos_low (n=93) and neg_low (n=99) had similar FCM levels (medians of 12.0, 10.3 and 10.4; 25-75% quartiles: 6.8-22.8, 4.9-22.0, and 4.2-18.5 ng/g FCM; $P=0.225-0.699$, Wilcoxon tests). In contrast, neg_high cows had significantly lower FCM levels (6.5 ng/g, 3.7-12.5, n=59) than all other groups ($P=0.001-0.029$). Neg_high cows presumably had higher fear levels towards humans than pos cows, and expressed this more actively than neg_low cows. Unexpectedly, however, the neg reaction was not related to a generally higher stress level. In this context, it must be taken into account that FCM medians and variation in the investigated sample were generally on a low level. Furthermore, other factors such as social rank or health status might have affected results more profoundly, deserving a more complex analysis including individual factors as a next step.

## Impact of handling frequency on drylot-housed heifer behavior

*Amanda Hubbard, Jason Sawyer, Reinaldo Cooke and Courtney Daigle*
*Texas A&M University, Animal Science, 2471 TAMU, College Station TX 77843, USA;*
*a.mathias@tamu.edu*

Feedlot cattle are rarely removed from pens due to concerns of handling events disrupting behavior. Research cattle are weighed at specific intervals to calculate growth and performance more precisely – yet impact of weighing events on behavior are unknown. The study objective was to quantify the impact of weighing events on behavior of feedlot heifers with the hypothesis that increased frequency of weighing alters behavior and performance. Angus heifers (n=60; ~8 months of age) were randomly assigned to one of 12 pens (n=5 heifers/pen). Pens were randomly assigned one of three treatments (n=4 pens/treatment) based upon frequency of handling applied across a 42-d feeding period. Power analysis determined the sample size used sufficient to detect effects. Heifers were weighed either weekly (7DAY), biweekly (14DAY), or at the beginning and end of the study (42DAY). Feed was delivered after all cattle had been weighed. Live behavior observations were conducted on all handling days from 08.00 to 12.00 and 13.00 to 17.00 by blind observers. Instantaneous scan sampling (n=48 scans/d; 10-min intervals) recorded number of heifers standing, walking, lying, eating and drinking per pen. Latency for each heifer to eat (LTE), drink (LTD), and lie down (LTL) after feed delivery, as well as latency for all 5 heifers in a pen to lie down simultaneously was recorded. Linear mixed models (PROC MIXED) evaluated impact of treatment and day on behavior. Orthogonal contrasts were used to analyze linear and quadratic effects of treatment on average daily gain (ADG) and average exit velocity (AEV). Treatment did not influence standing (P=0.11) or eating (P=0.89). Larger proportions of cattle were observed lying in 7DAY and 14DAY compared to 42DAY (P=0.02). Week influenced proportion of cattle observed standing (P<0.01), eating (P<0.01), and lying (P<0. 01). Proportion of cattle observed eating increased over time (P<0.01). Latency for an entire pen to lie down simultaneously after feed delivery varied throughout the study. Individual heifer LTE after feed delivery was longest (P<0.01) in week 1 (115.17±9.58 min) and 2 (41.11±2.97 min). There was a treatment × week interaction for individual heifer LTL (P=0.04) and to LTD after feed delivery (P=0.01) – no differences within week were observed for either measure. Neither linear (P>0.27) nor quadratic (P>0.23) effects of treatment were detected for ADG or AEV. Mean ADG per treatment was 0.9, 0.95, and 0.99 kg/d for 7DAY, 14DAY, and 42 DAY respectively. Results indicated cattle behaviorally adapt to new surroundings in approximately 3 weeks. In addition, while more frequent handling increased lying behavior, it appeared to negatively impact average daily gain, though the difference was insignificant in this study. Thus, feedlot cattle weighed more frequently in research trials will perform more resting behavior but may have reduced ADG compared to cattle handled infrequently.

## Human-directed behaviour of calves in a Novel Human and an Unsolvable Problem Test

*Sabine A. Meyer[1,2], Daniel M. Weary[2], Marina A.G. Von Keyserlingk[2] and Christian Nawroth[3]*
*[1]University of Kassel, Institute of Psychology, Holländische Str. 36-38, 34127 Kassel, Germany, [2]University of British Columbia, Faculty of Land and Food Systems, 248-2357 Main Mall, V6T 1Z4 Vancouver, Canada, [3]Leibniz Institute for Farm Animal Biology, Institute of Behavioural Physiology, Wilhelm-Stahl-Allee 2, 18196 Dummerstorf, Germany; sabine.busse@htp-tel.de*

Dairy cattle differ in their reaction towards humans and not accounting for these inter-individual differences can lead to decreased welfare and productivity. One test that is routinely used to assess the human-animal relationship in animals is the Novel Human Test. However, evidence that the human-directed behaviour of animals in this test is generalisable over different contexts is scarce. Our aim was to investigate whether the inclination of calves to interact with humans is stable across time and context. To this end, we assessed the human-directed behaviour of calves (n=29; 42-60 days old; 24 females, 5 males) in a Novel Human Test (NHT) and an Unsolvable Problem Test (UPT). In the NHT, subjects entered a test arena (5×5 m) with an unfamiliar human positioned in the middle. Subjects were free to move around the arena and interact with the human throughout the 10-min test. In the UPT, subjects were tested in a smaller arena (4×4 m) where they had been trained to receive milk from a dispenser located opposite to the entrance. After successful training, subjects were confronted with an unsolvable version of the task by positioning a mesh in front of the dispenser rendering the reward inaccessible. During training and test trials, each lasting 2 min, a familiar human was positioned in the middle of the arena. Subjects first received one trial in the NHT, followed by three trials on three consecutive days in the UPT in the subsequent week, followed by a second NHT trial approximately 3 weeks after the first one. We scored the time calves interacted with the human in the NHT and UPT trials. The time calves spent interacting with the human was correlated across the two NHT trials ($r_s$=0.46, P=0.011), indicating temporal consistency of this behaviour. Over the three UPT trials, the human-directed behaviour of the calves did not differ (trial 1: 11.0±3.84 sec; trial 2: 16.6±3.97 sec; trial 3=18.9±4.45 sec; data log-transformed, repeated measures ANOVA, F=2.07, P=0.136). Moreover, the time that calves spent interacting with the human in the NHT trials correlated with that during the UPT trials ($r_s$=0.54, P=0.002). These results indicate that human-directed behaviour in calves is relatively consistent across time and context.

## Artificial mothering during early life could boost the activity, exploration and human affinity of dystocia dairy calves

*Congcong Li, Chuang Zhang and Xianhong Gu*
*Institute of Animal Science, Chinese Academy of Agricultural Sciences, Yuanmingyuan West Road 2, Haidian District, 100089, Beijing, China, P.R.; congcongli1988@sina.com*

Dystocia calves are normally weak and need more care and attention. This study aimed to determine if artificial mothering during early life has effects on the behavior of dystocia calves. Dystocia level was evaluated based on a 5-point calving-ease score, with 1 = unassisted, 2 = easy pull, 3 = moderate pull, 4 = very hard pull, and 5 = Cesarian section. The study was conducted according to the guidelines of the Animal Welfare Committee of Institutes of Animal Science, Chinese Academy of Agricultural Science (IAS-CAAS). In total 32 Holstein calves (from primparous cows) with calving-ease score of 2 to 4 were collected and sequentially assigned to two groups (dystocia level was balanced in the two groups), namely Control (C, n=15) and Artificial Mothering (AM, n=17) group. Right after birth, calves were removed from their dams and placed in individual pens for the whole experimental period. Colostrum of 10% body weight were given to calves through esophageal tube within 30 min after birth (day 0). From day 1 onwards, calves were bucket fed three times of 2 l milk per day. The C calves were vigorously dried within 1-2 min by towel and left alone. Contact with the C calves including tactile and verbal stimulation were minimized during rearing. However, the AM calves were dried, groomed, and stimulated to stand for 30 min on day 0. In addition, 5 min of grooming were conducted 1 to 2 h after morning and noon milk feeding from day 1 to 6. On day 7, three 5 min test sessions were conducted for open field (OFT), novel human reactivity (NHT) and familiar human reactivity (FHT) test respectively 1 h after morning milk feeding. The calves were tested individually in an arena ($6\times7$ m$^2$). After the first 5 min OFT in the empty arena, a novel person entered the arena and stood immobile in the center for another 5 min (NHT). Afterwards, the novel person was replaced by a familiar person for 5 min (FHT). The latency, duration and frequency of behavior including standing, walking, jump-running, objective sniffing, self-grooming and human interacting were focally recorded and statistically analyzed by JMP 12.0. One-way ANOVA was conducted to compare the behavioral difference between C and AM calves in OFT, NHT and FHT tests respectively. The AM calves tended to be more active in the OFT and significantly more active in the NHT and FHT as they had longer and more bouts of jump-running behavior. The C calves had a tendency of longer and more bouts observing the novel person based on the results of staring at the person while standing still. The AM calves tended to lick-rubbing the novel person longer and had significantly more interaction towards the familiar person. The C calves tended to spent longer in heading towards the novel person. No difference was found in standing, walking, objective sniffing and self-grooming. In conclusion, artificial mothering during early life could increase the activity, exploration and human affinity of dystocia calves.

## Effects of dam bonded rearing on dairy calves' reactions towards humans vanish later in life

*Susanne Waiblinger[1], Kathrin Wagner[1] and Kerstin Barth[2]*
*[1]Institute of Animal Welfare Science, University of Veterinary Medicine Vienna, Department for Farm Animals and Veterinary Public Health, Veterinärplatz 1, 1210 Vienna, Austria, [2]Institute of Organic Farming, Johann Heinrich von Thünen Institute, Federal Research Institute for Rural Areas, Forestry and Fisheries, Trenthorst 32, 23847 Westerau, Germany; susanne.waiblinger@vetmeduni.ac.at*

Dairy calves in general are separated from their dam within hours after birth and reared artificially, but dam bonded rearing gets increased attention as welfare-friendly, more natural and, partly, work saving system, as bucket feeding of calves is often omitted. However, the first weeks of life are a sensitive period for developing relationships with humans. Human interactions with dam reared calves are reduced during this period and mostly negative in nature (ear tagging, disbudding) while association between feeding and humans falls away. This may negatively affect the animal-human relationship (AHR) also in the long-term although the AHR is dynamic. A study in artificially reared calves provided evidence that later experiences override former experiences, but there all calves were fed by humans during the postnatal sensitive period. Therefore, the aim of this study was to compare AHR of cows that had been reared differently during the first 12 weeks of life. Artificially reared animals (Artificial, n=7) had been separated from their mother within 12 h after birth, bottle-fed for 5 days and thereafter by automatic milk feeder (trained to suckle there by a human). Mother reared animals (Mother, n=12) were kept in the calving pen with their mothers for 5 days (human contact only if intervention such as treatment was necessary and when getting the cow for milking), and then had permanent access to their mothers by a selection gate (trained for using it by a human) connecting the calf area (where Artificial calves were housed as well) with the cow barn. Calves were weaned with 12 weeks and kept in young stock groups (mixed of both treatments) until integration into the cow herd before the first calving. We tested AHR by assessing the animals' responses towards an unfamiliar person in avoidance distance (AD) tests in the home environment at three ages: as calves, 4 weeks, heifers, 15 months and primiparous cows, 33±1.0 months old, i.e. 6±1.6 months after first calving. In calves, but not later in life, AD was also measured in a novel arena. Mother animals had higher AD as calves in the home pen (H=5.61, P=0.016, median: Mother: 0.9 m; Artificial: 0.4 m) and in the arena (H=4.85, P=0.026, Mother: 1.3 m, Artificial: 0.7 m), and tended to have higher ones as heifers (H=2.86, P=0.095, Mother: 1.2 m, Artificial: 0.8 m). However, there was no difference anymore in AD of the animals as primiparous cows between Artificial (median, min-max: 1.8, 1.0-2.3 m) and Mother (1.85, 0.9-2.7 m; Kruskal Wallis H=0.002, P=0.985). AD increased over time and was quite high in general, suggesting few positive and more negative experiences with humans. In conclusion under conditions similar to our investigated farm (i.e. relatively high AD in cows) there is no long-term disadvantage of mother bonded rearing on the AHR. Potential effects under different conditions regarding quantity and quality of human-animal interactions need further research.

## Validity aspects of behavioural measures to assess dairy cows´ responsiveness towards humans

*Asja Ebinghaus, Laura Schmitz, Silvia Ivemeyer, Leonie Domas and Ute Knierim*
*University of Kassel, Farm Animal Behaviour and Husbandry Section, Nordbahnhofstr. 1a, 37213 Witzenhausen, Germany; ebinghaus@uni-kasse.de*

For assessment of the human-animal relationship (HAR) in cows, different measures reflecting the animals´ behaviour towards humans are used. An established and widely tested measure is the avoidance distance (AD). This test records the distance a cow allows a person to approach. Further HAR measures applied relate to responses in more intense human-animal interactions. For instance, cows´ tolerance to tactile interaction (TTI), their behaviour during and after release from restraint (RB), and body language during the human-animal interaction by means of qualitative behaviour assessment (QBA) have been assessed in previous investigations, and tested with regard to reliability and criterion validity. However, uncertainties still exist regarding the measures´ construct validity, vulnerability to confounders or risks of observer bias. These aspects are addressed in the current study. AD, TTI, RB, and QBA were investigated on a research farm with 102 dairy cows regarding: (1) causal relationships with frequent positive human-animal interactions (6-days period of concentrate provision by hand); (2) potential confounding effects of lactation status (before/after calving); (3) potential short-term effects of management procedures (claw trimming (CT)); and (4) the occurrence of expectation bias. Concerning objectives 1-3, different samples of cows were assessed repeatedly and differences in behaviour before and after hand feeding of concentrate, calving, and CT were analysed by using paired Wilcoxon tests, for differences before and after CT using blinded observation of video records. For objective 4, non-blinded live and blinded video assessments before and after CT were carried out. Assessments strongly correlated regarding all measures (analysed before CT; AD: Spearman rank correlation ($r_s$)=0.93, n=29; TTI: Cohen´s Kappa (K)=0.88, n=29; RB: K=1.00, n=29; QBA: $r_s$=0.96, n=15, all P<0.001), but QBA scores ranged on different levels (paired Wilcoxon: P=0.027). Therefore, it was tested separately for live and blinded video assessments whether assessments before and after CT differed. After the period of hand feeding, the cows had lower ADs (P=0.013, n=27), responded less fearful regarding TTI (P=0.045, n=27) and QBA (P=0.026, n=14), and by tendency regarding RB (P=0.052, n=27), supporting the measures´ construct validity in the way that they reflect an improved HAR after frequent positive interactions. Neither (blinded) assessments before and one day after CT (P=0.489-1.0, n=29) nor before and after calving (P=0.244-1.0, n=13) differed, suggesting that single routine procedures and physiological changes around calving do not confound assessments. Also non-blinded live assessments did not yield any significant differences before and after CT (paired Wilcoxon: P=0.151-1.0, n=15-29). However, numerically observers assessed cows as less fearful after CT in live QBA compared to blinded QBA. Since observers were involved into the research question, they might have tried to avoid bias. Although, we could not detect a significant expectation bias, depending on the measures chosen, blinding should be considered in studies involving certain expectations.

## Reproducibility, similarity, and consistency of the flight responses of beef cattle

*Daisuke Kohari[1], Nahomi Ohtaka[1], Chikako Negoro[1] and Kyoko Kido[2]*
*[1]Ibaraki University, College of Agriculture, 4668-1, Ami, Inashiki-gun, Ibaraki, 300-0331, Japan, [2]National Institute of Agrobiological Sciences, 375-716, Miyota, Kitasaku-gun, Nagano, 389-0201, Japan; daisuke.kohari.abw@vc.ibaraki.ac.jp*

We compared the similarity and reproducibility of the flight responses of cows and calves by measuring flight distance (FD) to select human-friendly cattle. We also investigated the relationship between FD and avoidance reactions to clarify the consistency of these flight responses. With support from the Bio-Oriented Technology Research Advancement Institution (Research Project for Future Agricultural Production utilising Artificial Intelligence), 24 Japanese black (Bos Taurus) cow–calf pairs were studied in 2016 and 2017. These pairs included the same 18 cows in both years. The FD of each pair was measured by the same observer who approached the subject cow or calf in a pasture, approaching from the left or right, and measured the distance between the observer's feet and the nearest hoof of the animal at the moment it withdrew using a tape measure. This was repeated three times per animal. The avoidance reaction of some cows at a feeding rack at a barn before the grazing season was also recorded three times per cow in 2017. The median values were calculated and the correlations between cow–calf pairs, the same cows, the calves born in each year, and the avoidance reaction and FD of cows were analysed using R (ver.3.2.2, R Development Core Team). The median FD of each cow and calf was approximately $\pm 2$ m and was relatively constant. There was a significant correlation between the FD of cow–calf pairs in 2016 ($r=0.769$, $n=20$, $P<0.05$) and 2017 ($r=0.789$, $n=20$, $P<0.00005$). The correlation between the FD of the same cows for the two years was also significant ($r=0.794$, $n=18$, $P<0.0005$) and the FD of their calves born each year was also correlated between the two years ($r=0.581$, $n=18$, $P<0.05$). However, no relationship was observed between the avoidance reaction of cows at the feeding rack and the FD in the pasture ($r=-0.10$, $n=12$, NS). These findings imply that the flight response was similar and reproducible in cow–calf pairs and these responses might be inherited. However, the two flight responses (i.e. FD and avoidance reaction) were not correlated.

## Chute scoring as a potential method for assessing arousal state in ewes

*Kaleiah Schiller, Shayna Doyle and Kristina Horback*
*University of California Davis, Animal Science, One Shields Avenue, 95616, USA;*
*kmschiller@ucdavis.edu*

Chute scoring systems, and similar approaches, have been used among beef and dairy operations as a method to gauge the relative docility or temperament of animals based on response to physical restraint. While elements of the chute score, such as tail flicking, shuttering, exit speed and side kicking, have been validated as behavioral indicators of arousal in cattle, there is not a validated chute scoring system for sheep. The objectives of this study were (1) to develop an objective chute score scale that is specific for production ewes, (2) evaluate the consistency of chute score assessment over three repeated trials, and (3) determine if there are physiological measurements which validate chute score values. Twenty-eight multiparous, crossbred ewes from the same breeding group were restrained in a metal squeeze chute ($1.5 \times 0.6 \times 1.0$ m) for approximately 5 minutes, once a week for three weeks. Trained researchers, with previously established inter-rater reliability (Cronbach's Alpha>0.08), gave each ewe a chute score based on the behaviors observed. Chute scores were adapted from cattle literature and were as follows: 1 (Docile): Gentle, easily handled, stands calmly in chute with minimal movement, does not push against front or back gate, and producing few vocalizations; 2 (Restless): Quieter than average, but may be stubborn. May push against the front or back gate, and vocalize several times; 3 (Nervous): Impatient, with lots of movement within the chute as well as many vocalizations, may urinate or defecate within the chute, and push against the front or back gate. To validate these subjective chute scores, the heart rate, respiration rate and surface eye temperature were collected for each ewe while restrained. In addition, video data of restrained ewes were coded for latency to exit the chute, order of entry into the chute, step rate and vocal rate using Noldus Observer XT. Spearman's rho correlations revealed significant individual consistency in step rate ($P<0.001$), vocal rate ($P<0.001$), heart rate ($P<0.05$), chute score ($P<0.001$), eye temperature ($P<0.01$) between all trials of the study. The order in which sheep entered the chute and respiration rate did not produce significant within-subject correlations between all trials. Vocal rate maintained the most stable relationship with other physiological and behavioral variables for each trial, consistently correlating to step rate ($P<0.05$), chute score ($P<0.01$), eye temperature ($P<0.05$) and heart rate ($P<0.05$). Significant differences in mean ranks of heart rate [Wilcoxon signed-rank test; trial:1 ($\chi^2(2)=7.1$, $P=0.03$); trial: 2 ($\chi^2(2)=8.5$, $P=0.01$); and trial 3 ($\chi^2(2)=7.1$, $P=0.03$)], vocal rate [trial 1: $\chi^2(2)=8.6$, $P<0.01$); trial 2: $\chi^2(2)=7.9$, $P=0.02$); and trial 3: $\chi^2(2)=11.5$, $P<0.01$)] and step rate [trial 1: ($\chi^2(2)=14.3$, $P<0.01$); trial 2: ($\chi^2(2)=15.0$, $P\leq0.01$); and trail 3 ($\chi^2(2)=12.2$, $P<0.01$)] were found among chute score groups (1 vs 2 vs 3). These results suggest that chute scoring is a promising method for detecting consistent individual differences in ewes and could be improved by emphasizing species specific indicators of arousal such as vocalizations.

## Dairy and dwarf goats differ in their preference for familiar and unfamiliar humans

*Katrina Rosenberger[1], Nina Keil[1], Jan Langbein[2] and Christian Nawroth[2]*
*[1]Swiss Federal Food Safety and Veterinary Office, Agroscope, Centre for Proper Housing of Ruminants and Pigs, Tänikon 1, 8356 Ettenhausen, Switzerland, [2]Leibniz Institute for Farm Animal Biology, Institute of Behavioural Physiology, Wilhelm-Stahl-Allee 2, 18196 Dummerstorf, Germany; rosenberger@fbn-dummerstorf.de*

A detailed comprehension of the socio-cognitive abilities of domestic animals is necessary to improve human-animal interactions during management practices on farms. Goats are social animals and previous research has shown that they differentiate between conspecifics using various modalities. The aim of our study was to investigate if goats discriminate between humans by assessing if they show a preference for either a familiar or an unfamiliar experimenter. We were further interested in whether the selection of different production traits has altered the motivation to approach and interact with humans. We tested 60 female goats from different selection lines (30 Nigerian dwarf goats and 30 dairy goats (Saanen and Chamois coloured) at 16-17 months of age. All subjects were housed in similar pens and had similar experience with humans over the last 13 months prior test start. Subjects individually entered a familiar test arena (5×3.3 m, divided into 12 segments via a 3×4 matrix) from a start box (i.e. segment 2), with one familiar and one unfamiliar experimenter positioned at either the opposite left or right corner of the arena (i.e. segments 10 and 12, randomized between subjects). Goats were free to move around in the arena and interact with either human. All subjects received one single preference trial of 180 sec. We scored the number of goats that approached the segment of either the familiar or the unfamiliar human first. We also analysed the total duration each goat spent in the segment of either the familiar or unfamiliar human. Dairy goats and dwarf goats differed in their preference for approaching either the familiar or unfamiliar human first (Chi-Square test, $X^2$=4.35, P=0.037). More dairy goats approached the unfamiliar than the familiar human first (binomial test, 21 vs 9, P=0.043), whereas dwarf goats showed no preference for either human (binomial test, 12 vs 16, P=0.572). Dairy goats spent more time in the segment of the familiar and the segment of the unfamiliar human than dwarf goats (median ± SD; familiar human: dairy goats: 23.1±29.7 sec, dwarf goats 9.6±9.0 sec, Mann-Whitney U-test, U=274, P=0.009; unfamiliar human: dairy goats: 16.6±12.2 sec, dwarf goats 9.4±10.2 sec, Mann-Whitney U-test, U=314, P=0.045). In contrast to their initial approach behaviour, dairy goats tended to stay longer in the segment of the familiar person compared to the unfamiliar one (23.1±29.7 vs 16.6±12.2 sec, Wilcoxon rank test, W=321, P=0.07). No such difference was found for dwarf goats (9.6±90 vs 9.4±10.2 sec, Wilcoxon rank test, W=228, P=0.577). Our results show that dairy goats, but not dwarf goats discriminated between familiar and unfamiliar humans. These findings demonstrate that goats interacted with humans differently in relation to their familiarity with the human and to their selection line. These differences should be considered in goat-human interactions, e.g. during handling procedures.

## The provision of toys to pigs could improve the human-animal relationship: the use of an innovative test, the strange-person

*Míriam Marcet-Rius, Patrick Pageat, Cécile Bienboire-Frosini, Eva Teruel, Philippe Monneret, Julien Leclercq, Céline Lafont-Lecuelle and Alessandro Cozzi*
*IRSEA (Research Institute in Semiochemistry and Applied Ethology), Quartier Salignan, 84400 Apt, France; m.marcet@group-irsea.com*

A positive human-animal relationship is widely presumed to be beneficial for the production system, increasing not only animal welfare but also the productivity and quality of the product. Therefore, it exists an interest of finding feasible efficient strategies to use in farms to reach a positive human-animal relationship. Studies showed that mini-pigs can discriminate familiar from unfamiliar humans. Also, that each pig's own experience with humans allows it a recognition of familiar humans and a generalisation to others. Literature showed that play behaviour triggers positive emotions in pigs. The aim of the present study was to identify the valence of the interaction with a person (human-animal relationship), after a positive situation (object play) compared to a control situation (with no stimulus), using a 'Strange-person test'. Sixteen mini-pigs housed in an experimental setting were involved in the study. The 'Strange -person test', which is an innovative test measuring the trust towards humans as well as the emotional state of the pig, consisted of a disguised person with a colourful costume entering to the pen where two mini-pigs were housed, for a two-minutes video recording. The test was performed after two different situations: the provision of two middle-sized dog toys (Play session) and no provision (Control session). For the Play session, one operator entered the pen, provided the toys and left the room for 10 minutes. Each of the 16 animals participated in the longitudinal study for a total of two days in two consecutive situations: one Control session and one Play session, followed by the 'Strange-person test'. Behaviours were analysed by two independent observers. Results showed that the latency to approach the person was significantly lower after Play than after Control sessions (median Play=3.5 sec; median Control=26.5 sec; Wilcoxon *Signed-Rank Test*; $P<0.0001$); pigs spent more time near the person after Play than after Control sessions (mean Play=115.5±6.4 sec; mean Control=76.6±43.5 sec; *Paired Student t-test*; $P=0.001$); the direct contact with the person was significantly higher after Play than after Control session (mean Play=85.8±24.9 sec; mean Control=55.8±38.0 sec; *Paired Student t-test*; $P=0.002$); tail movement duration, a potential indicator of positive emotions, was significantly higher after Play than after Control session (mean Play=92.8±26.8 sec; mean Control=60.0±34.1 sec; *Paired Student t-test*; $P=0.003$). These results suggested an improvement of the human-animal relationship after Play than after Control sessions, as pigs seemed to be more confident and trusted more the strange person, showing a more positive emotional state. It suggested that toys provision could be a useful strategy to improve human-animal relationship and positive emotions, which, according to the literature, could also improve animal's health and welfare, human work conditions and overall pig productivity.

## Are human voices used by pigs (*Sus scrofa domestica*) when developing their relationship with humans?

*Céline Tallet, Sandy Bensoussan, Raphaëlle Tigeot and Marie-Christine Meunier-Salaün*
*Pegase INRA Agrocampus Ouest, 16 Le Clos, 35590 Saint-Gilles, France; celine.tallet@inra.fr*

The human-pig relationship develops through visual, tactile, olfactory, and auditory interactions. Our objective was to determine the effect of human voices on the development of the human-pig relationship. We hypothesised that human voice facilitates human-pig relationship development and handling. We studied the behaviour of 90 weaned female piglets divided into three treatments: human presence with voice (HPV), human presence without voice (HP), control (CTRL). For the HPV piglets the experimenter was present idle for 5 min/ day in the pen during three weeks and a female voice was broadcast from a speaker around the neck of the experimenter. The HP treatment was the same but a recorded background noise was broadcast from the speaker. For the CTRL piglets only routine husbandry care was provided. Piglets were then tested twice in a 3×3 m test area in the presence of the experimenter for 5 min. For test 1 the voice was broadcast for HPV piglets and the background noise for the others. For test 2 only the background noise was broadcast, HPV piglets were thus deprived of the human voice they were used to. Lastly, we recorded the time it took to move the animals from their pen to the truck (for transfer to the finishing pen), with a human voice broadcast for HPV piglets and a background noise broadcast for the others. We analysed the treatment effect with linear mixed models (R studio Version 1.1.453, means and sem are reported), except for the duration of transfer which was analysed with the non-parametric Kruskal-Wallis test (median and IQ are reported). Significance limit was fixed at $P<0.05$. In test 1 previous human presence decreased piglet fear reactions, but human voice broadcast had no effect. Indeed, the time spent gazing at the experimenter was lower in HPV and HP piglets ($17.1\pm0.9$ s and $22.6\pm1.5$ s respectively) than in CTRL piglets ($46.5\pm3.3$ s, $P<0.05$). They also investigated the experimenter earlier (HPV: $50.9\pm3.0$ s; HP: $44.7\pm2.6$ s vs CTRL: $170.7\pm3.5$ s, $P<0.05$). In test 2, HPV piglets deprived of a human voice expressed more stress reactions: their latency to move was longer compared to the others ($38.6\pm2.2$ s vs HP: $18.5\pm0.8$ and CTRL: $59.0\pm2.8$ s, $P<0.05$). HPV piglets also had more physical and vocal interactions: they stayed longer in the experimenter area than HP and CTRL piglets ($37.0\pm0.8\%$ vs HP: $29.5\pm0.7\%$ and CTRL: $25.9\pm0.7$, $P<0.05$), and grunted more ($47.6\pm1.6$ grunts vs HP: $40.5\pm1.1$ and CTRL: $27.6\pm1.2$ grunts, $P<0.05$). We found no effect of the treatment on the time taken to move animals from their pen to the truck ($P<0.05$). In conclusion, broadcasting a human voice did not modify pig response to human presence and handling in auditory conditions similar to treatment sessions (i.e. test 1 and moving). However, not broadcasting human voice (test 2) induced stress responses and increased interactive behaviour, which suggests that piglets identified human voice as part of the experimenter's necessary properties.

## Providing handling tools to move non-ambulatory pigs on-farm

*Ella Akin[1], Anna Johnson[1], Justin Brown[2], Cassandra Jass[3], Heather Kittrell[2], Suzanne Millman[4], Jason Ross[1], Kristin Skoland[2], Kenneth Stalder[1], John Stinn[3] and Locke Karriker[2]*
*[1]Iowa State University, Department of Animal Science, 806 Stange Rd, Ames, IA 50010, USA, [2]Iowa State University, Swine Medicine Education Center, 2412 Lloyd Veterinary Medicine Center, Ames, IA 50011, USA, [3]Iowa Select Farms, 811 S Oak Street, Iowa Falls, IA 50126, USA, [4]Iowa State University, Biomedical Sciences and Veterinary Diagnostic and Production Animal, 2412 Lloyd Veterinary Medicine Center 2201 VDPAM, Ames, IA 50010, USA; eeakin@iastate.edu*

The Common Swine Industry Audit (CSIA) was designed to meet U.S. customer swine welfare expectations. Established critical criteria that results in an automatic audit failure includes dragging a conscious animal. The objective was to determine if a Haz-Mat/Hospital sked (a commercial product for human rescue; SKED), revised deer sled (RDS) or ice fishing sled (IFS) were suitable humane handling tools for moving non-ambulatory grow-finish pigs on a commercial farm. Nine commercial crossbred genetic line pigs (104.3±18.3 kg [range 68.04 to 124.7 kg]) were randomly assigned to one handling tool. Each pig was moved to the start pen. A lidocaine epidural block was administered to induce a non-ambulatory state. Once confirmed non-ambulatory by a swine veterinarian, two production well-being employees positioned the pig onto the handling tool and moved them. Once inside the end pen, non-ambulatory pigs were monitored every 15 minutes by collecting their respiration rate and digital withdrawal reflex were collected until able to stand and walk unassisted. Outcomes included duration to move from start- to end pen (s), distance 20.6 m, change in pig temperature (°C), taken with an Infrared gun on the ventral plane, pig respiration rate (bpm), pig vocalization score (0 = none to 2 = continuous grunts/calls) and struggle score (0 = none to 2 = continuous movement of legs and/or head). Change in pig temperature, respiration rate and handling tool duration were analyzed using mixed model methods. Average total duration (± SE) and range (presented descriptively) to move the pig from start- to end pen was 123±31 s (96-157 s; SKED), 195±85 s (100-264 s; RDS) and 106±33 s (83-144 s; IFS). Average change in pig temperature was 1.6±3.8 °C (range of change -6.5-5.5 °C). Average change in pig respiration rate was 10.7±11.5 bpm (range of change -4-32 bpm). Average vocalization and struggle scores when placing the pig onto the handling tool was 1 (range 0-2). Average vocalization for restraining the pig was 1 (range 0-1) and struggle score was 0 (range 0-1). Finally, average vocalization score for moving the pig from start- to end pen was 1 (range 0-1) and struggle score was 0 (range 0-1). Based on the quickness of moving pigs from the start- to end location combined with pigs not displaying outward distress signs, this research supports the use of the HMH sked, revised deer sled and ice fishing sled as practical handling tools to move non-ambulatory grow-finish pigs on-farm.

## Animal handling and welfare-related behaviours in finishing pigs during transport loading and unloading

*Sofia Wilhelmsson[1], Paul H. Hemsworth[2], Jenny Yngvesson[1], Maria Andersson[1] and Jan Hultgren[1]*
*[1]Swedish University of Agricultural Sciences, Department of Animal Environment and Health, Box 234, 53223 Skara, Sweden, [2]University of Melbourne, Animal Welfare Science Centre, Faculty of Veterinary and Agricultural Sciences, Parkville, VIC 3052, Australia; sofia.wilhelmsson@slu.se*

Moving finishing pigs from farm to abattoir can be very stressful for both pigs and humans. Pig behaviour is mainly affected by previous on-farm handling, genetics, design of the loading facilities, and the handling behaviour of the animal transport drivers (TD) when loading and unloading the pigs. Taking a One-Welfare approach, we studied relationships between TD attitudes and handling behaviour, working conditions and welfare-related pig behaviour during loading and unloading of Swedish pigs. Data were collected between August 2018 and January 2019. Five trucking companies suggested farms based on location and scheduled deliveries. Eighteen TD were each observed handling pigs during one loading at farm and one unloading at an abattoir (range in pigs handled: 49 to 258 pigs per load). Farmers were contacted shortly before the intended delivery and asked about animal housing and management on farm. Direct observations were made of on-farm loading conditions; slope of ramp; duration of loading and unloading; pig behaviour, e.g. high-pitched vocalizations (HPV), slipping and crowding; and handling behaviour of TD, e.g. frequency of auditory and tactile contact with pigs. The participating TD responded to two questionnaires about their attitudes to pigs, pig handling and their working conditions. At loading, the slope of the loading ramp was from 5 to 22 (mean 12.2) degrees. The duration of loading varied from 16 to 114 (mean 43) min. Preliminary results show important differences between TD in handling behaviour, including the type of tool used. At loading, thirteen of the TD used both a paddle and a driving board and five used driving board only. At unloading, twelve used only a paddle, three used only a driving board and three used both. At loading, the number of oral interactions (shouting, whistling or talking) varied from 0 to 296 (mean 48). The number of tactile interactions (with board, paddle, hand or feet/leg) varied from 18 to 1,443 (mean 284). The most frequent tactile interaction was touching the rear of a pig with a paddle or hand. There were also variations in welfare-related pig behaviours during loading, e.g. the number of HPV (min. 0, max. 42, mean 12), crowding (min. 0, max. 45, mean 7) and slipping (min. 0, max. 62, mean 12). The association between TD behaviour involving tactile interactions and welfare-related pig behaviour (HPV, slipping, crowding, falling or climbing) was analysed by Spearman rank correlation, using the number of behaviours per pig handled by TD. The pigs showed significantly higher numbers of welfare-related behaviours at loadings where tactile TD interactions were more common (rho=0.54; P=0.021). These results highlight variation between TD in handling, with possible welfare implications on the pigs handled. Furthermore, they indicate a lack of awareness and training of TD to achieve conditions for good animal handling, working conditions and animal welfare at loading and unloading of finishing pigs.

# Comparison of a behavior test between does and bucks at different Brazilian rabbit farms

*Kassy Gomes Da Silva[1], Giovanna Polo[1], Tamara Duarte Borges[1], Leandro Batista Costa[1], Antoni Dalmau[2], Joaquim Pallisera[2] and Cristina Santos Sotomaior[1]*
*[1]Pontifícia Universidade Católica do Paraná, Graduate Program in Anima Science, Rua Imaculada Conceição 1155, Curitiba, Paraná, 80.215-901, Brazil, [2]IRTA, Program of Animal Welfare, Finca Camps i Armet, 17121, Monells, Spain; cristina.sotomaior@pucpr.br*

The maintenance of the natural behavior of farm species is a goal for good animal production. The management of male (buck) and female (doe) rabbits is different on the same farm. Does may have a more intense handling routine than bucks, because of their reproductive function and life time at the farm. This intense human interaction can modify the rabbit´s response to human contact by these two genders. The objective of this study was to evaluate if does and bucks have different responses in a behavior test, as well as to compare their responses according to different types of production. Ten Brazilian farms were evaluated and divided into three categories: university rabbit farms (U, n=3), meat rabbit farms (C, n=4) and farms with pet rabbit production (P, n=3). A total of 70 bucks and 248 does without kits of 25 different breeds were evaluated. The behavior test was a mix of novel object test and human-approach test, that consisted of holding a 10 cm stick near the rabbit's head (approximately 10 cm) and observing if it touched or sniffed the stick (score 0); if it stayed immobile or did not demonstrate interest in the stick (score 1); or if it ran away from the stick or attacked it with aggression (score 2). The 30-second observation period started when the stick was 10 cm away from the rabbit head. First, results for the scores 0, 1, and 2 from all does and bucks were compared, then between does and bucks of the farms U, C, and P, using score 0 and score 1+2 (score 1 and 2 together). Data were analyzed using Chi-squared test, with significance at 5%. The behavior test score varied according rabbit gender (P=0.0002). Bucks had more 0-scores (55.7%) than does (30.2%), with contrary results for 1-scores (does: 65.7%; bucks: 38.6%). Score 2 was given to 4.0% of does and 5.7% of bucks. In the comparison between farms, P farms showed no difference between does and bucks (P>0.05). At these farms, 42.7% of does and 45.8% of the bucks were 0-scored. At U rabbitries, bucks had more (P=0.001) 0-scores (58.3%) than does (22.1%). The same was observed at C farms, where 24.2% of does and 63.6% of bucks had a 0-score (P=0.0003). The results suggest that bucks were less timid than does in the overall evaluation, as well as at meat farms and university rabbitries. Does and bucks had similar behavior at pet farms. This may be a consequence of the management, as bucks and does may receive similar handling to pet rabbits. New studies may elucidate the relationship between the reproduction management, handling techniques and behaviour responses on rabbit farms.

## Beekeeper-honeybee interactions to measure personality at colony level: a pilot study of *Apis mellifera* in central Italy

*Rosaria Santoro[1], Jennifer Rowntree[1], Luca Modolo[2], Richard Preziosi[1] and Giovanni Quintavalle Pastorino[2]*

[1]*Manchester Metropolitan University, All Saints Building, M15 6BH, United Kingdom,* [2]*University of Milan, Department of Veterinary Medicine, Via dell'Università, 26900, Italy; rosaria.santoro@stu.mmu.ac.uk*

Personality differences play an important role in the lives of animals, influencing their actions. For organisms living in highly structured groups of related individuals such as colonies of social insects, personalities could also emerge at the group level. In this study, we apply the concept of personality to colonies of honeybees (*Apis mellifera*) to determine their behaviours across their lifespan. We collected information on 232 hives from 15 beekeepers in Lazio, central Italy, using a questionnaire developed by studies of keeper-animal interactions in zoos. We asked beekeepers to rate the behaviour of honeybee colonies in each of their hives. We listed five behaviours (aggressiveness, docility, cooperativeness, productiveness, moodiness) based on a previous study and each of them was rated on a scale of 1 (never exhibited) to 12 (always exhibited), with the intermediate score 6, by the participants for each hive that they managed. Beekeepers were also asked to report the year of introduction of the current queen and the age of the queen for each colony. In a subset of hives, we also carried out an empirical test of aggressivity by counting the number of stings on a piece of leather passed over the colony following disturbance. Correlations among behavioural traits were analysed using Spearman's rank non-parametric statistical tests. The influence of age of queen on the behavioural traits was measured by using general linear mixed effects models (lmm), where apiary was included as a random-effect and queen age as a fixed effect. The model was repeated for all behavioural traits listed in the questionnaire. The consistency of behaviours among beekeepers was assessed by estimating the proportion variance assigned to the random effects in the linear-mixed effects model and reported as a percentage of total variance in the data. We found a strong positive correlation between cooperativeness and docility (P<0.001), a positive correlation between productiveness and the aggressiveness of the bees towards beekeepers (P<0.001); a negative correlation between both docility/cooperativeness and aggressivity towards beekeepers (P<0.001); a weak negative relationship of cooperativeness and docile behaviors with moodiness (P<0.001). Regarding the results of aggressivity test, the number of stings counted in the patches was significantly positively correlated with the aggressiveness to beekeeper (P<0.001) and productiveness (P<0.001) and negatively correlated with both cooperative and docile behaviours (P<0.001). Also, our findings confirm that queen age might be a crucial factor having a positive significant effect on aggressiveness and productiveness behaviours (P<0.001) and aggressiveness as determined by the empirical test (P<0.001). 'Beekeeper' explained most of the variation of the data in hive productivity (>70%) and almost (>50%) of variation of data in docility. Finally, our results, based on a relatively simple questionnaire, provide evidence of the fact that beekeepers can consistently recognise different personality traits among hives.

## Tickling does not increase play in young male rats

*Tayla Hammond[1,2], Sarah Brown[1], Simone Meddle[1], Birte Nielsen[3], Vincent Bombail[3] and Alistair Lawrence[2]*
[1]*University of Edinburgh, Roslin Institute, EH25 9RG, United Kingdom,* [2]*SRUC, Roslin Institute, EH25 9RG, United Kingdom,* [3]*Institut National de la Recherche Agronomique (INRA), Université Paris-Saclay, 78350 Jouy en Josas, France; tayla.hammond@sruc.ac.uk*

There is abundant evidence that tickling can be used successfully to induce positive affect in rats. Play behaviour has been identified as one of the most promising indicators of positive welfare. However recently, Ahloy-Dallaire *et al.* have highlighted the lack of evidence to support that play increases with positive affect. In this work we tested the hypothesis that tickling would lead to an increase in rat play assumed to be mediated by an increase in positive affect. We also expected that tickled rats would show more anticipatory play in the home cage and the testing arena as a reflection of being motivated to receive tickling. Male Wistar rats (n=24; 45 days old) were housed in pairs and assigned by cage to receive tickling or control handling for 5 days. Home cage behaviour was recorded for 15 minutes before and after handling to allow investigation of the effect of tickling on spontaneous play. Rats were placed individually into the experimental arena and given a 30-s free-roam opportunity to express anticipatory or exploratory behaviour. Rats then experienced tickling or control conditions (hand moving on the side of the arena in the same 15 sec intervals as tickling) for a total of 2 minutes, followed by another 30-s free-roam opportunity. Within the arena, we recorded ultrasonic vocalisation (USV) production and scored behaviour during the free-roam tests. During the free-roam tests, tickled rats produced more positive USVs than control rats ($P=0.049$) and made more frequent investigations towards the static hand of the experimenter ($P=0.038$). There was no effect of treatment on home cage play (solitary $P=0.6$; social $P=0.4$). Play in the home cage only occurred before handling ($P>0.001$) and increased across test days ($P=0.002$) in all rats regardless of handling regime. Rats became inactive after handling ($P>0.001$). Thus, tickling did not induce play in rats despite some evidence that tickling enhanced positive affective state. However, predictable exposure to the arena and handling (either tickled or control) may have been sufficient enrichment to promote play. As such, the relationship between play and positive affect requires further exploration.

## Light spectrum for illuminating dark periods in reversed light cycles in laboratory rodent facilities

*Carlos Grau Paricio, Cécile Bienboire-Frosini and Patrick Pageat*

*IRSEA (Research Institute in Semiochemistry and Applied Ethology), IRSEA, Quartier Salignan Apt France, 84400, Apt, France; c.grau@group-irsea.com*

The human and murine diurnal rhythms are out of phase. Humans are considered a diurnal species and mice crepuscular/nocturnal. In standard artificial light cycles in laboratory rodent facilities (e.g.: 8am-8pm lights on/off) the mice's deep sleep is often disrupted because of routine husbandry and research procedures, welfare monitoring of the mice is limited by their inactivity, and scientific data obtained is from the naturally inactive period. Reversed light cycles allow to synchronize human and mice active periods, however they can suppose logistic difficulties for human work and depending on light spectrum and intensity, disturbance in mice/rodent circadian rhythms. In this literature review we have evaluated empirical and theoretical papers which describe the influence of light spectrum and intensity to illuminate rodent laboratory animal facilities during dark phases of light cycles. Four different perspectives were found, sensitivity to light intensity and spectrum, animal perspective in terms of their behaviour and activity, human perspective to evaluate feasibility of routine and research procedures, and animal-human interactions in terms of animal welfare evaluation and handling. Two lighting methods and associated spectrums have been described so far, red lights and sodium lamps. The first method, red lights are considered with a spectrum >630 nm, this wavelength is in the limits of mouse vision, but not 'invisible' as it has been described in some studies. Locomotor activity is significantly higher during dark periods illuminated with red lights and animal disturbance is minimal, however routine procedures and welfare evaluation has been described as challenging. The second method are sodium lamps with a spectrum of 589 and 589.6 nm. Locomotor activity was significantly higher during dark periods illuminated with sodium lamps, and routine and research procedures were feasible, including behavioural testing and welfare evaluation with some limitations in delicate procedures. Sensitivity to sodium lamps wavelengths and intensity has been described as theoretically capable of eliciting biological responses during the dark phase. We can conclude that actual lighting methods, red lights and sodium lamps have some limitations from human (red lights) or animal perspectives (sodium lamps). Led lights have been included during day light cycles but are still rare to illuminate dark cycles in laboratory rodent facilities. Future studies (including an ongoing evaluation in our facility) should evaluate the advantages/disadvantages of this technology as an alternative, which allows precise and modifiable wavelengths. Intermediate 600 nm wavelengths with led lighting seem a logical compromise between red lights and sodium lamps, but to our knowledge studies have not been published at this point.

**Heart rate – the effect of the handler on the dog after a potentially stressful situation**

*Helena Chaloupkova, Jitka Bartosova, Karel Novak, Nadezda Fiala Sebkova and Martina Bagiova*
*Czech University of Life Sciences Prague, Department of Ethology and Companion Animal Science, Kamycka 129, 16500 Prague, Czech Republic; chaloupkovah@af.czu.cz*

In working dogs, the relationship with the handler can significantly influence the dog´s performance and welfare. The aim of the study was to test whether unintentional signals from the handler after anticipated stress can influence the heart rate of the dog. Twelve dog-handler pairs were individually exposed to the same situation during routine obedience training lesson. Each handler received instructions from an assistant that included information that a person would be present, hidden behind a training building and performing an attack on the dog-handler pair as they would be passing along the building. However, in fact, there was no person hidden in the training grounds and no attack was taking place. Heart rate (HR) measured via beats per minute (BPM) of both the handler and the dog was measured with equipment SporttesterPolar® RS800CX Team 3.4. Data were divided into four-time periods that corresponded to the course of the test (start, control-measurement, attack notification, stress-measurement). Statistical analyses were provided using SAS program, version 9.4, Proc MIXED. The dogs' HR was 142.7±25.5 bpm and the handler HR was 120.2±20.0 bpm (mean ± S.D). The results of this pilot study shows that the dog´s HR was significantly related to the handler´s HR ($F_{3, 1869}$=11.93; P<0.001). We detected that at the start of the test, during control-measurements and during stress-measurement, with increased HR of the handler the dog's HR increased as well. On the other hand, during the attack notification, the handler´s HR was increased but the dog´s HR was decreased. The first data shows synchronization of HRs of both handler and dog during the obedience training regardless of whether the handler was exposed to a stressful stimulus. The opposite relationship of HRs of the dog-handler teams during the notification of the attack could have been caused by interrupting the handler with the dog when the handler's attention dedicated to the assistant.

## Small dogs display more aggressive behaviour than large dogs in social media videos

*Natalie Solheim Bernales and Ruth C. Newberry*

*Norwegian University of Life Sciences, Faculty of Biosciences, Department of Animal and Aquacultural Sciences, 1432 Ås, Norway; natalie.bernales@outlook.com*

Due to potentially greater vulnerability to accidents and attacks, the behaviour of small dogs may reflect greater threat sensitivity than that of large dogs. Based on this hypothesis, we predicted that dogs of small breeds (<10 kg) would be more likely to show signs of threat-perception and stress than dogs of large breeds (>20 kg). We extracted behavioural data from 310 videos posted on YouTube depicting adult dogs of four small dog breeds (Chihuahua, Jack Russell Terrier, Dachshund, Yorkshire Terrier) and four large dog breeds (German Shepherd, Border Collie, Labrador Retriever, Rottweiler; n=30-40 dogs/breed ). We searched for videos using the breed name alone (n=160), or the breed name with 'angry' (n=150). We included only one video per subscriber, and excluded videos that were compilations, depicting training, longer than 5 min or shorter than 10 s, showing play behaviour (in the 'angry' group), the wrong breed, with no handler present, or not showing the dog clearly. The observed videos were of comparable duration across dog size and breed (mean±SE: 86±3 s). Behaviours were scored as occurring or not occurring in each video (1-0 sampling), and generalised linear models were used to investigate associations between body size and behaviour accounting for breed, search strategy and location (indoors or outdoors). Results are presented as back-transformed least squares means. Snapping and/or biting were more commonly shown in small than large dog videos (22±5 vs 5±2%; $P<0.001$), as was growling (57±5 vs 35±5%; $P=0.004$) and, overall, small dogs exhibited a higher number of different signs of aggression (out of 8 possible behaviours) compared to large dogs (small: 1.9±0.1; large: 1.6±0.1; $P=0.034$). Compared to the large dogs, the small dogs were also more likely to show the stress-related behaviour, repeated lip licking: 44±5 vs 31±4%; $P=0.032$. Dogs were being held in their handler's arms or lap in 35±6% of the small dog videos compared to 8±3% of the large dog videos ($P<0.001$), and handlers were more likely to touch aggressive small than large dogs (20±5 vs 7±2% of videos, $P=0.002$). The findings support our hypothesis that small dogs are more threat sensitive than large dogs, and suggest that the observed behavioural differences between small and large dogs were influenced by differences in the behaviour of handlers towards the dogs. The results are consistent with greater tolerance of aggressive behaviour in small than large dogs.

## Efficacy of a 4-week training program for reducing pre-existing veterinary fear in companion dogs

*Anastasia Stellato, Sarah Jajou, Cate Dewey, Tina Widowski and Lee Niel*
*University of Guelph, 50 Stone Road E., Guelph, Ontario, N1G 2W1, Canada; astellat@uoguelph.ca*

Dogs often show signs of fear during routine veterinary appointments. Various strategies are recommended to clients to reduce this fear, including desensitization and counter-conditioning training, but no studies to date have examined its effectiveness at reducing pre-existing fear in companion dogs. We assessed the effect of a standardized 4-week training program involving desensitization and counter-conditioning to the clinic and a physical examination on behavioural and physiological signs of fear in dogs. Subjects were owned dogs rated as having mild-moderate fear. Responses were assessed during clinic entrance and during a physical examination, and assessments occurred before and after training. Sex (24 male, 22 female) and age (ranging from 1 to 13 years) were balanced across treatment groups and dogs were randomly allocated to receive training (n=24) or no training (n=22; Control). For dogs in the training group, the owner performed gradual, exam-style handling on their dog twice a week in their home and visited the veterinary clinic once a week, for a duration of 4-weeks. All handling was tailored to individual dog progress based on observed fear signs. Handling and visits were combined with food rewards reported to be highly palatable to each dog. Owners reported compliance with training requirements, and only dogs that received required training were included in the final analysis (3 excluded). No specific instructions were given to owners of control dogs. During assessments, dog behaviour was video recorded, and responses were assessed by a blinded observer for both clinic entrance (entrance, weigh-in), and examination. For behavioural measures, mixed Poisson and logistic regression models assessed the effects of training, testing phase (phases of clinic entrance or of examination), sex, and age, with dog as a random effect. Physiological measures assessed during examination were analyzed with linear regression models. Corresponding baseline responses for each measure taken from all phases of the first visit were included as covariates. Training did not significantly influence avoidance, posture, trembling, vocalizations, willingness and encouragement to step on the scale, or the physiological measures. During clinic entrance, the rate of lip licking showed an interaction with sex and phase ($F_{1,41}=6.19$, P<0.0008), and was lower for male, control dogs during entrance in comparison to female, controls. During examination, the rate of lip licking had an interaction with baseline lip licking ($F_{1,217}=5.99$, P=0.015); both treatment groups showed a reduction in lip licking from the baseline, and for trained dogs, the rate of lip licking decreased as the baseline rate increased. Results suggest that this particular 4-week training program was ineffective at reducing veterinary fear in dogs. Further research is necessary to explore alternative types of training programs (e.g. longer, more intensive and individualized), and to assess how individual dog characteristics and fear responses influence response to treatment.

## Cat pain and welfare: barriers to having good cat care management in Malaysian veterinary practices

*Syamira Zaini[1,2], Claire Phillips[2], Jill MacKay[2] and Fritha Langford[3]*
*[1]Faculty of Veterinary Medicine, Universiti Putra Malaysia, UPM Serdang, Selangor Darul Ehsan, 43400, Malaysia, [2]Royal (Dick) School of Veterinary Studies. The University of Edinburgh, Easter Bush Campus, Midlothian, EH25 9RG, United Kingdom, [3]Animal and Veterinary Sciences, Scotland's Rural College (SRUC), West Mains Road, Edinburgh, EH9 3JG, United Kingdom; s.s.zaini@sms.ed.ac.uk*

Environmental management in veterinary practices is crucial for cat welfare, however, there is a significant gap in the translation of optimal cat care into practices. In Malaysian veterinary practices, the post-operative environment often lacks enrichment (e.g. bedding, hiding places). Barren post-operative environments are known to be stressful for cats, with cats engaging in a variety of behaviour indicative of fear or stress. These stress-related behaviours can interfere with proper interpretation and use of established pain assessment scales. Therefore, this study used mixed methods approach to explore the current understanding of Malaysian vets about cat pain-behaviour and welfare, and the barriers to managing this in practices, in order to develop an effective educational intervention for use with Malaysian vets. In the first part of the study, an online survey was developed to ascertain what Malaysian veterinarians provided to cats' post-surgery and any barriers to the provision of enrichment. The survey consisted of 14-questions in three different sections. The three sections of the survey were: (1) Demographic details (e.g.: field of practice, availability of clinical protocol in the veterinary practice), (2) Attitudes to cat welfare and (3) Current management of post-operative cats and barriers to providing good care for the cat in practices. The survey was available for eight-weeks in Malaysia. Of 150 small animal practitioners contacted, 48 (32%) successfully completed the survey. Most respondents were senior veterinarians (54.2%, n=26) and 95.8% (n=46) were aware of Standard Operating Procedures (SOPs) within their veterinary practice. Most respondents were aware of the basic provisions that cats need post-surgery, yet identified barriers to that provision. The majority of respondents highlighted cost (45.5%, n=15) as the highest restriction to good care provision. In the second part of the study, interviews were carried out with 20 Malaysian veterinarians (selected from the survey sample). After thematic analysis, a number of key constructs emerged: firstly that the veterinarians recognized that comfortable post-surgery environments helped recovery and added a sense of security for the cat. Secondly, when asked directly, most veterinarians did not recognize the relationship between reducing stress and behavioural assessment of pain. Finally, they identified a lack of practice management skills as the other main reason for not providing better post-surgery recovery environments for cats. Therefore, any educational intervention should illustrate the tactics to improve cat welfare by suggesting cost-effective and efficient ways to have good cat care in the clinic and emphasize the need to reduce stress post-operatively in order to use behaviour as a method to assess pain.

## Does the personality of tigers influence the interaction with their keepers?

*Giulia Corbella[1], Giovanni Quintavalle Pastorino[1], Manuel Morici[2] and Silvia Michela Mazzola[1]*
*[1]Università degli Studi di Milano, Department of Veterinary Medicine, via Celoria 10, 20133 Milano, Italy, [2]Safary Park Pombia, Lago Maggiore, via Larino 3, 28050 Pombia (NO), Italy; silvia.mazzola@unimi.it*

When a zoo animal and a human interact on a regular basis, the quality of their interactions is influenced by various factors, also linked to the animal's history. But what about the animal's personality? Can the animals' expression of personality traits influence their relationships with their keepers? Through the determination of the personality profile of tigers and the analysis of their interactions with their keepers, we investigated the hypothesis that the different expression of personality traits may influence the relationship with their keepers. A group of six *Panthera tigris* sp., consisting of mother (1/2006) and 5 sons (7/2012), was studied. Personality trait expressions were evaluated (rating method): three keepers were asked to complete a previously published questionnaire, scoring each tiger against 31 traits of the animal's personality, rated on a scale of 1 (never exhibited) to 12 (always exhibited). Concordance between keepers was calculated by inter-rater reliability (Intraclass correlation coefficient). Keeper-animal interactions were evaluated through the analysis of video recorded behaviors of tigers expressed during evening re-entry procedures in the internal enclosure. The behaviors were analyzed, with Boris Software, on the basis of a previously published ethogram. The frequency (events) and duration (states) of tiger's behaviors in the presence of the different keepers were evaluated, and then related to the personality traits. The considered variables were analyzed through descriptive statistics, and the differences between animals were calculated using the Kruskal-Wallis nonparametric test. The data from the questionnaires allowed drawing of a personality profile for each tiger, which was compared with the Interaction plots analysis of the tiger's behavior. The results provided evidence that each animal interacted differently with the three keepers. All the tigers showed a significantly higher expression of affiliative behavior with the female keeper, and this was correlated with the expression of the *friendly* trait. Each tiger showed a significantly different level of *compliance with the keepers' calls*, and a high expression of the trait *dominant* was related to a lower level of compliance. Other interaction-related behavior analyzed, such as *approach* behaviors, were significantly different for each tiger, and were related to the expression of *cooperative* and *timid* traits. In conclusion, the results of this pilot study have highlighted the possibility that the expression of some personality traits could be predictive of the interaction that the animal will have with the keeper. Further studies are needed to confirm this hypothesis, and it would also be useful to deepen analysis of the personality of the keepers.

## The variable effect of visitor number and noise levels on behaviour in three zoo-housed primate species

*Anita Hashmi[1], Matthew Sullivan[1], Silvia Michela Mazzola[2] and Giovanni Quintavalle Pastorino[1,2]*
*[1]Manchester Metropolitan University, School of Science and the Environment, John Dalton, Chester Street, Manchester, M1 5GD, United Kingdom, [2]University of Milan, Veterinary Medicine, Via Celoria 10, 20133 Milano, Italy; anita_hashmi@yahoo.co.uk*

Visitors may impact on the welfare of zoo-housed animals, and this has become an established research area in the modern zoo, with diverse results, particularly in primates. This study aimed to further the current literature and tested the hypothesis that visitor number and noise levels significantly affected the behaviour of three ape species housed at Blackpool Zoo, UK, chosen to compare reactions across apes: Western lowland gorilla (*Gorilla gorilla gorilla*; 3 male, 3 female), Bornean orangutan (*Pongo pygmaeus*; 1 male, 5 female) and pileated gibbon (*Hylobates pileatus*; 2 male, 2 female). This was studied through 10-minute focal observations and measurements of visitor numbers and noise levels around enclosures. Observations were conducted twice per week from April-August 2018 between 10:00-16:00, with three observations per individual per session, species observed on a rotating basis and order of individuals chosen with a random number generator, creating data across quiet and busy times of day and season. Changes in behaviour were analysed on species and individual levels, using RStudio version 1.1. Noise levels had a significant positive correlation with visitor number ($r^2$=0.27, P<0.001) and both factors had significant positive or negative associations with named behaviours: known stress indicators or evidentiary of a visitor effect in previous studies. Inactivity showed negative correlation with visitor number as expected (gorillas: estimate=-0.028, z-value=-3.089, P<0.001; orangutans: estimate=-0.025, z-value=-2.902, P<0.001) but no significant effect of noise was found. Stereotypic behaviour was unexpectedly not significantly associated with visitor number or noise levels in gorillas but was in orangutans: positive association was seen with visitor number, negative with visitor noise. Different visitor avoidance behaviours also showed varying results: levels of individuals sitting with their back to the window was not correlated with visitor number or noise (P>0.05), but the orangutan-specific behaviour of using substrate or sacks to cover the head displayed a significant negative correlation with noise level (estimate=-0.027, z-value=-1.976, P<0.05). There was a corresponding rise in the visitor attention behaviour with rising noise levels (estimate=0.040, z-value=2.807, P<0.01), displaying an increase in human-directed vigilance behaviour in response to increased noise levels. Individual and species differences were seen in this study, emphasising the complexities of personality and studying the visitor effect. Increases in stereotyping and clinging, and decreases in inactivity suggest a potential negative effect on the welfare of these apes. The mixed results reinforce that the visitor effect is influenced by many factors, such as husbandry and personality. This study highlights the need for off show areas for captive apes, and the importance of considering individual differences when attempting mitigation of unwanted behaviours.

## The influence of human interaction on guinea pigs: behavioural and thermographic changes during animal assisted therapy

*Sandra Wirth[1,2], Jakob Zinsstag[1,3] and Karin Hediger[3,4]*
*[1]Swiss Tropical and Public Health Institute, Socinstrasse 57, 4051 Basel, Switzerland, [2]University of Zurich, Veterinary Medicine, Winterthurerstrasse 260, 8057 Zürich, Switzerland, [3]University of Basel, Missionsstrasse 60, 4055 Basel, Switzerland, [4]REHAB Basel, Im Burgfelderhof 40, 4055 Basel, Switzerland; sandra.wirth@uzh.ch*

This study examines effects of retreat possibility, human contact and presence of conspecifics on guinea pigs involved in animal-assisted therapy (AAT) at a rehabilitation clinic for people with brain injuries and/or paraplegia. Therapies include cutting vegetables, filling the vegetables in wooden pet puzzle toys, trying to encourage the guinea pigs to approach by feeding them and attempt to pet them. The aim was to increase knowledge of guinea pig's well-being and stress behaviour in order to carry out AAT in a long-term, ethical and 'one-health' manner. Guinea pigs (n=20) were assigned to a randomised, controlled within-subject trial with repeated measures. Each guinea pig took part in four different settings: (1) therapy with retreat possibility, (2) therapy without retreat possibility, (3) therapy with retreat possibility without presence of conspecifics, (4) control without human interaction. Setting 1, 3 and 4 took place on a 1.2 m$^2$ table cage specially designed for therapies. In setting 2, the guinea pig was located in a pet bed on the patient's lap. Guinea pigs were involved in AAT for a maximum of 35 min, of which a maximum of 5 min took place without possibility to retreat or without the presence of conspecifics. All sessions were video recorded and each guinea pig's behaviour was coded with Noldus Observer XT 12.5 using continuous recording and focal animal sampling. Behaviour was coded according to a specifically designed ethogram including individual behaviour, conspecifics' social behaviour, active and passive human-animal interactions, interactions with the environment and vocalizations. Moreover, changes in body temperature (represented by eye temperature) were measured in 5-s intervals with a thermography camera (FLIR T530). Statistical analysis was performed using generalized linear mixed models, with therapy setting as a fixed effect and individual guinea pig as a random effect. Preliminary results from 10 guinea pigs indicate a significant influence of settings on thermographic results (P=0.027), with the slope of the temperature trend line in setting 3 (mean ± SD, 0.0082±0.0197) differing from those in setting 1 (-0.0004±0.0017, P=0.012) and setting 4 (-0.0004±0.0018, P=0.012) but not from that in setting 2 (0.0050±0.0074, P=0.347). Furthermore, the slopes of the temperature trend lines did not differ between settings 1 and 4. Our findings contribute to the identification of conditions under which the involvement of guinea pigs in animal-assisted interventions can benefit both animals and humans.

## The use and well-being of cats in elder care in Sweden

*Lena Lidfors, Ida Jonson and Ellen Nyberg*
*Swedish University of Agricultural Sciences, Department of Animal Environment and Health, P.O. Box 234, 532 23 Skara, Sweden; lena.lidfors@slu.se*

The use of animals for therapeutic benefits is common in elder care in Sweden, however there are no information on how common cats and other animals are. Earlier research has shown that the presence of cats can lead to better overall health, lower stress levels and more social interactions in elderly. The aim of this study was to investigate how common cats are in relation to other therapy animals in elder care in Sweden and to assess the behaviour and health of cats within these nursing homes. A questionnaire was e-mailed to 434 nursing homes and units in 57 municipalities in two regions of Sweden regarding the use and perception of animals within the elderly care. The response rate was 62% (n=268), and 56% (n=243) answered the whole questionnaire. Of the 194 nursing homes that answered the question about which animals they have now or have had answered; dog 79%, cat 43%, birds 24%, fish 20%, horse 11%, farm animals (cattle, sheep, goats) 6% and other animals (for ex. rabbits) 6%. One answered using a robot cat 'Justocat'. Of those having a cat 77% had resident cats and 23% had cats coming in during the day. In a follow up study the well-being of 18 cats in 13 nursing homes was investigated by interviews, clinical examinations (n=17) and behavioural observations (n=9) in three regions of Sweden. The cats age ranged from 2-15 years. Focal animal observations were carried out directly after the clinical examination during 60 minutes. The cats placement in the room, body position and behaviour was recorded with 0-1 recording in 3 minutes intervals. Descriptive data are shown as medians, Q1 and Q3 of the percentages of recordings. The cats were found on an elevated place, i.e. furniture or window shelf (median 61.9%, Q1=23.53, Q3=76.19), the floor (median 23.8%, Q1=14.29, Q3=60) or the feeding place (median 4.8%, Q1=0, Q3=14.29). One cat was lying in the lap of an elderly 14% of the observations. The cats were lying a median of 38.1% of the observations (Q1=20, Q3=52.38), sitting 28.6% (Q1=17.65, Q3=33.33), standing 14.3% (Q1=9.52, Q3=19) and walking 14.3% (Q1=4.48, Q3=20). Behaviours performed by the cats were resting (median 19%, Q1=4.76, Q3=38.1), looking (median 35.3%, Q1=28.57, Q3=42.86), attention seeking from elderly/personal (median 14.29%, Q1=4.76, Q3=19.05), being petted (median 9.5%, Q1=4.76, Q3=13.33) and eating/drinking water (median 0, Q1=0, Q3=4.76). Licking its fur was recorded 4-14% in four individual cats, but median over all cats became 0. Play was not recorded in any of the cats. No cats showed fear or aggression during the behavioural observation. Body condition score (BCS) with a scale from 1-9 showed that 6 cats (35%) had over 5 (overweight), 9 cats had BCS=5 and 2 cats had BCS 4 vs 3 (underweight). Teeth health had the highest number of remarks, where 10 of 16 cats (63%) had moderate to severe tartar. Older cats had more tartar. Four cats had remarks on the fur/skin. Health checks on lungs, heart, mouth, lymph nodes, nose, eyes, ears and back/tail showed that all 17 cats were healthy. In conclusion, cat was the second most common animal in elder care, the observed cats could perform natural behaviour, some were obese and had bad teeth health, but were otherwise healthy.

## Camels in animal-assisted interventions – survey of practical experiences with children, youth and adults

*Malin Larsson[1] and Dana Brothers[2]*
*[1]Swedish University of Agricultural Sciences, Department of Animal Environment and Health, Box 234, 53223 Skara, Sweden, [2]Hospice of the Northwest, Community Outreach and Education Manager, 227 Freeway Drive, Suite A Mount Vernon, WA 98273, USA; malinlarsson@yahoo.com*

Horses have been used in physiotherapy at least since the 1950s, and more recently in other forms of equine-assisted interventions (EAI). Another large domestic animal can also be used in animal-assisted interventions (AAI), including riding – the camel. We here introduce a new field within AAI: Camel-assisted interventions (CAI). Camels in AAI fill similar functions as horses – they are big, soft, warm, and they mirror emotions. Camel riding provides sensory, motor and vestibular stimulation, which may make camel riding suitable in physiotherapy. Camels and horses have different locomotion and may provide different motor stimuli for riders. Camels are less flighty than horses. The humps of two-humped camels provide tactile stimulus and support. Camels with their unusual appearance enhance alertness and interest, while their calm behaviour transmits to people. This duality makes camels potentially suitable in interventions for people with ADHD or autism. A camel in recumbent position is on eye-level with children, allowing safe, direct contact. The camel responds immediately to the child's behaviour. If the child becomes restless, the camel stops ruminating or gets up: authentic, non-judgmental feedback. The purpose of this study was to locate practitioners of CAI around the world and collect information about experiences from CAI. CAI was evaluated in a survey to practitioners, through the Facebook group Cameleers. The survey was sent by e-mail to 17 practitioners in 5 countries, with 6 answers from 4 countries (Austria, Australia, Germany and USA). The survey included questions about kinds of CAI performed, how the camels react, diagnoses of participants, their reactions and effects of CAI. All respondents had worked with young people (15-24 years), and 5 with children or adults, respectively. All had worked with people on the autism spectrum, 5 with depressed people, 4 with people with ADHD and 4 with physically disabled people. 5 respondents had worked with CAI for more than 10 years. All respondents had worked with ground-based activities, 5 with educational activities, 4 with leisure riding, 3 with CAI together with a therapist and 2 with physiotherapeutic riding. According to all respondents, a typical first time reaction to camels was surprise and joy. 4 respondents also reported curiosity, and 3 also fear or withdrawal as reactions when meeting camels first time. No respondent reported aggressiveness or indifference as typical first reactions. Improved well-being, attention, social engagement, physical coordination and self-esteem were some of the reported effects of CAI. Camels used in CAI were selected for the tasks and acclimated to the work. They were reported to be calm and relaxed and they liked being groomed. There are quite few CAI practitioners around the world. No scientific studies on CAI are available yet. The survey showed that camels have a potential in AAI and may be a complement to horses. More research is needed to evaluate effects of CAI.

## Optimal dog visits for nursing home residents with varying cognitive abilities – an interdisciplinary study

*Karen Thodberg[1], Poul Videbech[2], Tia Hansen[3] and Janne W. Christensen[1]*

*[1]Aarhus University, AU-Foulum, Department of Animal Science, Blichers Allé 20, 8830 Tjele, Denmark, [2]Centre for Neuropsychiatric Depression Research, Mental Health Centre, Glostrup, 2600 Glostrup, Denmark, [3]Aalborg University, Department of Communication and Psychology, Kroghstræde 3, 9220 Aalborg Ø, Denmark; karen.thodberg@anis.au.dk*

Visiting dogs are popular in nursing homes, and studies indicate several benefits from these animal-assisted activities. However, the more specific effects of the dog and the residents' cognitive impairment level on immediate responses are unknown. Use of ethological methodology could be an important tool in inter-disciplinary research to obtain evidence-based knowledge in this field. The aim of this inter-disciplinary project was to study: (1) the immediate response of nursing home residents to dog visits with different opportunities to interact with the dog; and (2) whether the residents' cognitive impairment level affected their behaviour. We randomly assigned 186 nursing home residents to three treatment groups that either received 12 bi-weekly 10-minute visits accompanied by a dog (D); 12 visits accompanied by a dog, including a new activity with the dog (E.g. brushing, 'training', doing tricks) during each visit (DA) or 12 visits without a dog, but with a new activity during each visit (A). Apart from the visitor, an observer was present during all visits, recording the frequency and duration of the residents' verbal and physical interaction with the dog and person. We divided the 12 visits into 3 periods of 4 visits for data analysis. For each period, data were analysed either as binomial variables in a generalized linear mixed model or as durations with Wilcoxon Rank Sum/Kruskal-Wallis tests. Comparing the three visit types, we found that residents talked less to the persons in DA compared to D and A in all periods ($P<0.01$). During visits with an activity, but no dog, the residents took more part in the activity (period 1, $P=0.07$; period 2: odds=3.4 [1.5-8.0], $P<0.01$; period 3: odds=18.9 [4.9-71.4] $P<0.001$), and talked more about the activity, compared to DA (Period 1: odds=6.9 [1.8-19.2], $P<0.01$; period 2: odds=11.1 [2.9-41.7], $P<0.001$; period 3: odds=10.6 [2.9-38.5] $P<0.001$). In both DA and A, the odds of taking part in the activity and talking about it were lower with higher levels of cognitive impairment ($P<0.001$-0.07 and $P<0.001$-0.07, respectively). In period 1, the residents in DA had lowered odds of touching the dog with increasing level of cognitive impairment ($F_{1,85}=5.2$; $P<0.05$). The likelihood of talking to the dog was affected by an interaction between visit type and level of cognitive impairment (Period 1: $F_{1,90}=4.6$; $P<0.05$; Period 2: $F_{1,87}=5.3$, $P<0.05$), and the most severely impaired residents were less likely to talk to the dog during DA visits compared to D. In conclusion, nursing home residents vary in their immediate response to visits, both due to the type of visit and because of differences in the level of cognitive impairment. Adding an activity might not be optimal for all, especially not for the most cognitively impaired residents. The results from this study can help optimise use of animal-assisted activities, and ensure the best interventions for a broad range of nursing home residents.

## Geography affects assistance dog availability in the United States and Canada

*Sandra Walther[1], Marko Yamamoto[2], Abigail P. Thigpen[1], Neil H. Willits[1] and Lynette A. Hart[1]*
*[1]Univ. Calif., Davis, 1 Shields, Davis, CA 95616, USA, [2]Teikyo Univ. Sci., Adachi, Yamanishi, Japan; lahart@ucdavis.edu*

Regarding assistance dogs, this study's objective was to characterize numbers and types of assistance dogs placed in 2013/2014 with persons having disabilities, by U.S./Canada non-profit facilities with Assistance Dogs International (ADI) or International Guide Dog Federation (IGDF), and non-accredited, sometimes for-profit, U.S. facilities. This revealed facilities' geography and dog types placed (guide, hearing, and service--mobility, autism, psychiatric, diabetes, seizure). ADI/IGDF facilities responded--22 states and 3 provinces: 55% (55/100). Non-accredited facilities in 16 states responded, 17% (22/133). ADI/IGDF facilities had 2,374 total placements; there were 797 non-accredited placements. Guiding and mobility placements were similar; autism dogs were third most. Psychiatric dogs were fourth most accredited, and most non-accredited, placements. Twenty states had no responses; 17 of these states lacked ADI/IGDF facilities. Accredited facilities sourced dogs from breeding programs or outside breeders, not other sources; in contrast, most responding non-accredited facilities used other sources. Non-accredited facilities used many breeds. Dog types placed differed by geographic regions, by facility accreditation status, and by all dog sources (all Chi squares: P<0.0001). In regions lacking facilities, people with disabilities face challenges, living far from supportive facilities to provide dogs or lend support. Despite increasing dog placements, availability of well-trained assistance dogs is geographically limited. Accredited facilities provide traditional dog types, and for autism, using standardized dog sources. Non-accredited facilities utilize diversified dog sources, often placing newer types of dogs (psychiatric), but they lack a unified standard of training, and their dogs vary in cost, now increasing confusion for potential service dog handlers.

## Horses and ponies differ in their human-related behaviour

*Lina S.V. Roth and Josefine Henriksson*
*Linköping University, IFM-Biology, AVIAN Behavioural Genomics and Physiology Group, 58183 Linköping, Sweden; lina.roth@liu.se*

The domestic horse has been living in close association with humans for thousands of years and while they probably were important as a food source and means of transport in the beginning of the domestication, horses today are mostly used for sport and leisure. In this study we investigated whether horses would seek contact with a passive human when presented with an unsolvable problem, consisting of pieces of carrot in a closed bucket, and if they could make use of the information from an active human demonstrator in a detour task, resembling the shape of V. The human-directed behaviours of 22 horses/ponies where studied during a problem-solving experiment and the same horses/ponies participated in a detour experiment where their success and interest-related behaviours (proximity and attention towards detour set-up/demonstrator) were recorded. The age of the horses/ponies (n=22) ranged from 7-25 years (13.5±0.9) and included equal numbers of horses and ponies, mares and geldings. All experiments were performed in accordance with relevant guidelines and regulations. Interestingly, the human-directed behaviours differed between ponies and horses, where the horses sought more human contact when presented with a problem (n=11, Z=-2.13, P=0.033), compared to before. The ponies, however, were in close proximity to the human regardless whether there was a problem present or not. In the detour experiment, one horse succeeded in the initial control without demonstration, while five succeeded after human demonstration. This did, however, not reach statistical significance (n=11, $\chi^2(1)=2.25$, P=0.13). Six of the eleven ponies succeeded in both initial control and during test. The horses/ponies that succeeded in the detour test after human demonstration did not show more human-directed behaviours when presented with an unsolvable problem, than the unsuccessful horses/ponies. Hence, if some individuals did use interspecies social learning in the detour experiment this social ability was not related to their human-directed behaviour according to our results. However, comparing the behaviours from the two experiments revealed that horses showed a positive correlation between their interest-related behaviours in the detour test and, when presented with an unsolvable problem, human proximity (n=11, rs=0.83, P=0.002), eye contact-seeking behaviour with human (n=11, rs=0.69, P=0.020), and looking behaviour towards the unsolvable problem (n=11, rs=0.61, P=0.047). No correlations were found for the ponies (n=11, P>0.1). Hence, this study show that size (horse/pony) have an impact on the human-related behaviours in both experiments.

## Practice what you preach – the discrepancy in knowing and doing based on moral values to farm animals

*Elske N. De Haas[1,2] and Frank A. Tuyttens[2]*
*[1]Department of Animals in Science and Society, University of Utrecht, Yalelaan 2, Utrecht, the Netherlands, [2]Institute for Agriculture and Fisheries research, Scheldeweg 68, Melle, Belgium; e.n.dehaas@uu.nl*

Most of us agree that animals should have a life that is worth living. Most of us agree that animals should be treated good, and consider animals as sentient beings. This is formulated under the EU treaty of Amsterdam and Lisbon; in which we should pay full regard to their welfare requirements. For our moral consideration, an animals capacity to feel (experience) appears more important than its capacity to think (agency). Sentience is therefore a crucial element which forms our moral attitude. The conditions under which farm animals are kept and treated are often far from perfect. As applied ethologists we know this, we are aware of animal mistreatment and animal suffering, yet this knowledge and our concern is often not reflected in our behaviour or consumption choices. Why is it so difficult to stick to believes, and why do humans sometimes fail to act on their moral values? There are different psychological processes whereby people undermine their own moral values. One of these is cognitive dissonance. Caring for animals and consumption of animal based products are in moral conflict, as the treatment of animals does not reason with moral concerns about sentient beings. As a result, studies showed that people who eat meat judge the sentience of farm animals less than of our pet animals. Hereby, people can be viewed as irrational or moral hypocrites under certain circumstances/food choices. This phenomenon can be explained, partly, by our emotions. Our moral attitude is based on our own emotions, and thereby provides the values we have about animals. But people appear to be very bad at monitoring their emotions, and thus we act contrary to what we think we ought to do. However, will power and self-control under cognitive dissonant dilemmas can be trained. Strong will helps us in being better in self-control. We highlight the difficulty in thinking right and doing right based on moral psychology. The information we provide is based on scientific literature on human psychology of meat eating, consumption of animal products and human attitudes to animals. As animal welfare researchers, we critically evaluate whether we practice what we preach, and whether we should have a more opinionated viewpoint in relation to care for farm animals. Our goal is to encourage animal welfare researchers in introspection of their own moral values regarding how animals are treated, and in which way they could influence others.

## North American stakeholder perceptions of the issues affecting the management, performance and well-being of pigs

*Sarah Ison[1,2], Kaitlin Wurtz[2], Carly O'Malley[2] and Janice Siegford[2]*
*[1]World Animal Protection, Farming, 5th Floor, 222 Grays Inn Road, WC1X8HB, London, United Kingdom, [2]Michigan State University, Animal Science, 274 S. Shaw Lane, 48824, East Lansing, MI, USA; sarahison@worldanimalprotection.org*

Changes to production practices that improve animal welfare are often slow to be implemented. This could, in part, be due to methods of communication or understanding between stakeholders. Online surveys, implemented using convenience sampling, investigated pork producers' ($n=360$), veterinarians ($n=162$), and researchers and outreach/extension educators ($n=129$) perceptions of the issues affecting the management, performance and well-being of pigs. All respondents were asked two open-ended questions: 'what do you consider the top three most important issues affecting the management, performance, and well-being of breeding pigs' and 'market hogs'? Three scorers independently created qualitative coding systems to categorise responses into a general, then specific theme. Results from scorers were combined to create the final coding scheme that included the same/similar identified themes or chose themes that best fitted the responses. Data analysis was conducted in R. Chi-square tests compared frequencies of responses attributed to themes among stakeholders. Raw text was visualised to investigate differences in the use of language. Five general themes were chosen: basic health and functioning (BHF, 46% of responses), human inputs (HI, 21%), environmental inputs (EI, 15%), human issues (HIS, 11%), and behaviour/welfare (BW, 7%). Producers differed in the balance of responses among general themes from research/extension professionals ($P<0.001$) and veterinarians ($P<0.001$), where no difference was found ($P=0.08$). Most notably, producers mentioned HIS less (4%) than research/extension professionals (17%) or vets (15%). HIS included the sub-themes 'control', 'profit' and 'sustainability'. Control included responses related to outside threats that change production practices on-farm (e.g. sow housing changes, restrictions on antibiotics), which, interestingly was more often seen as an issue for vets and research/extension professionals than producers. Most commonly mentioned specific themes across all respondents were health and nutrition. In terms of the raw text, research/extension professionals frequently used the word 'welfare', whereas vets used both welfare and care, and producers used the word care, but less frequently. In addition, producers were more positive in their use of language, mentioning what pigs should have (e.g. quality/proper feed, health), compared with research/extension professionals and vets who more often indicated what should be avoided (e.g. disease, issues). The stakeholder surveys had a different focus, which could contribute towards differences shown here. Regardless, by understanding differences in language used and issues identified among stakeholder groups, targeted communications can be developed that should be more powerful at driving improvements in pig management, performance, and welfare.

## What's in a name – the role of education and rhetoric in improving laying hen welfare

*Huw Nolan[1], Lauren Hemsworth[2] and Peta S. Taylor[3]*
*[1]University of New England, Faculty of Humanities, Arts, Social Sciences and Education, Armidale, NSW, 2351, Australia, [2]University of Melbourne, Welfare Science Centre, Veterinary Clinical Sciences, Parkville, VIC, 3010, Australia, [3]University of New England, Faculty of Science, Agriculture, Business and Law, Armidale, NSW, 2351, Australia; hnolan3@une.edu.au*

The rhetoric of laying hen welfare influences people's emotions. Terms like 'cage-free' and 'caged' connotes 'liberty' or 'imprisonment', or simply, 'good' or 'bad' for the hen. Science can determine the risk that the chicken egg industry (hereafter industry) practices pose to hen welfare, but a social license to operate will ultimately determine whether these practices are acceptable. Science and social licence do not always align and can lead to serious negative welfare consequences for hens. Furnished cages were designed as a compromise between the welfare implications of conventional cage and free-range systems, but societal concerns may still occur on the rhetoric that any cage is still a 'cage'. Hence, we are investigating the relationship between education and social licence using the 'furnished cage' system as a case study. We hypothesise that support for furnished cages will not occur so long as the rhetoric of 'cage' persists. Furthermore, if emotive language influences the effectiveness of education campaigns then objective science will not change attitudes towards housing systems. We designed an online experiment in which a random sample of the Australian public (n=851) were assigned to one of three treatments (T1, T2 and a control, (CT)). The experiment surveyed the public's knowledge of, and attitudes towards industry and hen welfare using a mix of open-ended questions, Likert scale responses and true/false statements. Participants were surveyed before and re-surveyed again, after an educational video intervention. T1 and T2 were shown a video containing objective facts about industry practices and welfare pros and cons of conventional cage and free-range systems before they were introduced to a compromise, i.e. the 'furnished cage' (T1) or the 'furnished coop' (T2). The Control video contained general information about chickens and no information about industry or welfare. The post-video survey then re-tested participant's knowledge of the egg-laying industry and included a question about their support for furnished cages (T1+ 50% of CT) or furnished coops (T2 + 50% of CT). Preliminary results were analysed using Chi square comparisons. The proportion of correct true/false responses indicated no difference between treatments pre-video ($X^2$=1.22, df 4, P=0.87) but there was a significant difference post-video ($X^2$=260.39, df 4, P<0.001). This difference is explained by the 38% increase in correct responses in both T1 and T2 compared to 4% increase in CT. Post-video support for furnished systems differed between treatment groups ($X^2$=57.32, df 6, P<0.001) with 45% of T1, 46% of T2 and 21% of CT indicating they would support furnished systems. Support for furnished systems did not differ in relation to language ($X^2$=1.47, df 3, P=0.69). This study suggests the rhetoric of 'cage' may not be as pervasive as previously thought and we found no evidence that emotive language impacts education or support for alternative housing systems.

## A food and agriculture course to raise awareness of animal welfare in university students majoring in pre-school education

*Yuki Koba[1], Aira Seo[2] and Hajime Tanida[2]*

*[1]Teikyo University of Science, Faculty of Education & Human Sciences, 2-2-1, Senjusakuragi, Adachi-ku, Tokyo, 1200045, Japan, [2]Hiroshima University, Graduate School of Integrated Sciences for Life, 1-4-4, Kagamiyama, Higashi-Hiroshima, 7398528, Japan; yukikoba@ntu.ac.jp*

Food products derived from farm animals are popular in the Japanese diet, but awareness of animal welfare in Japan is relatively low. This stems from the fact that consumers' daily life is alienated from the production sites for food animals. Some kindergarten children, who have had no opportunity to visit livestock farms, tend to believe that milk and meat are synthetically produced. Childhood experience with farm animals would heighten their awareness of living creatures and would spur them to think about the humane treatment of animals used for various purposes. Kindergarten teachers should play a key role in animal-related awareness and education for children; however, most teachers have no educational background in animal science. Our university farm, which houses mainly dairy cows, started a course on food and agriculture education for university students majoring in pre-school education (PE) in 2015. This course, the first of its kind in Japan, is a residential training program spanning four days. The purpose of the course is to offer students the opportunity to learn by experience (feeding, milking, handling, etc.) how livestock animals are reared. The objective of the present study was to clarify the impact of the course on the awareness of animal welfare among PE students. The study targeted 38 PE students from four universities who participated in the course in summer, 2018. Thirty-four agricultural (AG) students from six universities, who should have clear opinions toward animal welfare and who participated in a similar course, were chosen for comparison. We gave the students a farm animal quiz and a questionnaire on awareness of animal welfare before and after the course. Before the course, the students also took the Yatabe-Guilford personality inventory, which is one of the most popular psychological tests in Japan. The data were analyzed with nonparametric statistics. The quiz score of the PE students significantly ($P<0.05$) improved upon the completion of the course, but the score was significantly ($P<0.01$) lower than that of AG students. Awareness of animal welfare in PE students and AG students differed. Both before and after the course, PE students were significantly ($P<0.05$) more compassionate to animals than were AG students, who had a more positive attitude to animal use for humans. The awareness of PE students changed after the course. They became significantly ($P<0.02$) more positive to using animals for food or clothing but became significantly ($P<0.001$) more negative to 'raising farm animals to be more profitable because such animals get little stimulation and have low desire to be active.' Meanwhile, the awareness of AG students was unchanged by the course. As awareness of animal welfare and the personality of PE students showed significant correlations ($P<0.05$), incorporating individual learning support system would improve the effectiveness of the course. In conclusion, this course might help equip future teachers to improve the awareness of pre-school children about animal welfare.

## New perspectives for assessing the valence of social interactions

*Irene Camerlink*
*University of Veterinary Medicine Vienna, Institute of Animal Welfare Science, Veterinarplatz 1,*
*1210 Vienna, Austria; irene.camerlink@vetmeduni.ac.at*

Ethology has long underestimated the paramount importance of the social base that functions as a container out of which all other behaviours arise. In humans it is well known that having a social base (family and friends) has a positive impact on health, well-being and longevity, but also that relationship quality is amongst others related to problem behaviour and emotional stability. In animals, social relationships have hardly been acknowledged, and even less so investigated. Primate studies show that, in line with the 'relationship quality' hypothesis, groups with strong ties show fewer and less intense agonistic behaviour and more reconciliation after conflicts. Animal husbandry systems generally inhibit the formation of long-term relationships through early life separation of dam and offspring and by frequent regrouping. To provide animals a 'life worth living' it is therefore needed to focus on the individual within the context of its social structure. To do so it is important to study social interactions not only in terms of quantity but also in quality. For example, little is known about the importance of behaviours performed jointly, such as sleeping in close proximity or behavioural synchrony. Thus, not only 'what' the animal is doing, but also 'how' the animal is doing this and with whom. Overall, the quality of interactions may have a major role in the direction and magnitude of valence (i.e. positive / negative emotional state) and arousal (i.e. high / low activity) in any behavioural expression. Current methods are largely unable to detect the subtle differences in physiology and emotional state that happen before, during, or as a consequence of an interaction. What we have been researching are predominantly macro behaviours; well defined behaviours of substantial and measurable duration. Instead, ethograms could be refined to include micro behaviours: behaviours, gestures or signals that typically last only a few seconds. These occur frequently and may, in combination with other techniques, be a valuable method to further our understanding of the valence of social interactions. Moreover, with the increasing use of precision technology, new possibilities arise for non-invasive measures of greater quantity and quality. Methods, some based on human studies, are gradually explored in animal science and include amongst others infrared thermography to detect emotional response, social network analysis to assess social structure, body demeanour for an overall qualitative behavioural assessment, and facial expression to obtain information on signals and emotional state. To study quality of life in animals, collaboration on the development of techniques and the publication of methodological standards is needed for the progress and soundness of research on positive welfare in group-living animals.

# Ranging patterns and social associations in laying hens

*Yamenah Gómez[1], John Berezowski[2], Filipe Maximiano Sousa[2], Sabine Gabriele Gebhardt-Henrich[1], Michael Jeffrey Toscano[1], Ariane Stratmann[1], Sabine Vögeli[1] and Bernhard Völkl[3]*
*[1]Center for Proper Housing: Poultry and Rabbits (ZTHZ), Division of Animal Welfare, VPH Institute, University of Bern, Burgerweg 22, 3052 Zollikofen, Switzerland, [2]Veterinary Public Health Institute, University of Bern, Schwarzenburgstrasse 155, 3097 Liebefeld, Switzerland, [3]Division of Animal Welfare, VPH Institute, University of Bern, Länggassstrasse 120, 3012 Bern, Switzerland; yamenah.gomez@vetsuisse.unibe.ch*

In an exploratory study, the ranging pattern of laying hens in a semi-commercial system was recorded using an RFID tracking system. We monitored the location of 421 hens within four different areas (indoor, wintergarden, yard and free-range) of a commercial laying hen house sub-divided into four pens containing 355 hens/pen. The monitoring was conducted continuously for a period of four months (167 days) with 72 days of free access to all four areas. Our analysis and methods focused exclusively on the 72 days of free access. The two aims of the study were first, to characterize spatio-temporal ranging patterns based on movement between areas, and second, to identify non-random social associations based on simultaneous co-occurrence at the gates and pop-holes. To group hens based on the level of similarity of hen-individual ranging patterns, dissimilarity matrices were generated by Dynamic Time Warping (DTW), a method of time series analysis. In a second step, hierarchical clustering from the DTW dissimilarities between time series and the corresponding dendrograms were used for visual grouping of hens. For identification of social associations at the gates, hens passing between areas within five seconds of each other were identified for each day and a social network developed to investigate the social association between hens. There seem to be 4-5 groups of hens defined by distinct ranging pattern, depending on the pen. Two distinct groups were found in all four pens. These two groups were correctly classified between 69 and 87% (Linear Discriminant Analysis). Hens of one of the two groups showed a low number of transitions and spent a large amount of time indoors, while those of the other group had a high number of transitions, spent little time indoors, and more time in the yard and/or free range. The remaining groups spent either a lot of time indoors and in the wintergarden or yard and differed on pen-level. Furthermore, the median of the DTW distances of within-hens between-day comparisons was 1.53 to 1.84 times smaller than the median of the between-hens across day comparisons (Wilcox Signed-Ranks Test: all $V<8$; all $P<0.0001$). This means that each hen's ranging pattern was consistent over time and varied more between hens than within hens. Moreover, the average between-hen DTW distance per day decreased slightly with time. Based on the social network analysis, hens that were closely associated by their movement patterns between areas became more similar over time. Cluster-specific ranging patterns (= hens with similar time series cluster together) correlated with social associations (Pearson's product moment-correlation with 1000 permutations: all $r<-0.25$). We believe this is the first study that has detailed hen-individual ranging patterns with social association between hens.

## Socio-positive interactions in goats: prevalence and social network

*Claire Toinon, Susanne Waiblinger and Jean-Loup Rault*
*Institute of Animal Welfare Science, University of Veterinary Medicine Vienna, Veterinaerplatz 1, 1210 Vienna, Austria; claire.toinon@vetmeduni.ac.at*

Little is known about socio-positive behaviour in goats. Our goal was to assess the prevalence and the distribution of socio-positive behaviours between individuals. A stable herd of 15 horned dairy goats was studied. Goats had free access to an outdoor run connected to an indoor deep litter pen. One observer recorded the presence or absence of each goat in the field of view through five min scan sampling. Simultaneously, she performed continuous behaviour sampling during 25 min recording sessions, alternating between indoor and outdoor, with 15 sessions per day during week one (three days), week two (two days), week three (three days) and week six (two days), hence for a total of 62.5 h of observation with 40 days in between the first and the last observations. The frequency of five socio-positive behaviours was recorded: passive touching, rubbing, grooming, licking, social play; and three socio-negative or neutral behaviours: agonistic with and without physical contact, and third-party interventions in an agonistic interaction. For each socio-positive interaction, the identities of the donor and receiver and their posture (standing or lying) were recorded. For each agonistic interaction, the loser and winner were recorded to calculate the index of success (IS) of each goat (IS = number of interactions won/number of agonistic interactions). Social network analyses were performed to study edges, defined as relationships between individuals. On average, each goat was observed for 52.3±6.5 h. During this time 2,387 interactions were recorded: 39% were passive touching, 21% agonistic behaviours with physical contact, 13% social play, 12% agonistic behaviour without physical contact, 11% rubbing and other social behaviours accounted for the remaining 4%. The social network based on socio-positive interactions showed that six dyads interacted particularly frequently (182 interactions/dyad vs 7 for other dyads, $\chi^2$<0.001). These preferential associations were consistent whether the goats were observed inside or outside and over the 40-day study length. Furthermore, social network analysis revealed that goats were more selective regarding individuals for rubbing than for passive touching (respectively 31 and 67 edges, $\chi^2$<0.001). Goats also tended to be more selective about the peers they touched while lying than while standing (36 vs 54 edges, respectively, $\chi^2$=0.058). Therefore, goats were more selective for some types of socio-positive behaviours over others. Interestingly, the socio-positive network was distributed completely differently from the socio-negative network. Hence, goats interacted with different individuals depending on the valence of the interaction. As four of the dyads that interacted the most included one high ranking (IS>0.65) and one submissive individual (IS<0.33), socio-positive interactions do not seem related to hierarchy. Moreover, because touching while lying was frequently observed and selective, it would be a good indicator of the socio-positive network.

## Play fighting social network position does not predict injuries from later aggression between pigs

*Simon P. Turner[1], Jennifer E. Weller[2], Irene Camerlink[3], Gareth Arnott[2], Taegyu Choi[4], Andrea Doeschl-Wilson[4], Marianne Farish[1] and Simone Foister[1]*

*[1]SRUC, Edinburgh, EH25 9RG, United Kingdom, [2]Queen's University, Belfast, BT9 7BL, United Kingdom, [3]University for Veterinary Medicine Vienna, Vienna, 1210, Austria, [4]University of Edinburgh, Edinburgh, EH25 9RG, United Kingdom; simon.turner@sruc.ac.uk*

Play fighting is common in young mammals. Proposed functional benefits include enhanced motor, social and cognitive development whereby, in some species, improved physical skill and opponent assessment are expected. This could reduce social conflict later in life and may contribute to improved welfare in managed animals. Using pigs as a model, we examined whether variation in play fighting involvement before weaning was predictive of the number of injuries received from aggressive interactions later in life. The work was approved by the UK government. Piglets (n=239) experienced either pre-weaning socialisation (n=135; joining two neighbouring litters at 2 weeks old; 'socialised') or remained in their litter group ('control'; n=104). Sows were kept in farrowing crates. Play fighting interactions were extracted from video between the ages of 14-26 d. Social network analysis quantified the centrality position of pigs in the play fighting network of their pen. Centrality (degree centrality, Eigenvector, betweenness, clustering coefficient) in play fighting networks quantified metrics such as a pig's tendency to select play partners who themselves had experienced play with many partners. The effect of social network parameters on the number of skin lesions at 11 weeks of age was assessed 24 h after pigs were mixed into new groups (n=8 socialised groups; 94 pigs; 5 control groups; 63 pigs) and at 3 weeks post-mixing (14 weeks of age). Prior experience of other tests which may have led to variation in fighting experience did not affect the test outcome and was omitted from subsequent analysis. Assortment analysis showed that in 2 out of 6 socialised groups, piglets positively assorted by litter (P<0.001), indicating a strong preference to play with their own littermates. At the pen level, play network structures were similar in socialised and control groups. Large variation existed in play fighting engagement and individual network position between pigs (e.g. 1-19 play partners; median clustering coefficient 0.78, range 0.19-1.00 which is expressed on a 0 to 1 scale). Mixed model analysis showed that pigs with many play partners had more lesions at 14 ($F_{1,46}$=4.11; P=0.05) but not 11 wk. Centrality metrics that described the animal's position within its play network (e.g. whether it preferentially played with experienced play partners) did not affect skin lesions ($F_{1,88}$=0.06-1.76, P=0.18-0.95). These results suggest that play fighting does not reduce costs of fights in immature pigs. Future work should explore if it promotes social cohesion or reduces conflict between adults where fitness costs of aggression may be higher.

## Quantifying play contagion: low playing calves may depress play behaviour of pen-mates

*Verena Größbacher[1], Katarína Bučková[2,3], Alistair Lawrence[4], Marek Špinka[2,3] and Christoph Winckler[1]*

*[1]University of Natural Resources and Life Sciences, BOKU, Department of Sustainable Agricultural Systems, Gregor-Mendel-Straße 33, 1180 Vienna, Austria, [2]Institute of Animal Science, Department of Ethology, Přátelství 815, 104 00 Prague, Czech Republic, [3]Czech University of Life Sciences, Department of Ethology and Companion Animal Science, Kamýcká 129, 165 00 Prague, Czech Republic, [4]SRUC, Animal Behaviour and Welfare, West Mains Road, EH9 3JG Edinburgh, United Kingdom; verena.groessbacher@boku.ac.at*

Emotional contagion is the transfer of an emotional state from one animal to another. Play behaviour has been hypothesized to act as vector of emotional contagion with one playing animal generating and amplifying play activity in conspecifics. This would create the potential to induce positive behaviour and emotions and improve the well-being of group-living animals. The current study aimed to quantify the strength of play contagion in dairy calves. We modelled low and high play levels in response to different milk amounts and recorded play in low-milk-calves cohabitating high-milk-calves. Holstein Frisian dairy calves (n=66) were housed in groups of three and submitted to one of three treatments: In the Negative control (N6, n=5 groups) and Positive control (P12, n=5), all three calves received either 6 L or 12 L of milk/day. In the Contagion group (n=12), two calves received 12 L of milk/day (C12) and one calf received 6 L of milk/day (C6). Behaviour was assessed using accelerometers sampling at 1 Hz on the vertical axis only from 05.00 to 23.00 on two consecutive days at both four and eight weeks of age. Sets of ten accelerometer readings, representing periods of 10 s, were combined and classified into lying, standing or locomotor play using quadratic discriminant analysis. Durations of locomotor play were then calculated as minutes per 18 hours. Volume of residual milk was recorded per meal (RESID) and temperature inside the barn was recorded at 12.00 and 18.00 (TEMP). Calf health was scored weekly for presence of diarrhoeic faeces and coughing (HEALTH). Data were analysed using a mixed model (fixed effects: treatment, week, RESID, TEMP, HEALTH; random effects: day nested in week, calf and group; date). Locomotor play decreased with higher RESID (estimate=-0.09; $F_{1,196}$=8.18; P<0.01), with higher TEMP (estimate=-0.12; $F_{1,196}$=7.18; P<0.01) and when calves showed signs of impaired HEALTH (estimate=-2.18; $F_{1,196}$=11.75; P>0.01), thus indicating the sensitivity of play as a welfare indicator. Play was not affected by age ($F_{1,196}$=1.26; P=0.26) but by treatment ($F_{3,196}$=7.56; P<0.01). P12-calves performed more locomotor play than N6-calves (Mean ± SD=16.6±5.9 vs 12.0±4.2; estimate=3.9). However, play of C6-calves (11.7±3.7; estimate=-0.4) and C12-calves (13.1±5.2; estimate=1.08) was not different from play of N6-calves. Contrary to expectation, play was neither increased in C6-calves nor in C12-calves in contrast to the Negative control. One interpretation is that the contagion effect was reversed relative to our expectations due to the presence of the low-milk-calf depressing play behaviour of its high-milk pen-mates.

## Salivary oxytocin is associated with ewe-lamb contact but not suckling in lactating ewes

*Elinor Muir[1], Joanne Donbavand[2] and Cathy Dwyer[1,2]*
*[1]Royal (Dick) School of Veterinary Studies, University of Edinburgh, Easter Bush Campus, EH25 9RE, Edinburgh, United Kingdom, [2]SRUC, Animal Behaviour and Welfare, Roslin Institute Building, Easter Bush Campus, Edinburgh, EH25 9RE, United Kingdom; cathy.dwyer@sruc.ac.uk*

Maternal care in sheep is dependent on a strong selective ewe-lamb bond. The formation of this bond is facilitated by release of central oxytocin at parturition, and the immediate onset of affiliative maternal behaviour in the ewe such as licking and grooming the neonate, close contact, low-pitch vocalisations and suckling activity. This study assessed whether affiliative behaviour and oxytocin release is also associated with maternal care expressed later in the lactation period. The behaviour of 15 ewes (7 primiparous and 8 multiparous) and their 6 week-old twin lambs was measured by: (1) focal animal observation during a 3-minute separation-reunion test; and (2) scan sampling for 12 days of undisturbed behaviour at pasture. Pre and post-test saliva samples were collected from ewes and analysed for oxytocin using an ELISA. Data were compared by paired t-tests, Mann Whitney tests and Spearman's Rank Correlation. There was a significant increase in oxytocin concentration in the ewes after the separation-reunion test compared to before (mean increase=43.47 pg/ml, paired-t=-2.51, P=0.02). Separation from the lambs induced high frequencies of distress behaviours in the ewes and lambs (e.g. high pitched vocalisations P<0.001), and a motivation to reunite with the lambs. Reunion promoted close ewe-lamb contact, nosing and suckling behaviour (P=0.04). Oxytocin concentration was positively associated with duration of close contact in the separation-reunion test (rs=0.767, P<0.001) and ewe-lamb spatial proximity in the field (rs=0.596, P=0.01), but not with frequency of suckling (rs=0.330, P>0.05). There were no significant effects of parity on affiliative behaviours expressed in the test, behaviours in the field or salivary oxytocin concentration. However primiparous ewe exhibited behaviours indicative of greater distress and anxiety in the separation phase compared to multiparous ewes (activity during separation: T=3.40, P=0.005), and travelled further to reunite with their lambs on release (median distance (with IQR): Primiparous=7.0 m (4.9), Multiparous=2.37 m (2.5), W=46, P=0.039). The results indicate that ewes maintain a motivation to be in close contact with their lambs during lactation, and this contact is associated with oxytocin release. Whether oxytocin is causal for the motivation to maintain close contact with lambs, or a consequence of the contact, is still to be elucidated.

## Maternal protection behavior in Hereford cows

*Franciely Costa[1], Tiago Valente[1,2], Mateus Paranhos Da Costa[1] and Marcia Del Campo[3]*
*[1]FCAV-UNESP, Grupo de Estudos e Pesquisas em Etologia e Ecologia Animal (ETCO), FCAV-UNESP, 14884-900, Jaboticabal, SP, Brazil, [2]University of Alberta, Livestock Gentec, 116 St & 85 Ave, Edmonton, AB T6G 2R3, Canadá, [3]Instituto Nacional de Investigación Agropecuaria, INIA, Programa Nacional de Carne y Lana, Ruta 5, km 386, 45000 Tacuarembo, Uruguay; mdelcampo@inia.org.uy*

Maternal protection behavior (MPB) has a paramount importance in extensive conditions, because of its effect on calf survival and workers safety. The Hereford breed is characterized by a genetically quiet temperament. However, there is not enough scientific information regarding the behavior of Hereford cows after the calf birth. The aim of this study was to characterize the maternal protection behavior (MPB) variability of Hereford cows and its effect on calf performance. The experiment was carried out at the National Institute of Agricultural Research, INIA Tacuarembó, Uruguay. The animals were kept in an extensive rearing system, under grazing. MPB was determined on 145 cows around four days after calving, while the newborn calf weighing procedure was being performed. MPB was assessed by observing the cow reaction towards the handlers when working with its offspring, using a 6-score scale, ranging from 1 (cow tries to escape) to 6 (cow tries to attack handlers). Calves were weighed 4 days after birth and on day 45. To test the effect of MPB on calves' average daily gain (ADG), the PROC GLIMMIX of SAS was used. MPB, cow body condition score, cow category (primiparous or multiparous) and the calf sex were considered as fixed effects. The model included cow age at calving and calf birth weight as covariables. Extreme MPB scores were not registered, with no cow showing abandonment behavior towards the calf (MPB1 = try to escape), or aggressive behavior towards the handlers (MPB5 = threating or MPB6 = attacking). The results showed that 35.2% of the evaluated cows were distant and indifferent to the calf management (MPB2), 42.7% remained distant but attentive to the calf management (MPB3) and 22.1% were close and attentive to the calf management (MPB4). MPB was not associated with calf ADG (F=0.78; P=0.46). In short, 100% of the evaluated Hereford cows exhibited a desirable score of MPB (2, 3 or 4).

### Do stocking density or a barrier affect calving location and labor length in group-housed Holstein dairy cows?

*Katherine Creutzinger[1], Heather Dann[2], Peter Krawczel[3] and Kathryn Proudfoot[1]*
*[1]Ohio State University, 1920 Coffey Rd, Columbus, OH 43210, USA, [2]William H. Miner Agricultural Research Institute, 1034 Miner Farm Road, P.O. Box 90, Chazy, NY 12921, USA, [3]University of Tennessee, 258 Brehm Animal Science Building, Knoxville, TN 37996, USA; creutzinger.1@osu.edu*

Understanding maternal behavior can aid in the design of suitable maternity areas for dairy cows. Cows seek seclusion at calving when kept on pasture or individual pens. However, cows are often housed in group pens of variable stocking densities before calving without space for seclusion. Our objectives were to determine the impact of two stocking densities and the provision of a barrier to allow seclusion on the choice of calving location and the length of stage II labor. We predicted that cows will use a barrier to give birth regardless of stocking density, and will have lower labor length when housed at low stocking density with a barrier. Cows were cared for according to a protocol approved by an Animal Care and Use Committee at the Miner Institute. Holstein dairy cows with unassisted calvings (n=211) were included in the analysis. Pens were dynamic as cows entered 21±3 days before their expected calving date and were removed after calving. At enrollment, cows were assigned randomly to one of four treatments using a 2×2 factorial arrangement: (1) high stocking density without a barrier (9.7 to 12.9 $m^2$ per cow; 'non-barrier'); (2) high stocking density with a barrier ('barrier'); (3) low stocking density without a barrier (19.4 to 25.8 $m^2$ per cow); and (4) low stocking density with a barrier. The 'barrier' was created using plastic road barriers and plywood (3.6×0.6×1.5 m). Treatment pens were created using gates separating four areas within a sawdust bedded pack. Pens were replicated four times so that all treatments were in all positions in the larger bedded pack. Video data were used to determine the duration of stage II labor (from the first abdominal contraction to the calf's hips fully expelled) and if cows calved in the area containing the barrier or in the same relative area in non-barrier pens. Treatment differences were determined using PROC GLM (SAS 9.4) including fixed effects of barrier, stocking density, and their interaction. Pen was the experimental unit (n=16). A total of 50% of cows calved in the area containing a barrier or the same relative area in non-barrier pens. A higher percentage of cows calved next to a barrier in barrier pens compared to those that calved in the same relative area in non-barrier pens (37 vs 13%; P=0.002). There was no effect of stocking density (P=0.8), however, there was a tendency for interaction between barrier and stocking density (P=0.06); cows were more likely to calve next to a barrier in low stocking density pens (42 vs 6%; P=0.001) but not in high stocking density pens (31 vs 20%; P=0.21). There was no impact of stocking density on the duration of stage II labor (79 vs 80±4 min; P=0.93), but there was a tendency for cows in barrier pens to have shorter labor compared to those in non-barrier pens (74 vs 83±4 min; P=0.08). Group-housed cows preferred to calve next to a barrier when one was provided, but limited space allowance may limit this choice. Providing a barrier in a group pen has the potential to reduce the duration of stage II labor.

## Sociability is associated with pre-weaning feeding behaviour and growth in Norwegian Red calves

*Laura Whalin[1], Heather W. Neave[1], Kristian Ellingsen-Dalskau[2], Cecilie Mejdell[2] and Julie Føske Johnsen[2]*
*[1]University of British Columbia, Animal Welfare Program, 2357 Main Mall, Vancouver, BC V6T 1Z4, Canada, [2]Norwegian Veterinary Institute, Ullevålsveien 68, P.O. Box 750 Sentrum, 0106 Oslo, Norway; laura.whalin@alumni.ubc.ca*

Pre-weaning feed intake and growth of calves are highly variable. Personality has been suggested as a possible influence, but the focus is often on fearfulness. Other personality traits such as sociability may also influence pre-weaning feeding behaviour, perhaps because social facilitation encourages increased intakes. The aim of our study was to describe the relationship between personality traits, feeding behaviour, and growth in pre-weaned Norwegian Red dairy calves. Twenty group housed calves (3-5 calves/group, 6 groups) had access to automatic milk and concentrate feeders, and *ad libitum* access to water, hay, and silage. For the first 30 d all calves were offered 12 L of milk/d, after which milk was reduced by 25%. This study measured milk and concentrate intakes, drinking speed, visits to the milk feeder, and average daily gain until d 42 of age. Each calf was tested in three standardized personality tests at d 21: a novel environment test (10 min), a novel object test (10 min with a 10 L bucket), and a social motivation test (time taken to return to the group following separation). A Principal Component Analysis revealed three factors that explained 73% of the variance in behaviours expressed during these tests. Factor 1 (labeled as 'Playful/Exploratory', explaining 39%) had high positive loadings for activity (number of quadrants crossed), locomotor play, and exploration, and a high negative loading for time spent standing inactive. Factor 2 (labeled as 'Social', explaining 20%) had a high positive loading for number of vocalizations, and high negative loadings for the time taken to return to the group and time spent looking at the novel object. Factor 3 (labeled as 'Fearfulness', explaining 13%) had a high positive loading for the latency (time taken) to touch the object and a high negative loading (less time spent) for touching the object. Using a mixed multiple regression analysis, calf scores on each of the three factors (continuous variables) were tested for their effects on feeding behaviours and growth, with birth weight and sex as covariates and group as a random effect. Calves scoring highly on the 'Social' factor drank milk at a faster speed (P=0.02), had more unrewarded visits to the milk feeder (P<0.01), and had higher average daily gains (P<0.05); these calves also tended to consume more concentrate (P=0.08), perhaps due to social facilitation. Calves scoring highly on the 'Playful/Exploratory' factor tended to have fewer unrewarded visits to the milk feeder (P=0.09), and lower average daily gains (P=0.10), perhaps due to greater energy expenditure on activity. We found no associations between the 'Fearfulness' factor and pre-weaning feeding behaviour or average daily gain. These results suggest that personality, particularly sociability, is associated with feeding behaviour and may have impacts on growth.

# Effect of cow-calf contact on motivation of dairy cows to access their calf

*Margret L. Wenker[1,2], Marina A.G. Von Keyserlingk[1], Eddie A.M. Bokkers[2], Benjamin Lecorps[1], Cornelis G. Van Reenen[3], Cynthia M. Verwer[4] and Daniel M. Weary[1]*
*[1]University of British Columbia, Animal Welfare Program, 2357 Main Mall, BC V6T 1Z6 Vancouver, Canada, [2]Wageningen University and Research, Animal Production Systems group, De Elst 1, 6708 WD Wageningen, the Netherlands, [3]Wageningen Livestock Research, De Elst 1, 6708 WD Wageningen, the Netherlands, [4]Louis Bolk Institute, Kosterijland 3-5, 3981 AJ Bunnik, the Netherlands; margret.wenker@wur.nl*

It is common practice to separate dairy cows and calves within a few hours after parturition. When cow-calf contact is allowed a bond develops, even in absence of suckling, but little is known about how suckling affects the strength of this bond. The aim of this study was to assess the motivation of dairy cows with different levels of cow-calf contact to access their calf. We hypothesised that cows directly separated from their calf would work less hard to get access to their calf compared to non-suckled and suckled cows routinely kept with their calf. Thirty-four Holstein Friesian cows were randomly assigned to one of three treatments: (1) separated from their calf within 2 hours postpartum (n=11); (2) allowed to spend nights with their calf but fitted with an udder net to prevent any suckling (n=11); or (3) allowed to spend nights with their calf and the calf was able to suckle (n=12). Nightly cow-calf contact lasted from approximately 18:30 h until 06:30 h. Cows were trained to push a weighted gate to access their calf. Testing was undertaken once daily immediately following the afternoon milking at approximately 18:00 h. The weight on the gate was increased by 9 kg each day until the cow failed to push open the gate or reached the maximum weight of 90 kg. If a cow was unsuccessful in pushing the weighted gate, she was retested at that same weight for 2 additional days. Kaplan–Meier survival curves were used to analyse the maximum weight pushed by the cows in each treatment group. Video was used to record latency to make nose contact with the calf and duration of licking the calf during the test. Behaviour was analysed using a mixed model accounting for repeated measures. Separated cows worked as hard (median=18 kg) as cows with an udder net (median=27 kg, $\chi2=0.98$, P=0.32) to access their calf, but worked less hard than suckled cows (median 45 kg, $\chi2=6.30$, P=0.01). The latency to make nose contact with the calf was not different between separated cows (mean±SE, 29±16 s) and udder net cows (30±9 s, F=0.00, P=0.97) or suckled cows (20±6 s, F=0.53, P=0.47). Once reunited, separated cows licked their calf (36±10 s) more than suckled cows (15±2 s, F=7.56, P=0.01), but did not differ from the udder net group (21±7 s, F=2.90, P=0.10). These results indicate that dairy cows' motivation to access their calves is greatest when cows are suckled, and that cows are still motivated to access their calf even when fitted with an udder net or separated at birth. The greater motivation of suckled cows to access their calf could be explained as a stronger bond between cow and calf rather than urge for relief of udder pressure, as all cows were tested right after milking. Overall, the calf is a valuable resource that dairy cows are willing to work for.

## Vomeronasal organ alterations and their effects on social behaviour: results and perspectives after 10 years of research

*Pietro Asproni, Alessandro Cozzi, Cécile Bienboire-Frosini, Violaine Méchin, Céline Lafont-Lecuelle and Patrick Pageat*
*IRSEA, Research Institute in Semiochemistry and Applied Ethology, Quartier Salignan, 84400 Apt, France; p.asproni@group-irsea.com*

The vomeronasal organ (VNO) plays a key role in animal behaviour since it is responsible for intra- and interspecific chemical communication. Despite this well-known role, until recently, no VNO spontaneous pathologies were reported in the literature, as well as their effects on animal behaviour. We present our research focused on VNO spontaneous changes and their association with behavioural disorders. From 2009 to 2019, we studied by histopathological analysis VNOs from cat, pig, mouse, rabbit, sheep and dog. The studies were conducted according to ISAE ethical guidelines, IRSEA Ethics Committee and French law. In a study conducted on 20 owned cats, we observed that 70% of the cats were affected by chronic inflammation of the VNO. Moreover, the chi-square test revealed that, when affecting the vomeronasal sensory epithelium, this condition was associated with the presence of intraspecific aggression (P=0.038). In a study on 38 farm pigs sampled at the slaughterhouse, we observed that chronic vomeronasalitis was associated with the number of skin lesions, expressed as linear values and collected according to the Welfare Quality® protocol (P<0.001, Wilcoxon two-sample test). In a second study (34 pigs), we observed that this condition was also associated with the presence of tail biting (P=0.002, $\chi2$ test). More precisely, the post-hoc comparison revealed that pigs affected by vomeronasalitis were more frequently victim (P=0.0208) or biter (P=0.011) than neutral. Interestingly, aggression was enhanced in pigs presenting more intense inflammation and in those affected in both the VNOs. Immunohistochemical and morphometrical analyses showed that an inflamed VNO has fewer pheromone receptor neurons than normal (420 vs 802 neurons/1 mm$^2$, respectively), confirming that an inflamed VNO cannot assure normal chemical communication among conspecifics, contributing to the presence of aggressive behaviours. Other preliminary studies demonstrated that vomeronasalitis can affect rabbit, sheep and dog. Recently, we focused our research on VNO alterations other than inflammatory changes. Analysing by histology 20 two-year-old mice (10 RjOrl:SWISS and 10 C57BL/6JRj), we observed that 100% of the VNOs presented signs of degenerative changes compatible with the ageing of the sensorial epithelium, such as vacuolar degeneration, nuclear swelling and chromatin margination, affecting sensory neurons. These preliminary data suggest that also the VNO can be subjected to pathological changes due to the ageing process, as other sensory organs. In conclusion, the VNO can be affected by pathological changes that can alter its functionality impacting animal behaviour and welfare. As the VNO is crucial for various behaviours, its alterations may also influence behaviours other than social relations, such as maternal, sexual and predator-prey interactions. Future studies will aim to investigate these perspectives in various species.

## Sweaty secrets: plantar gland secretions influence male mouse social behavior in the home cage

*Amanda Barabas[1], Helena Soini[2], Milos Novotny[2] and Brianna N. Gaskill[1]*
*[1]Purdue Univ., Animal Science, 270 S. Russell St, West Lafayette, IN 47907, USA, [2]Indiana Univ., 800 E. Kirkwood Ave, Bloomington, IN 47405, USA; abarabas@purdue.edu*

Severe wounding, resulting from excessive home cage aggression, is a common reason for premature euthanasia in male laboratory mice. Aggression can be reduced by transferring the old nest during cage cleaning which is thought to contain aggression appeasing odors from the plantar sweat glands. However, neither the deposits on used nesting material nor the contents of plantar sweat have been evaluated. The aims of this study were to determine if (1) strain (and known aggression levels) influences nest and sweat odor profiles and (2) if odor profiles correlate with social behavior. Home cage social behavior was recorded for 7 days from the following 3 strains: SJL (high aggression), C57BL/6 (moderate aggression), and A/J (low aggression; n=8; 24 cages×5 mice/cage = 120 mice). Escalated aggression, allo-grooming and group sleep were recorded using 1,0 sampling every 5 min (00:00-12:00) on days 1, 2, and 7. Welfare checks for severe wounding were conducted daily and if any mice met our humane endpoint criteria (wound 1 cm$^2$), they would be euthanized. However, none met this criteria. Cage hierarchy was assessed via the tube test on days 5 & 6. Sweat from 2 mice per cage (dominant and subordinate) and nest samples were collected on day 7 for volatile organic compound (VOC) analysis using stir bar sorptive extraction and gas chromatography-mass spectroscopy. Nest (N-) and sweat (S-) VOC proportion data were analyzed with separate principal component analyses. Each component (PC) represents a set of VOCs with loading values greater than 0.4. For aim 1, significant nest and sweat PCs were analyzed in a GLM with strain as an independent variable. For aim 2, significant PCs were run in a stepwise regression to determine the best model for each behavior. Models with the lowest AIC value were analyzed using a GLM. The strain significantly affected two nest components (N-PC1: $F_{2,18}$=5.71; P=0.012 and N-PC3: $F_{2,18}$=5.13; P=0.017) and one sweat component (S-PC1: $F_{2,18}$=22.88; P<0.01). Several social behaviors were influenced by sweat and nest VOCs. Cages with high S-PC1 scores displayed low escalated aggression ($F_{1,19}$=6.68; P=0.02), high allo-grooming ($F_{1,20}$=17.99; P<0.01) and high amounts of group sleep ($F_{1,21}$=13.13; P<0.01). Similarly, for N-PC1 scores, cages with high values also engaged in low levels of escalated aggression ($F_{1,19}$=20.69; P<0.01) and high allo-grooming ($F_{1,20}$=6.41; P=0.02). Cages with high N-PC3 scores engaged in more aggression ($F_{1,19}$=6.42; P=0.02). Overall, C57BL/6 cages displayed the most allo-grooming ($F_{2,18}$=32.36; P<0.01), group sleep ($F_{2,18}$=9.56; P<0.01), and had high S-PC1 scores. SJLs displayed high overall aggression ($F_{2,18}$=117.30; P<0.01), had high N-PC3 scores, and had low N-PC1 scores. Previously, nest transfer has been shown to reduce aggression at cage change and this data shows that PCs representing VOCs in nesting material and sweat correlate with low escalated aggression and high affiliative behaviors. VOCs found in these significant PCs will be candidates for future work to determine their direct influence on mouse social behavior.

## Can live with 'em, can live without 'em: implications of pair vs single housing for the welfare of male C57BL/6J mice

*Luca Melotti[1], Niklas Kästner[1], Anna Katharina Eick[1], Anna Lisa Schnelle[1], Rupert Palme[2], Norbert Sachser[1], Sylvia Kaiser[1] and S. Helene Richter[1]*
*[1]University of Münster, Department of Behavioural Biology, Badestrasse 13, 48149 Münster, Germany, [2]University of Veterinary Medicine Vienna, Department of Biomedical Sciences, Veterinärplatz 1, 1210 Vienna, Austria; melotti@uni-muenster.de*

The basic question as to whether male laboratory mice should be singly or group housed represents a major animal welfare concern within current laboratory animal legislation and husbandry. To better understand the behavioural and physiological mechanisms underlying this issue, we conducted two longitudinal experiments using C57BL/6J mice. In the first experiment (n=32), we explored social behaviour of pair housed males from weaning to adulthood, over a period of nine weeks. Males were randomly assigned to pairs without knowing their litter of origin, in order to reflect standard lab practice, and cage enrichment consisted of a red plastic house and a gnawing stick. We took weekly measures of agonistic, socio-exploratory and affiliative behaviours within two different contexts, i.e. in the undisturbed home cage (one hour per week) and during the 15 minutes immediately after cage cleaning. In the second experiment (n=36), we investigated whether separation of male pairs into single housing at different ages (35, 56 or 77 days of age; 12 mice for each separation age) affected welfare-related measures such as faecal corticosterone metabolites (FCMs; collected six days before and one day after each separation event) and anxiety-like behaviours (assessed in elevated plus maze, novel cage, and open field tests between 83 and 87 days of age). All procedures complied with the EU regulations on animal experimentation and were approved by the federal authorities (LANUV). In the first experiment, levels of agonistic behaviour were low overall (no skin wounds observed); they were higher after cage cleaning than in the undisturbed cage as expected (rates per minute; Median=0.292, IQR=0.463 vs Median=0.030, IQR=0.058, respectively; Wilcoxon signed rank test, z=3.11, P=0.002), but did not significantly change with age in either context (Friedman´s tests; $\chi^2(8)=13.20$, P=0.10 and $\chi^2(8)=12.76$, P=0.12, respectively). Instead, affiliative behaviour increased with age in the undisturbed home cage (Friedman´s test, $\chi^2(8)=54.17$, P<0.001) but did not change over time after cage cleaning ($\chi^2(8)=10,05$, P=0.26). In the second experiment, age of social separation did not affect levels of FCMs (no significant interaction between age of separation and faecal sampling day; mixed design ANOVA: F(3.66,27.42)=0.86, P=0.49) or anxiety-like behaviours (one-way ANOVAs: all P>0.24). Taken together, this study shows that pair housed male mice can maintain low levels of aggression across a long period of their life and perform increasing levels of sociopositive behaviours which may serve to promote stable social relations. At the same time, our results suggest that male mice can quickly adapt to separation into single housing at different ages, from adolescence to adulthood. These findings are in line with the behavioural ecology of wild male mice, which suggests that both solitary and group living represent two alternative strategies.

## The effect of LPS and ketoprofen on social behaviour and stress physiology in group-housed pigs

*Christina Veit[1], Andrew M. Janczak[1], Birgit Ranheim[1], Anna Valros[2], Dale A. Sandercock[3] and Janicke Nordgreen[1]*

[1]*Norwegian University of Life Sciences, Faculty of Veterinary Medicine, Ullevålsveien 72, 0454 Oslo, Norway,* [2]*University of Helsinki, Production Animal Medicine, Yliopistonkatu 4, 00014 Helsinki, Finland,* [3]*Scotland's Rural College, Animal and Veterinary Science Research Group, Roslin Institute, Midlothian, EH25 9RG, United Kingdom; christina.maria.veit@nmbu.no*

Poor health is a risk factor for tail biting but the mechanism behind this link is still unclear. An injection with lipopolysaccharide (LPS) can be used to model aspects of sickness. Recent studies showed more tail- and ear- directed behaviour in LPS injected pigs compared to saline injected pigs. The aim of this study was to evaluate social behaviour and stress physiology of pigs after LPS injection [1.2 µg/kg], and to test the effect of a nonsteroidal anti-inflammatory drug (NSAID) on the effects of LPS. The experiment was approved by the national animal research authority (FOTS id 15232). Fifty-two female pigs (11-12 weeks) were randomly allocated to four treatments all comprising two injections: Saline-Saline (SS), Saline-LPS (SL), Ketoprofen-Saline (KS), Ketoprofen-LPS (KL). The first substance was administered intramuscularly (i.m.). The latter was administered intravenously (i.v.) on average 60±14 min afterwards in an ear vein during fixation with a mouth snare. Pigs were marked on the back for individual identification and video recorded. Activity was scan sampled every 5 min for 6 h after the last i.v. injection in the pen. Social behaviour was observed continuously in 10×15 min intervals between 8 am and 5 pm at baseline and one and two days after i.v. injection with focus on ear and tail exploration as well as manipulation of other body parts. Data were analysed by a mixed model to account for repeated measures and baseline behaviour. Saliva was sampled at baseline and at 4, 24, 48, 72 h after the i.v. injection and analysed for cortisol. Salivary cortisol was significantly higher in SL pigs (1,067.6±77.8 pg/ml) compared to KL pigs (641.4±77.8 pg/ml; student's t-test: P<0.001), SS pigs (418.0±84.6 pg/ml; P<0.001) and KS pigs (258.0±77.8 pg/ml; P<0.001) at 4 h post injection (F(treatment×time point)$_{12,182.8}$=4.65; P<0.001). SL pigs showed more sternal recumbence (296 counts) and were less active (52) than SS pigs (lying inactive sternal: 149 | active: 186, Student's t-test for both models respectively: P<0.001), KL pigs (174 | 217, P<0.001), and KS pigs (151 | 215, P<0.001) during 6 h post injection (lying inactive sternal F(treatment)$_{3,48}$=9.26; P<0.001; active F(treatment)$_{3,48}$=14.66; P<0.001). Treatment effects on social behaviour were not consistent. SL pigs performed longer ear exploration (sum per day: 1702.0 s) than SS pigs (465.9 s) two days after i.v. injection, (student's t-test: P=0.02; F(treatment×day)$_{6,88}$=1.49, P=0.19). They did not differ at baseline. The duration of tail exploration (F(treatment×day)$_{6,87.58}$=0.23, P=0.97) and manipulation of other body parts (F(treatment×day)$_{6,88}$=0.36, P=0.90) did not differ between treatments. LPS activated the HPA-axis (measured 4 h after injection) and elicited so-called sickness behaviour within 6 h after injection as indicated by lower activity in LPS injected pigs.

## Housing gilts in crates prior to mating compromises welfare and exacerbates sickness behavior in response to LPS challenge

*Luana Alves[1], Thiago Bernardino[1], Lays Verona Miranda[1], Gina Polo[2], Patricia Tatemoto[1], Sharacely De Souza Farias[1] and Adroaldo José Zanella[1]*
*[1]Center for Comparative Studies in Sustainability, Health and Welfare, University of São Paulo, Preventive Veterinary Medicine and Animal Health, VPS, Av. Duque de Caxias Norte 225, Jardim Elite, 13635-900 Pirassununga, SP, Brazil, [2]Research Consulting, Bogotá, Colombia; adroaldo.zanella@usp.br*

Previous studies have shown the impact of restrictive housing for pregnant sows in their welfare, for example supressing the immune system and making them more susceptible to disease. We hypothesized that restrictive housing prior to mating could exacerbates the impact of diseases on cortisol, behavior and rectal temperature. The aim of this study (reviewed and approved by the Ethics and Animal Use Committee – CEUA 902100816, and supported by the São Paulo Research Foundation FAPESP/CAPES grant 2017/03818-5) was to assess the consequences of housing gilts, previously synchronized (Regumate, MSD®), in crates (C) (n=14), group housing (GH) (n=14) and outdoor (O) (n=14) systems, on their responses to a disease simulation. During estrous, eighteen animals (n=6/housing system) were challenged intravenously with lipopolysaccharide (LPS – *E. coli* O111:B4, Sigma Aldrich®, 2μg/kg) whereas 24 animals (n=8/housing system) were given sterile saline solution. Data collected included twice daily salivary sampling (6:00 am and 6:00 pm), rectal temperature (right after inoculation – T0 until 7 hours after inoculation – T7) and behavioral observation (12 min/animal morning and afternoon) for five days prior and for four days after estrous. Salivary cortisol was measured prior LPS inoculation and four times post-inoculation. Data were analyzed with 'R' using Shapiro-Wilk test for residual normality and Bonferroni or Kruskal Wallis post-hoc test. Cortisol levels were higher in LPS challenged gilts than in saline injected animals, as in temperature assessment. LPS challenged gilts slept longer than saline animals (P<0.001), as well as lying ventrally (LV) (P=0.003). C housed gilts spent more time in LV than GH gilts (P<0.01) but did not differ from O animals. For lying laterally behavior (LL), O and C gilts were in LL longer in the 6:00am observation than during the 6:00pm observation (P=0.03). On the LPS challenge day, salivary cortisol levels were highest at the first measure for C gilts (P=0.03). We also found higher cortisol level in crate housed gilts inoculated with LPS (P=0.03) than saline. O and C animals appeared to respond faster to the challenge. Moreover, on T6 and T7 the recovery of basal cortisol levels in O and C gilts, were faster than GH animals. As to corroborate our results we observed low activity and high LV in LPS gilts, particularly when housed in crates, both measures are indicative of sickness behavior. Animals housed in C had higher cortisol levels than GH and O gilts since the beginning of the study showing that restrictive housing is a major challenge to animal welfare. However, GH gilts appeared to recover from the challenge slowly than others, demonstrating that barren group housing can be challenging as well. We showed that keeping animals in crates exacerbates the impact of LPS challenge on welfare indicators.

## Behaviour and sociability of yaks among different regions in Bhutan

*Nedup Dorji[1,2], Peter Groot Koerkamp[2], Marjolein Derks[2] and Eddie Bokkers[1]*
*[1]Wageningen University and Research, Animal Production Systems group, Wageningen, P.O. Box 338, 6700AH, Netherlands Antilles, [2]Wageningen University and Research, Farm Technology group, Wageningen, P.O. Box 16, 6700AA, Netherlands Antilles; nedup.dorji@wur.nl*

Highland pastoralists in Bhutan are highly depend on their yaks because these animals deliver food, fur and products to sell, but they are also used as pack animals and for draught power. Yak-based communities in Bhutan are rather isolated and practice seasonal migration in response to forage availability. Often they experience labour shortage to herd yaks, forage shortage in rangeland, poor extension and veterinary services and high yak mortality. Since traditions and beliefs of yak-based communities differ between regions, opinions and perceptions on how to manage yaks may also differ. Therefore, it was hypothesized that the welfare of yaks might vary among different yak farming regions. As a first step, we studied the behaviour and sociability of yaks in three regions (west, central and east) of Bhutan. Observation were done between 5:20-12:00 hr and 15:30-17:40 hr depending on the weather conditions and location of herds during grazing. Cows (354) and calves (272) of 51 herds (17 herds per region) were assessed for flight distance before milking. Also, behaviour of these herds and two additional herds was observed by counting the number of animals eating, lying, standing idle and walking during 5 min, thereafter for 15 min counting the number of events of agonistic behaviour (head butt, chasing, fighting), allo-grooming, rubbing and scratching, sexual behaviour, and playful behaviour. Both observations were repeated 6 times in a row (6×20 min). For each herd, the total number of events was summed for the different time periods, and the medians were compared between regions using the Kruskal-Wallis test. The test person could approach and touch 42.0% of the calves in west, 12.5% in central and 6.5% in east, which was significantly different ($P<0.05$). However, no difference was found for touching the lactating cows: 20.4% in west, 13.9% in east and 4.8% in central. The median flight distance for calves was 49 cm (east), followed by 17 cm (west) and 6 cm (central). In similar order, the median flight distance for lactating cows was 38 cm (east), 31 cm (west) and 13 cm (central). The number of animals eating (east: 78%; west: 79%; central: 70%), lying (east: 7%; west: 10%; central: 15%), standing (east: 8%; west: 7%; central: 10%) and walking (east: 7%; west: 4%; central: 4%) was not different among the regions. The median number of counts for head butts (east: 2; west: 0; central: 0), chasing (east: 6.5; west: 3; central: 3.5), self-licking (east: 50; west: 59; central: 104), sexual behaviour (east: 9.5; west: 3; central: 0), and play behaviour (east: 1.5; west: 1; central: 0) was significantly different ($P<0.05$). Scratching and rubbing (east: 19.5; west: 29; central: 24.5), however, was not different. This study shows that there are differences in behaviour and sociability between herds of yaks living in three regions of Bhutan, which might indicate different management practices and animal welfare status.

## Effects of herd, housing and management conditions on horn-induced altercations in cows

*Julia Johns and Ute Knierim*
*University of Kassel, Farm Animal Behaviour and Husbandry Section, Nordbahnhofstr. 1a, 37213*
*Witzenhausen, Germany; johns@uni-kassel.de*

The keeping of horned cows in loose housing systems carries an increased risk that situations of competition (e.g. during feeding) and consequently of increased agonistic interactions lead to skin alterations including hairless areas, superficial scratches or deeper wounds and swellings. The aim of this study was to identify potential risk factors regarding horn-induced altercations. In total, 36 dairy farms that either kept fully horned herds or switched to horned herds were investigated during three winter periods from 2014 to 2017. Herd sizes ranged from 18 to 135 cows and the percentage of horned cows varied between 2 and 100. Horn-induced altercations were expressed as mean frequency per cow and herd and recorded in both horned and dehorned cows. During farm visits, 68 factors regarding herd, housing and management conditions were collected. Since number of farms and their conditions varied between years, three separate sets of analyses (one per year) were calculated. Univariable pre-selection of factors (at $P \leq 0.2$) was carried out using non-parametric analyses. Subsequently, multivariable linear regression models were calculated with bidirectional stepwise selection of factors according to the Aike information criterion (AIC). The mean proportion of horned cows in the herds increased from 66% in 2014/15 to 72% in 2015/16 to 76% in 2016/17. The mean number of altercations increased on 7 farms with an increasing proportion of horned cows, did not change on 11 farms, decreased on 8 farms despite increasing proportions of horned cows, and stayed at a constant low level ($\leq 5$ altercations) on 10 farms that included two fully horned herds. Overall, this led to mean numbers of 8.3±5.9 altercations in 2014/15, 7.3±4.9 in 2015/16 and 8.6±6.1 in 2016/17 with mean proportions of 73% hairless areas, 18% superficial scratches and 9% deeper wounds and swellings. The following factors showed the same direction of association with horn-induced altercations in all 3 years and were statistically significant in at least one year. More altercations were found in herds with higher proportions of horned cows (2014/15: t=2.64, P=0.02; 2015/16: t=4.95; P<0.01; 2016/17: t=4.31, P<0.01), with limited forage availability (2014/15: t=2.20, P=0.04), silage feeding (vs hay feeding) (2015/16: t=2.35, P=0.03), TMR feeding (vs hay feeding) (2015/16: t=2.83, P=0.01), increased concentrate amounts per cow and day (2016/17: t=2.49, P=0.02), group (vs individual) heifer integration into the herd (2015/16: t=3.69, P=0.01), without additional observation of the herd after heifer integration (2015/16: t=2.37, P=0.03), with unevenly distributed water troughs, brushes, minerals, etc. (2016/17: t=3.41, P=0.01), without optimal water trough height (2016/17: t=2.78, P=0.01), and in herds with Holstein-Friesians (2014/15: t=2.94, P=0.01). Our results underline the importance of feeding and social management for the prevention of horn-induced altercations that may be higher than certain aspects of design or dimensions of the housing system.

## The relations between the individual and social behaviour of dairy cows and their performance, health and medical history

*Miroslav Radeski and Vlatko Ilieski*

*Faculty of Veterinary Medicine, Animal Welfare Center, st. Lazar Pop Trajkov 5-7, 1000 Skopje, Macedonia; miro@fvm.ukim.edu.mk*

Observing domestic herd animals kept in a confined area, such as dairy cows, gives us the opportunity for a detailed behavioural analysis at the individual and herd level. The objective of this study was to determine whether and how dairy cows' performance, health and medical 'treatment history' is related to, and predictive of, behavioural and social differences at the individual and herd level. Specifically, this study aims to generate hypotheses through series of unsupervised learning analyses of short term behavioural observations. In a free stall confinement area of 900 $m^2$, 91 dairy cows were observed for 14 hours/day (07.00-21.00) on two consecutive days using four cameras covering the whole area. For the subsequent video analysis, the cow catalogue was created with five photos of each individual cow (front, back, left and right side). The created ethogram consisted of 16 different behaviours, including state events – lying, moving, standing, feeding; and point events – headbutts, displacements, chasing and licking. Furthermore, the spatial distribution of cows was observed by dividing the whole confinement area into eight zones (lying area ZL1 – ZL4 and feeding area ZR1 – ZR4). The video observations with coded ethogram were done using BORIS® software. The individual characteristics were represented by 30 different variables, including age (4.33±1.84 years), origin (33 purchased and 58 cows born on farm), body condition score (BCS, 3.60±0.70), pregnancy state (108.51±71.54 days), milk production (9,216.24±1,494.73 liters), but also lameness, mastitis, reproductive disease and other medical conditions in their lifetime. These data were generated by the InterHerd® software, kept and updated at the farm, for each animal from its birth up to the observation period. After the data pre-processing and reducing of the variables, beside the descriptive statistics, an unsupervised learning clustering (i.e. nonhierarchical K-means cluster analysis) was used to determine possible links between the observed behaviours and individual features. Clusters with older cows were linked with longer standing duration (Cluster mean age = 8.15 and standing 0.21±0.09 vs Cluster mean age = 3.04 andstanding 0.09±0.04); cows with lower BCS lay down less (BCS 2-3.5 with lying proportion 0.28±0.06 vs higher BCS with minimum lying 0.34±0.06), moved less (BCS 2-3 had lowest moving of 0.02) and had lowest headbutts, displacements, chasing and fights (-0.18, -0.20, -0.20 and -0.24, respectively). The number of mastitis cases in a cow's lifetime was linked with longer lying duration (minimum lying 0.40±0.03 vs clusters without mastitis and maximum lying 0.23±0.02), and lameness was clustered with standing duration (no lame cows cluster's, standing<0.11 vs standing of lame cows >0.11) and the lying location (cows that had lameness ≥4 times lay at the outer side 1.35±0.98 vs mid side cubicles, -1.06±0.48). In total, we identified 58 significant clusters that link certain individual characteristics with different types of behaviours. These types of analyses might help in understanding the influence of animal performance and medical history/health on the behaviour at individual and herd level.

## Extracting herd dynamics from milking order data: preliminary insights comparing social network and manifold-based approaches

*Catherine McVey[1,2], Diego Manriquez[1] and Pablo Pinedo[1]*
*[1]Colorado State University, Animal Science, Fort Collins, CO 80521, USA, [2]University of California Davis, Center for Animal Welfare, 1 Shields Ave, Davis, CA 95616, USA; mcveyc13@gmail.com*

Emerging research suggests that genetic selection for improved dairy health and longevity traits may indirectly select for social dominance and aggression. If unintended welfare complications arising from the subsequent disruption of herd dynamics are to be avoided, efficient strategies for quantifying and monitoring social structures will be needed. Entry order into the milking parlor represents a massive untapped reservoir of information on social interactions within dairy herds. The aim of this project was to explore what features of a herd's social organization could be recovered from this data using machine learning and social network algorithms. Data was repurposed from a closely supervised feed trial. RFID logs from morning milkings were collected from a closed pen of 182 post-freshening organic dairy cows of mixed parity over a 100 day observation window that included the transition to spring pasture. Preliminary data visualization via loess curve fitting revealed the entry orders of some cows to be fairly consistent over time, but others to be highly dynamic. Subsequent PCA embeddings revealed this data to be low dimensional but quite noisy. Both visualizations indicated significant reordering of animals entering the parlor following pasture access. The *Perc* package in *R* was utilized to estimate rank via a social network approach. Directed dyads were generated by parsing the parlor entry data and recording each time a given cow entered immediately ahead of another. For both pen and pasture subsets, the resulting adjacency matrices were extremely sparse, with 91 and 81% of dyadic relationships unobserved. An annealing algorithm was then used to estimate rank order, utilizing beta probabilities to account for uncertainty in the pairwise dominance relationships. Visualization of these results revealed a highly linear hierarchy with no distinct subhierarchies, but confidence in these rank estimates was fairly low. Kendal Tau analysis subsequently revealed no correlation between rank order estimates generated separately from pen and pasture periods, nor were they correlated to age. An unsupervised machine learning approach was next explored to avoid extensive data reformatting and the risk of artifacts it carried. It was hypothesized that, if herd dominance hierarchy could be conceptualized as having a geometric shape, then high dimensional parlor entry data might actually exist on a low dimensional manifold: an abstract topological feature that frequently arises in complex nonlinear systems. A Laplacian Eigen Embedding implemented in Matlab revealed that cows occupied a low dimensional predominantly linear geometric form with greater ambiguity in the middle ranks and with some clustering patterns to suggest the presence of subcommunities. Future research aims to explore if this geometric feature is temporally consistent across a wider range of herds, how it relates to direct observations of social behaviors, and if deviations from this form may be used to identify abnormal social dynamics that precede welfare issues.

## The effect of coping style on behaviour and autonomic reactions of domestic pigs during dyadic encounters – a pilot study

*Annika Krause[1], Jan Langbein[1] and Birger Puppe[1,2]*
*[1]Leibniz Institute for Farm Animal Biology (FBN), Institute of Behavioural Physiology, Wilhelm-Stahl-Allee 2, 18196 Dummerstorf, Germany, [2]University of Rostock, Faculty of Agricultural and Environmental Sciences, Justus-von-Liebig-Weg 6B, 18059 Rostock, Germany; krause@fbn-dummerstorf.de*

The objective assessment of affective states of pigs is increasingly important in terms of understanding and improving pig welfare. It has to be considered that individual pigs differ in their general adaptive response to environmental and social demands according to their emotional perception. The effect of different coping styles on behaviour and autonomic activity was studied during 2-h dyadic encounters of familiar pigs with special focus on agonistic interactions. The pigs were selected according to their coping style based on the behavioural reactivity in the backtest and underwent a surgical implantation of a telemetric device which allowed the real-time measurement of electrocardiogram and intra-arterial blood pressure (BP) in order to assess the activity of both branches of the autonomic nervous system simultaneously (Ethic approval by the LALLF, No. 7221.3-1.1-037/12). Based on 6 pigs (3 proactive, 3 reactive) a total of 15 dyadic encounters were carried out. Social and non-social behaviours were analysed (resting, feeding, active behaviour, snout contact, grooming, fighting) as well as the respective initiator and success after fighting (win, defeat). Heart rate (HR), BP and their respective variabilities (HRV, BPV), indicative for vagal (HRV) and sympathetic (BPV) activity, were analysed in 5-min time intervals (TI) during dyadic encounters. More specifically, to evaluate affective appraisal in the pigs during agonistic interaction, the autonomic activity was analysed in 10-s TI immediately before, during and after the first decisive fight in each dyadic encounter. All parameters were analysed by repeated measures analyses of variance and multiple pairwise Tukey-Kramer-tests. Our results indicated an impact of coping style on behaviour and autonomic activity during dyadic encounters. Proactive coping was accompanied by increased active behaviour ($F_{1,4}=2.5$, $P<0.05$) and higher HR ($F_{1,4}=11.8$, $P<0.05$), whereas reactive pigs showed shorter contact latencies ($F_{1,4}=6.3$, $P<0.05$) and initiated the first positive social contact more often than proactive pigs ($F_{1,4}=3.6$, $P<0.05$). In the context of fighting, coping style was not found to have an effect, but rather fighting success seemed to play a major role in modulating autonomic responses. Winning pigs showed lower HR ($F_{1,23}=2.9$, $P<0.1$) during and immediately after fighting, which was accompanied by higher vagal activation ($F_{1,60}=6.4$, $P<0.05$) and lower sympathetic activation ($F_{1,4}=6.9$, $P<0.05$) compared to defeated individuals. The present investigation reveals, for the very first time, autonomic responses of pigs during dyadic encounters in real-time and provides evidence suggesting that the coping style may play an essential role in affecting behaviour and autonomic reactions during social interaction. In the specific context of agonistic interaction, however, we found that fighting success, rather than the individual coping style, modulated the autonomic reaction in domestic pigs indicating differences in affective appraisal.

## Focussing on significant dyads in agonistic interactions and their impact on dominance indices in pigs

*Kathrin Büttner, Irena Czycholl, Katharina Mees and Joachim Krieter*
*Institute of Animal Breeding and Husbandry, Christian-Alberechts-University, Olshausenstr. 40, 24098, Germany; kbuettner@tierzucht.uni-kiel.de*

Dominance is defined as a pattern found in repeated agonistic interactions between two individuals characterised by the consistent outcome of the fight to the advantage of one animal. Thus, it is proposed that the consistent asymmetric outcome of dyadic interactions has to be tested for significance before further sociometric measures are calculated to avoid misinterpretations of the results. If and to what extent dyadic interactions should be tested for a significant asymmetric outcome has not yet been clarified. Thus, in the present study two different calculation methods for the limits of significant dyads were carried out by a sign test based on the differences in won and lost fights considering all dyadic interactions in the pen (PEN: pen individual limits) and a sign test focussing on each individual dyad (DYAD: dyad individual limits) which are compared to the data set containing all dyadic interactions (ALL). Data of animals in three mixing events (93 pens with 8.91±0.58 weaned piglets; 26 pens with 20.88±1.66 fattening pigs; 12 pens with 20.75±3.39 gilts) were used. All animals were video recorded for three days (weaned piglets) or two days (fattening pigs, gilts), respectively, after mixing and all agonistic interactions (initiator, receiver, winner, loser) were recorded. Dominance indices (DI) were calculated for all data sets (ALL, PEN, DYAD) which take the number of won and lost fights as well as the number of opponents and the group size into account. Considering all dyadic interactions, fattening pigs and gilts showed a significant decrease in the number of fights after the first day of video observation ($P<0.05$), whereas weaned piglets had in total a higher level of fights with more fluctuation. In order to answer the question how many days have to be analysed to get reliable results of the DIs, Spearman rank correlation coefficients (rs) were calculated at the end of each observation day. In each comparison, the data set ALL showed slightly lower rs values compared to the data sets PEN and DYAD independent of the considered age group. For weaned piglets, the highest rs values could be obtained for the comparison between day 2 and day 3. Here, the rs values ranged between 0.93 and 0.98 independent of the used data set. This implies that, although the number of fights showed a temporal fluctuation with no significant decrease during the observation period, focussing on only 2 days of video observation might be sufficient in order to get reliable DIs. Furthermore, comparing the DIs between the three used data sets in the temporal course, rs values above 0.70 could be obtained for all age groups. This implies that focussing on significant dyads has no substantial impact on the results. Comparing both calculation methods (PEN, DYAD), it has to be kept in mind that dyad individual levels depend on the total number of fights in the pen which is probably only appropriate in groups with higher frequency of agonistic interactions (e.g. weaned piglets). Otherwise, no DI can be calculated for too many animals.

## Maternal serum cortisol and serotonin response to a short separation from foal

*Daniela Alberghina[1], Maria Rizzo[1], Francesca Arfuso[1], Giuseppe Piccione[1], Irene Vazzana[2] and Michele Panzera[1]*

*[1]University of Messina, Veterinary Sciences, Polo Universitario dell'Annunziata, 98168 Messina, Italy, [2]Experimental Zooprophylactic Institute of Sicily, Via Gino Marinuzzi 3, 90100 Palermo, Italy; dalberghina@unime.it*

Separation from the mother is known to negatively impact behavioral, neuronal and hormonal responses in mammalian offspring. Numerous studies have shown that even brief separation elevates circulating cortisol in the foal. Furthermore, gender-related differences in response to weaning have been described: fillies seemed more affected by weaning stress than colts. However, less attention has been paid to separation impact in mothers. The role of central serotonin in response to acute and chronic stress is well documented. Peripherally serotonin is mainly stored in platelets. This study was designed to investigate cortisol and serotonin response to separation from 6-to-7-month-old foals in serum of multiparous mares. The protocol and the study design were approved by the Ethical Committee of the Department of Veterinary Sciences of the University of Messina. Fourteen Thoroughbred mother-foal pairs (10±4 years of age) were recruited in this study, six were used as control and eight equally distributed for gender of foals) were subjected to a short separation (30 min) in order to assess serum cortisol and serotonin responses in mares. During short-term separation, there was no visual contact between mare and foal, but vocal contact could not be prevented. After 30 min, the mares were returned to the box. Blood sampling was performed via jugular venipuncture 30 min prior to separation, at 30 after separation and 30 min after the return of the mother in the box. Similar blood sampling protocol were applied to control mares. The samples were scheduled to be collected between 9.00 and 11.00 am. Blood samples for cortisol and serotonin were collected into 10 ml serum vacuum tubes with clot activator. The samples were allowed to clot followed by centrifugation at 1.750 g for 10 minutes. The obtained sera were stored at -20 °C until analysis. The serum aliquots obtained from each sample were stored frozen at −20 °C. Samples were analyzed within 2 weeks of collection. Cortisol concentration was measured by chemo-luminescent assay and serotonin by enzyme-linked immunosorbent assay (ELISA). Differences between samples were analyzed by two-way ANOVA with Bonferroni's Multiple Comparison Test. Separation significantly increased cortisol levels ($F_{2,12}$=5.89, P=0.016) and significantly decreased serotonin levels ($F_{2,12}$=3.92, P=0.049), no changes were detected in control mares. Interestingly mares of colts showed cortisol levels higher than mares of fillies (ANOVA $F_{1,6}$=7.25, P=0.035,). These preliminary findings suggest that serotonin also might be considered as a peripheral marker of acute stress in equines and that different cortisol responses in mothers could be related to gender of foal.

## Understanding stereotypic pacing: why is it so difficult?

*Colline Poirier, Catlin J. Oliver, Janire Castellano Bueno, Paul Flecknell and Melissa Bateson*
*Newcastle University, Framlington Place, NE1 7RU Newcastle-upon-Tyne, United Kingdom;*
*colline.poirier@ncl.ac.uk*

Stereotypic behaviours are commonly observed in captive animals and are usually interpreted as a sign of poor welfare. However, the causal factors underlying stereotypies are still unclear and mounting evidence suggests that, even within one species, different stereotypies can have different causes. Understanding the cause of stereotypies is particularly important in laboratory animals, since the use of stereotyping animals in research might compromise the validity, reliability and replicability of scientific findings. Despite this lack of understanding, stereotypies are often used as indicator of chronic or acute stress. In this study, we investigated whether pacing, the most frequent stereotypy displayed by laboratory macaques, is reliable at detecting acute stress. To answer this question, we measured the reaction of rhesus macaques to the stressful event of witnessing spontaneous agonistic interactions between conspecifics housed in the same room but in a different cage. The behaviour of 13 socially-housed adult males was quantified before, during and after agonistic interaction exposure, based on video recordings. Displacement behaviours, which are pharmacologically validated indicators of stress or anxiety, increased after agonistic interaction exposure, confirming that the events were experienced as stressful by the focal individuals ($\chi^2(3)=31.1$, P<0.001). The occurrence of pacing did not increase during or after the agonistic interactions ($\chi^2(3)=1.1$, P=0.79). Instead, agitated locomotion, a non-stereotyped locomotor behaviour defined as 'moving rapidly between locations, with a stiff un-relaxed gait', increased during the agonistic interactions ($\chi^2(3)=154$, P<0.001). These results suggest either, that pacing as an indicator of acute stress is prone to false negative results, increasing in some stressful situations but not others, or that agitated locomotion has been mistaken for pacing in previous studies and that pacing is in fact unrelated to current acute stress. Both interpretations lead to the conclusion that pacing is unreliable as an indicator of acute stress in laboratory rhesus macaques. Our data also suggest that agitated locomotion might potentially be a useful indicator of acute stress, or at least arousal, but more studies are needed to confirm this. Combined with previous studies, these results highlight the current lack of understanding of the causal factors underlying pacing behaviour in laboratory macaques, especially whether it is due to stress (chronic or acute, current or from the past) or to something else. We will discuss the different potential causes of pacing and their implications in term of welfare and validity of neuroscientific findings coming from pacers and non-pacers.

## Identification of seven call types produced by pre-weaning gilts during isolation

*Mariah Olson, Maggie Creamer and Kristina Horback*
*University of California Davis, Department of Animal Science, Center for Animal Welfare, One Shields Ave, Davis, CA 95616, USA; kmhorback@ucdavis.edu*

There is a current boom in the development and use of real-time computer monitoring of animal vocalization to evaluate animal health and safety (i.e. precision livestock farming). The domestic pig is an ideal candidate for this type of assessment as it is a highly vocal mammal with anatomical structures to support the source-filter theory of speech production. Previous studies on swine vocalizations predominately describe calls as being either high frequency (>1000 Hz) or low frequency (<1000 Hz) within a specific context, but a standardized description of the vocal repertoire of domestic swine has yet to be established. The objectives of this study were to (1) identify and categorize vocalizations produced by 3-week-old gilts (n=40) during a 5 minute social isolation test, and (2) identify the behavioral state of gilts when producing each vocalization. Video data of isolation tests were coded for duration gilt was in each behavior state (walk, freeze, and nose environment) as well as the number of escape attempts (jumps at walls of test box). The teat position for each gilt at nursing was recorded two times a day for the week of the experiment. Piglet weight was recorded prior to placement in the isolation test (2.4×1.5 m plastic kiddie pool). Audio data was extracted from video files and subsequently analyzed in Raven Pro 1.5. Calls were classified according to acoustic and spectral characteristics such as peak frequency, tonality, frequency range, and harmonics. Based on this analysis of over 15,000 vocalizations, seven call types (A-G) were identified. The most frequent vocalizations identified were classified as call type E or 'squeal' (4,180 calls), C or 'croak' (3,177 calls), D or 'high-grunt' (2,664 calls), and A or 'low-grunt' (2,611 calls). The majority of E calls (43.6%) and C calls (50.8%) occurred when the gilt was walking in the isolation test, while the majority of A calls (54.8%) and G calls, or 'moderate-grunts' (64.7%), occurred when the gilt was nosing the environment. There were significantly positive correlations between teat rank and the rate ($r_s$=0.42, P<0.01) and duration ($r_s$=0.48, P<0.01) of vocalization during isolation, as well as a significantly negative correlation between teat rank and weight ($r_s$=-0.40, P<0.01). There was a significantly negative correlation between the number of C calls a gilt produced and bodyweight ($r_s$=-0.42, P<0.01), as well as a significantly positive correlation between C call production and the number of escape attempts ($r_s$=0.38, P<0.01). Taken together, these results indicate that the smallest gilts in this study suckled at the most posterior teats of the sow and vocalized significantly more often, for longer durations, and produced more C calls ('croaks') and escape attempts when isolated from their natal pen. Acoustic features which may reflect the emotional state of animals could have far-reaching implications for the use of real-time monitoring to assess the welfare of pre-weaning piglets. The results of this study indicate that grouping individual calls into two broad categories of high and low frequency may not provide sufficient specificity when it comes to identifying individuals in need.

# The use of pig appeasing pheromone decreases negative social behavior and improves performance in post-weaning piglets

*Paula Ramírez Huenchullán[1] and Cinthia Carrasco[2]*
*[1]NGEN BA, Las Cortezas Poniente 1813, 7911655, Chile, [2]Pontificia Universidad Católica de Chile, Facultad de Agronomía e In, Av. Vicuña Mackena 4860, 7820436, Chile; prhuenchullan@gmail.com*

The pig appeasing pheromone (PAP) has been effective in reducing stress responses and aggressive behaviour in different trials. Our objective was to evaluate the effect of PAP on the welfare and performance of weaned piglets in a commercial system. We used 800 weaned pigs divided over 4 identical rooms with 20 pens/room and 20 pigs/pen. The rooms were washed and disinfected 48 hours before the newly weaned pigs arrived. PAP blocks were installed at 1.50 meters from the ground (1 block/25 $m^2$) in the rooms. Behavioural observations were performed at 24, 39 and 49 days post-weaning on randomly selected pens and pigs. Behaviour was evaluated in 10 pens/treatment through direct observation (focal: 3 pigs/pen, n=30; group: 20 piglets/pen, n=200) for four 5- min blocks a day: twice between 09:30 am and 12:30 and twice between 14:30 pm and 16:30. Between each focal observation was 2 min of group observation, leading to 28 min of observation per pen in each record. In addition, behaviour was observed from continuous video recordings through scan-sampling every 10 minutes for 24 hours at 7 and 15 days post-weaning. Production performance (weight, daily weight gain and feed conversion) were measured additionally to day 21. The statistical unit for this was the pen (n=40). The Shapiro-Wilk test was applied to verify the normal distribution of the data. For the variables without normal distribution the Mann-Whitney Wilcoxon test was used to compare groups. Normally distributed data were analysed using a general linear model (GLM) where the following factors were considered: treatment, day and treatment × day. The differences were considered significant with P<0.05. The statistical program was GraphPad PRISM 7.0 During the 24 hours post-mixing, there was a significant effect of pheromone on the frequency of negative social behavior: bite (Control=702.75 vs PAP=406; P=0.0016) fight (Control=125.25 vs PAP=67.5; P=0.0094) which remained very significant on day 15 post-mixing. On day 15, an increase in the frequency of ear biting was recorded among the piglets of the control group as compared to the PAP group (P=0.0012). There was a difference in the number of piglets that huddled during the first 24 hours and 15 days after mixing (Control=302.63 vs PAP=99.4; P=0.0008 and Control=395.6 vs PAP=467.2; P=0.04 respectively). At 24 hours, a higher activity (walking and exploring) was observed in the control group (P=0.0012). Belly-nosing was seen more in the control group than the PAP group at 24 hrs and day 15 (Control=18.63 vs PAP=9.38; P=0.0068 and Control=71.6 vs PAP=54.4; P=0.01 respectively). The pigs in the control group drank more frequently at 24 hours and day 7 after weaning (P=0.0097 and P=0.021 respectively). In live observation (focal and group), no significant differences were found between the pigs in the control and treatment group. The PAP and control group differed in the feed conversion at 45 and 70 days of age (P<0.05), in the total feed conversion of the period and in the adjusted conversion (P<0.05). The PAP block is a tool that helps the adaptation of newly weaned piglets

## The influence of sex on dog behaviour in an intraspecific Ainsworth Strange Situation Test

*Chiara Mariti[1], Beatrice Carlone[1], Emilia Votta[1], Marco Campera[2] and Angelo Gazzano[1]*
*[1]University of Pisa, Dept. of Veterinary Sciences, Viale delle Piagge, 2, 56124 Pisa, Italy, [2]Oxford Brookes University, Dept. of Social Sciences, Gibbs Building, Gipsy Lane, OX3 0BP Oxford, United Kingdom; chiara.mariti@unipi.it*

The Ainsworth Strange Situation Test has been largely used to investigate dog-human bond. Only recently, its use has been extended to the study of dog-dog bond. The aim of the current research was to investigate the influence of sex on dog behaviour in an intraspecific Ainsworth Strange Situation Test. Fifty-five dyads of adult dogs were involved in a modified Ainsworth Strange Situation Test in which a dog is tested with a cohabitant dog, the latter playing the role of the presumed attachment figure and an unfamiliar person plays the role of the stranger. Following the protocol described in Mariti *et al.*, seven 2-min episodes were carried out in an experimental room, unfamiliar to all dogs. In episode 1, dogs are led into the room and left free to explore it. As recommended in the ASST, the procedure included two separations from the presumed attachment figure (episode 3, dog with stranger; episode 5, dog in complete isolation; and episode 6, dog with stranger) and two reunions with him/her (episode 4 and 7). Dogs were all pets, they had been living in the same household for at least one year, they were more than 14 months old, and were healthy. The behaviour of both dogs was observed and the durations of 17 social and non-social behaviours, including signs of stress (hypersalivation, trembling, fore hind lifting, self-licking, self-scratching, shaking, yawning, and nose licking) were measured in seconds. Data were analysed using multiple Linear Mixed Models considering single behaviour types as dependent variables. Investigated factors were: episodes (1, 4, and 7, in which both dogs were present), kinds of relationship (mother-offspring couples n=18, non-blood relatives n=37), sex (female-female n=27, male-female n=24, and male-male n=4), and age difference. Fisher's Least Significant Distances post hoc tests have been performed to allow pairwise comparison (P<0.05). In this study, only results relative to the sex of dogs involved (considered as fixed factors) will be reported. When two female dogs were involved in the test, they spent more time in physical contact compared to dyads formed by two males (6.4 vs 2.0; P=0.048) or one female and one male dog (3.9; P=0.033). Female dogs also spent less time close to the door compared to both males (1.3 vs 4.8; P=0.020) and male-female dyads (3.0; P=0.031). Female-female dyads differed from female-male dyads displaying less barking (1.3 vs 4.2; P=0.025), fewer signs of stress (5.5 vs 7.2; P=0.033), and more whining (23.0 vs 15.7; P=0.044). These results suggest that the sex of dogs has a strong impact on the way they relate to other dogs and they, as a dyad, cope with potentially stressful conditions. More in detail, female dogs seem to find in their female companion a buffer against stress.

## Sow response to sibling competition during nursing in 2 housing systems in early and late lactation

*Gudrun Illmann[1,2], Sébastien Goumon[2], Iva Leszkowová[2] and Marie Šimečková[2]*
*[1]Faculty of Agrobiology, Food and Natural Resources, Czech University of Life Sciences Prague, 160 21 Praha-Suchdol, Czech Republic, [2]Animal Science, Ethology, Přátelství 815, 10400 Prague, Czech Republic; gudrun.illmann@vuzv.cz*

Litter competition at the udder may lead to an increase of non-nutritive nursings in order to encourage weaning in late lactation. Sows have been shown to respond to increased litter competition during the first two days postpartum (pp) by terminating nursing only during the post-massage phase, after milk has been released. Before weaning, litter competition is expected to increase again due to physical constraints (increase in piglet body size) and reduced space due to presence of protective bars in crates compared to pens. Sow response to litter competition, if any, would be expected to be different than the one found in early lactation because the sows can 'afford' to terminate nursing before milk ejection as piglets of that age are no longer completely dependent on milk in both crates and pens. The aim of the study was to assess the intensity of sibling competition during nursings in early and late lactation (days 5 and 25 pp) before and after milk ejection in pens compared to crates and to assess whether sows' response to litter competition differs between both periods and housing systems. Sows and their piglets were housed either in farrowing crates (n=14) or in farrowing pens (n=13). Behavioural assessment was done for 4 h through video recording on days 5 and 25 pp. Using scan sampling, presence of piglets at the udder was scored every 15 s interval, commencing 5 intervals (i.e. 75 s) prior to milk ejection, and continuing for 9 intervals (i.e. 135 s) after milk ejection. The mean number of piglets exhibiting fights was calculated for both before (pre-massage) and after milk ejection (post-massage). Sow postural changes and nursing status ((nutritive and non-nutritive) were also noted. Data were analysed using PROC GLM and PROC GENMOD of SAS including housing, age and litter size as fixed effects. Litter size did not differ between pens and crates (13.5 vs 13.7 piglets; SEM: 0.7). There was a higher number of piglets fighting during post-massages but not during pre-massages ($\chi^2=0.81$; NS) on day 5 than on day 25 ($\chi^2=4.25$; P<0.05) with no effect of housing ($\chi^2=0.24$; NS). A higher number of piglets fighting during pre-massages was associated with a higher proportion of piglets missing milk ejection on days 5 ($\chi^2=6.08$; P<0.05) and 25 ($\chi^2=4.69$; P<0.05) with no effect of housing ($\chi^2=1.65$; NS). On both days, sows tended to respond to increased number of fighting piglets with non-nutritive nursings during pre-massages ($\chi^2=2.91$; P<0.1) but not during post-massages ($\chi^2=0.76$; NS). No effect of housing was found ($\chi^2=0.03$; NS). In conclusion, housing had only a very limited effect on litter competition and on the sow's response. Our results indicated that a higher number of piglets fighting during pre-massages reflected an increased number of piglets missing milk ejection like previously found during the first two days pp. The tendency of sows to respond to higher sibling competition with non-nutritive nursings might indicate that they suppress litter competition by not releasing milk. Non-nutritive nursings might prolong the inter-nursing interval and decrease milk output, hence encouraging weaning in later lactation (parent- offspring conflict theory).

## Influence of bull biostimulation on puberty and estrus behaviour of Sahiwal breed (*Bos indicus*) heifers

*Sanjay Choudhary, Madan Lal Kamboj and Pawan Singh*
*National Dairy Research Institute, Livestock Production Management, Room no.175 Narmada hostel NDRI, Karnal, 132001, India; rajsaya07@gmail.com*

Efficient, accurate and timely detection of estrus is a major factor in reproductive success of dairy cows. Biostimulation by exposing the growing females to the males has been successfully demonstrated to prime the females for early onset of puberty through priming pheromones and aid in accurate and efficient detection of estrus through signaling pheromones in sheep, goat and pigs. It has also been reported to improve estrus detection and reproductive performance in some beef breeds of cattle through direct contact during post-partum period. The hypotheses of the present study were to investigate the effect of biostimulation on growing heifers through only fenceline bull contact and both fenceline and direct bull contact on age at puberty and estrus behaviour. For this, a total of 24 prepubertal Sahiwal heifers were allotted to 3 groups of 8 each on the basis of age (14.44±0.28 months) and body weight (152±6.12 kg). In no bull exposure (NBE) group, the heifers were not exposed to bull; in fenceline bull exposure (FBE) group, the heifers were exposed to a bull through a fenceline contact and in FBE+DBE (direct bull exposure) group, the heifers were housed in a fenceline contact with bull round-the-clock along with direct contact for a period of 6 hours through another bull. Heifers were confirmed to have attained puberty if $P_4$ concentrations were >1 ng/ml. The estrus behaviours were recorded on day -3, -2 and -1 (prior to estrus), d 0 (on the day of estrus) and on day +3, +2 and +1 (post estrus) using 24 hours CCTV camera recording. The overall mean daily gain was significantly (P<0.05) higher in FBE+DBE (508.11±14.57 gm) as compared to FBE (493.57±14.94 g) and NBE (462.35±16.28 gm). The mean leptin and growth hormone concentrations were significantly (P<0.05) higher in FBE (3.11±0.23 and 6.46±0.55 ng/ml) and FBE+DBE (3.25±0.25 and 6.47±0.55 ng/ml) groups than in NBE (3.25±0.25 and 5.08±0.24 ng/ml). The mean age and weight at puberty in FBE (19.33±0.36 m & 226.20±6.35 kg) and FBE+DBE (19.11±0.58 m and 224.19±4.54 kg) were similar but significantly (P<0.05) lower than in NBE (24.00±0.12 m and 258.35±13.13 kg). In first estrus the mean frequencies of estrus behaviours viz., sniffing/licking, tail raising, micturition, chin resting, and allowing mounting attempts were similar in FBE and FBE+DBE but significantly (P<0.05) higher than in NBE heifers from d-2 and d0 of estrus and then declined from d+1. These frequencies of estrus behaviors increased in second estrus as compared to first estrus in all three groups. Percent reduction in daily times spent on eating in NBE, FBE and FBE+DBE heifers (29.41, 42.92 and 43.64%), rumination (44.44, 44.77 and 44.07%) and resting (39.02, 47.96 and 44.12%) on the day of estrus from the reference days. We concluded that biostimulation of Sahiwal heifers both by fenceline bull contact or fenceline plus direct bull contact appeared to advance the age at puberty and elicited greater expression of estrus behaviour. Fenceline plus direct bull exposed heifers showed higher frequencies and intensities of estrus behaviours than only fenceline bull exposed heifers.

## Observation of ram-estrous ewe interaction enhances mating efficiency in subordinate but not in dominant rams

*Rodolfo Ungerfeld[1], Agustín Orihuela[2] and Raquel Pérez-Clariget[3]*
*[1]Facultad de Veterinaria, Universidad de la República, Departamento de Fisiología, Lasplaces 1620, Montevideo 11600, Uruguay, [2]Facultad de Ciencias Agropecuarias, Universidad Autónoma del Estado de Morelos, Av. Universidad 1001, Colonia Chamilpa, Cuernavaca, Morelos 62210, Mexico, [3]Facultad de Agronomía, Universidad de la República, Departamento de Producción Animal y Pasturas, Garzón 780, Montevideo 12400, Uruguay; rungerfeld@gmail.com*

Sexual performance of rams does not improve after observing hetero-sexual behavior of other rams, as happens in bucks and bulls. According to dominance position, rams may differ in their response, masking the possible differences within both categories. Thus, the objective of the present experiment was to determine if rams' sexual behavior is modified differently in dominant (DOM) and subordinate (SUB) rams after observing the sexual behavior of each other. The dominance relationship of nine dyads of Corriedale rams was determined using a food competition test. Next, each ram was tested in two different situations in different days with an overcross design; a control test (CT), and a test in which the other ram was joined with two estrous ewes in the contiguous pen (sexual stimulation before; VSST). Estrous ewes were placed in one pen, and each ram was located for the first 20 min in the contiguous pen. Then, the ram was moved into the pen with the estrous ewes, staying there for other 20 min. During the CT, the ewes remained alone for the first 20 min. During the VSST, the other ram of the dyad was in the pen with the ewes, but was taken out before introducing the tested ram. There were no differences in the frequency of any behavior during the first 20 min between DOM or SUB rams. The DOM rams did not modify any behavior according to the type of test. On the other hand, SUB rams increased the number of ejaculations and the ejaculation efficiency [ejaculations: 4.2 vs 2.8, pooled SEM=0.4; P=0.04; Arcsine (root ejaculates/total mounts): 0.65 vs 0.40, pooled SEM=0.08; P=0.03] when tested after viewing the DOM rams, and tended to achieve their 1st ejaculation earlier in the VSST test [log time to first ejaculation (s): 1.5 vs 1.9, pooled SEM=0.2; P=0.058]. It was concluded that while visual pre-stimulation of a SUB ram courting and mating estrous ewes had no effect on DOM ram' sexual behavior, the observation of a DOM ram, enhanced mating behavior of SUB rams. This expands the concept that DOM and SUB rams respond with different strategies to visual stimulation.

# The phase of oestrous cycle influences the stress response to social isolation in Corriedale ewes

Aline Freitas-De-Melo[1], Camila Crosa[1], Irene Alzugaray[1], María Pía Sánchez[1], Mariana Garcia Kako Rodriguez[2], Belén López-Pérez[1] and Rodolfo Ungerfeld[1]

[1]Facultad de Veterinaria, Universidad de la República, Lasplaces 1620, 11600 Montevideo, Uruguay, [2]Faculdade de Ciências Agrárias e Veterinárias, Universidade Estadual Paulista, São Paulo, Prof. Paulo Donato Castelane, s/n, Jaboticabal, 14884-900, Brazil; alinefreitasdemelo@hotmail.com

In ruminants, sexual steroid concentration during hormonal treatments and/or gestation affect the response to different stressors. As sheep are gregarious animals, individual isolation from the flock provokes a stress response. The aim of this study was to compare the behavioural and physiological responses to social isolation in oestrous and dioestrous ewes. Oestrous (day of oestrus, n=7) or dioestrous (9-10 days after oestrus, n=8) ewes were individually isolated in a novel place (4×4 m² pen) during 10 min, alternating one ewe from each group. The pen had white walls, and the floor was marked every 1 m generating 16 squares. During the social isolation test, the ewes had no visual, olfactory or auditory communication with other animals or humans, and all the ewes' activities were video recorded with two cameras. Total protein, albumin, and globulin serum concentrations, as well as surface temperature on the skin of the abdominal area, were determined before and after the tests. The number of occurrences of the following behaviours were analysed from the videos: lines crossed, vocalizations, freezing, as well as the time that each ewe remained on the peripheral squares. The variables were compared with a mixed model; the model included the group, the time as repeated data when corresponded as well as the interaction between group and time. Results were considered significantly different when P≤0.05, and a tendency when 0.05<P≤0.1. Oestrous ewes tended to vocalize more times during social isolation than dioestrous ewes (7.0±3.3 vs 1.1±0.5, respectively; P=0.1), and had greater concentrations of total protein and globulin than dioestrous ewes (total protein: 7.0±0.08 vs 6.7±0.05 g/dl; P=0.05; globulin: 2.7±0.1 vs 2.4±0.1 g/dl; P=0.04, respectively). Surface temperature on the skin of the abdominal area tended to be greater in oestrous than dioestrous ewes (36.3±0.4 °C vs 35.2 °C±0.7; P=0.07, respectively). The other variables did not differ according to the phases of the oestrus cycle. In conclusion, oestrous ewes were more sensitive to social isolation than dioestrous ewes. Thus, the stress response to different stressful farm practices may vary according to the phase of oestrous cycle.

## Role of animal behaviour in addressing future challenges for animal production

*Laura Boyle[1], Matthias Gauly[2,3] and Hans Spoolder[4]*
*[1]Teagasc, Pig Development Department, Animal and Grassland Research and Innovation Centre, Moorepark, Fermoy P61 C996, Co. Cork, Ireland, [2]The European Federation of Animal Science (EAAP), Via Tomassetti 3, 00161 Roma, Italy, [3]Piazza Università, Animal Science, Faculty of Science and Technology, Piazza Università 51, 39100 Bozen-Bolzano, Italy, [4]Wageningen Livestock Research, P.O. Box 338, 6700 AH Wageningen, the Netherlands; laura.boyle@teagasc.ie*

Climate change, population growth, dwindling resources, pollution, antimicrobial (AM) resistance (AMR) and the unequal distribution of food (obesity vs starvation) are challenges with major implications for animal agriculture. The United Nation's Global Goals for Sustainable Development acts as a roadmap to aid in addressing these challenges. The present paper explores how a good understanding of animal behaviour may help to achieve some of these goals, and make animal production more sustainable. The world population is growing, and there will be an increased demand for intensively and efficiently produced animal protein for human consumption. Although technically possible, there are several obstacles that challenge sustainable growth. These include threats to animal health through climatic changes and new emerging diseases. There is also a need to reduce the global use of AM to combat AMR. Meanwhile, 'livestock's long shadow' reflects the important negative role that animal production has on pollution, climate change and loss of habitats, etc. Applied ethology can play a modest, but valuable role in addressing some of these issues. Behavioural traits associated with resilience to heat and cold stress and disease challenges will play a major role in maintaining animal welfare and production efficiency. Changes in behaviour also play a crucial role in the early detection and prevention of disease, reducing the need for AM. The advantage of animal behaviour as the first line of defence against a threat makes it the main driver of precision livestock farming technologies, thus improving resource use (labour, feed, etc.) and reducing emissions. The inclusion of animal behaviour and animal based measures in quality assurance schemes facilitate transparency and accountability, and improves consumer acceptance of animal products. Finally, a better understanding of animals and their behaviour can be used to valorise resources that humans cannot use, such as marginal lands and food waste, to address problems of hunger and poverty. An important caveat to the role of animal production in achieving the sustainability goals is not only that animal welfare must not be compromised but that they must, in line with the main theme of this conference, experience 'a life worth living'. Indeed the concept of One Welfare recognises that animal welfare, biodiversity and the environment are interconnected with better human welfare. Hence in cases where animal welfare conflicts with environmental sustainability important compromises will have to be reached. The ISAE, in collaboration with the EAAP, can play a vital role in this regard. Their combined knowledge can aid the wider animal sciences to better understand and interpret animal behaviour such that it can make a valuable contribution to achieving sustainability of animal production systems.

## Association between locomotion behavior and *Campylobacter* load in broilers

*Sabine G. Gebhardt-Henrich[1], Ivan Rychlik[2] and Ariane Stratmann[1]*
*[1]University of Bern, ZTHZ, Division of Animal Welfare, VPH Institute, Burgerweg 22, 3052 Zollikofen, Switzerland, [2]Veterinary Research Institute, Hudcova 70, 621 00 Brno, Czech Republic; sabine.gebhardt@vetsuisse.unibe.ch*

Movement patterns of broiler flocks (optical flow) have been shown to predict welfare parameters and colonization by *Campylobacter*. The mechanism and causation between optical flow and health and welfare of broilers are unclear. The aim of this study was to assess the social motivation for movement in individual birds with known *Campylobacter* load (CL). Furthermore, we expected associations of speed with body mass, pododermatitis, and hockburn. In each of 6 *Campylobacter* positive flocks from 3 farms, 16 broiler chicks (Ross 308) between 25 and 28 days of age were randomly chosen and separated from the flock in a catching frame (114×114×60 cm) inside the barn with visual contact to the flock. One chick at a time was carried to the end of a 342 cm long runway with opaque sides that led to the chicks in the catching frame and then released. The time to reach the flock mates in the catching frame was recorded with a timeout of 5 min. After the test, chicks were weighed, sexed, and scored for pododermatitis and hockburn while the observer was unaware of the speed in the social reinstatement test. Individual faecal samples were collected and the percentage of CL was determined by next generation sequencing. Two general linear mixed models with the fixed factors CL, test sequence, sex, body mass and the random factors farm and flock nested in farm were calculated with all birds and excluding birds with a timeout. Associations between pododermatis or hockburn with CL or with speed were tested with Kruskal Wallis Tests. An association between speed, pododermatitis, and hockburn with body mass was tested in a larger dataset of 20 *Campylobacter* negative or positive flocks with a general linear mixed model with the random factors farm and flock nested in farm. Chicks with a high CL reached the flock mates faster than chicks with a low CL depending on body mass (all birds, CL: $F_{1,71}=3.95$, P=0.05, interaction between CL and body mass: $F_{1,71}=3.82$, P=0.054, n=81, without timeout, CL: $F_{1,58}=7.8$, P=0.007, interaction between CL and body mass: $F_{1,58}=8.1$, P=0.006, n=67). *Campylobacter* in chickens is not subclinical so an effect of CL on speed in either direction was expected. It is unclear whether the test measures sociality or anxiety. Various degrees of pododermatitis were found in 61% of the chicks but were not associated with CL ($x^2=1.3$, P=0.26, n=93) or speed ($x^2=2.6$, P=0.10, n=93) and did not vary with body mass ($F_{1,313}=1.57$, P=0.21, n=319). The prevalence of hockburn was low (18%) and not associated with CL ($x^2=0.06$, P=0.81, n=93). However, chicks without hockburn reached the flock mates faster than chicks with hockburn ($x^2=4.7$, P=0.03, n=93). Chicks without hockburn were lighter than chicks with hockburn ($F_{1,313}=10.6$, P=0.001, n=319). We conclude that individual variation in movement was associated with CL interacting with body mass and could lead to differences in optical flow in *Campylobacter* positive and negative flocks.

## Changes in activity and feeding behaviour as early-warning signs of respiratory disease in dairy calves

*Marie J. Haskell, Jenna M. Bowen, Gemma A. Miller, David J. Bell, Colin Mason and Carol-Anne Duthie*
*SRUC, Animal and Veterinary Sciences, West Mains Road, Edinburgh EH9 3JG, United Kingdom; marie.haskell@sruc.ac.uk*

Respiratory disease is a major cause of calf mortality early in life. In surviving calves, the resulting lung damage can result in poorer health, fertility and growth in later life. Early detection of disease in calves is important, as it allows for early treatment which reduces the severity and duration of the disease for the affected individual and reduces disease spread to group-members. Disease on-set is associated with changes in behaviour. These changes are often termed sickness behaviour, and can be defined as the behavioural response to infection or injury. Behavioural responses to disease include decreased appetite, increased lethargy and reductions in social and exploratory behaviour. The use of technology offers a means of detecting these behavioural responses on commercial farms, as behaviour can be monitored remotely and continuously over long periods. This study aimed to determine whether changes in activity and feeding behaviour can be used as early-warning indicators of respiratory disease in dairy calves. One hundred entire male Holstein pre-weaned calves (age ~8-42 days) were used. The calves were housed in groups of 14 in an 'igloo' + yard system, with a total floor area of ~62.5 m. Each calf was fitted with a tri-axial accelerometer on the hind leg to record standing and lying time. Feeding behaviour was recorded through automatic computerised milk feeders. To detect the presence of respiratory disease, each calf was assessed daily using a modified version of the 'Wisconsin Scoring System', which involves recording rectal temperature, coughing, nasal and ocular discharge. The peak day of the most extreme illness event (i.e. day of disease with the highest Wisconsin score) was identified for each calf experiencing illness. Each event and the days before and after were paired with the corresponding days from a healthy calf of similar age and weight. Data for 22 disease events were analysed to assess differences in behaviour (PROC MIXED, SAS v9.3). The results showed that compared to healthy calves, ill calves lay for longer and tended to have longer lying bouts (daily lying: 17.6±0.3 vs 16.7±0.2 h, P<0.01; bout length: 74.8±10.6 vs 56.0±3.7 min, P=0.09 for ill and healthy calves respectively). Ill calves fed for a shorter time and had fewer feeder visits (with intake) each day compared to healthy calves (feeding time (min): 19.3±1.4 vs 22.8±1.5; P<0.05; visits: 2.1±0.2 vs 3.2±0.4; P<0.05). Importantly, differences between ill and healthy calves were evident in both activity and feeding behaviour on days prior to illness being detected through clinical signs. Lying bout length was higher in ill calves for 2 days prior to illness being detected (P<0.05), lying time was longer on Day -1 (P<0.05) and feeder visits with consumption were less frequent 3 days prior to clinical detection (P<0.05). The results show that assessments of both activity and feeding behaviour can be used as early warning indicators of respiratory disease in calves, and that sensors and precision farming technologies are effective methods of capturing this data.

## Using calves' behavioural differences to design future control measures for important zoonotic pathogens

*Lena-Mari Tamminen[1], Ulf Emanuelson[1] and Linda Keeling[2]*
*[1]Swedish University of Agricultural Sciences, Department of Clinical Sciences, Box 7054, 75007 Uppsala, Sweden, [2]Swedish University of Agricultural Sciences, Department of Animal Environment and Health, Box 7068, 75007 Uppsala, Sweden; lena.mari.tamminen@slu.se*

Within cattle herds, the transmission and resilience of VTEC O157:H7 (a serotype of verotoxin-producing *Escherichia coli* – known for its life threatening complications in humans) is dependent on a small proportion of cattle who become colonised and shed high numbers of the bacteria. Reducing the proportion of these animals is considered key for decreasing the prevalence of VTEC O157 and preventing transmission to humans. However, as the animals do not show any clinical symptoms this is difficult. The aim of this observational study was to analyse differences between colonised and non-colonised animals with a view to increasing understanding of the drivers in the colonisation process and potentially enabling identification of colonised animals. The study included calves between 7 and 306 days of age from 12 Swedish dairy farms (n=26 pens) where the presence of this highly virulent strain of VTEC O157 had been established. Behavioural observations of undisturbed animals, fearfulness to human approach and other external clinical assessments were followed by sampling at the individual level using recto-anal mucosal swabs for assessing if animals were colonized with VTEC O157. During restraining for sampling, animal reactivity was scored. Behavioural observations and animal-based indicators for colonised animals (n=56) were compared with non-colonised calves (n=135) housed in the same pens (and therefore of similar age) using elastic net regression. The model was run with 10-fold cross validation and the model that generated the smallest mean square error was selected. Logistic regression was performed on the variables included in the elastic net model and backwards selection using Akaike information criterion used to reduce the model further. Colonisation was associated with more oral behaviours, such as self-licking (P<0.01), tongue rolling (P<0.05) and higher frequency of oral manipulation of the environment (P<0.05). Coughing, having diarrhoea and poor ruminal fill (P<0.01) was negatively associated with colonisation. Non-colonised calves were more likely to be observed being displaced and displacing others (P<0.001) but the risk of being colonised increased with the number of displacement events in the group (P<0.01). These results indicate that colonised calves are not found among the sick, stressed and so potentially immunocompromised calves in a group, as we had predicted, but among the animals that are active, grooming and exploring. The association between oral behaviours and colonisation may indicate that the main driver of colonisation is simply increased exposure to the bacteria by some individuals. Thus, improved hygiene measures to reduce oral exposure on infected farms could be an important method to reduce the number of colonised animals. Using this approach for studying how individual differences influence susceptibility and transmission of zoonotic pathogens can provide important information for designing control-measures and interventions in the future.

## Dairy cows with Johne's disease spend less time lying and increase feed intake around peak lactation

*Gemma Charlton[1], Vivi Thorup[2], Dimitris Dimitrelos[2] and S. Mark Rutter[1]*
*[1]Harper Adams University, Edgmond, Shropshire, TF10 8NB, United Kingdom, [2]IceRobotics, Bankhead Steading, South Queensferry, Edinburgh, EH30 9TF, United Kingdom; smrutter@harper-adams.ac.uk*

Johne's disease (JD), also known as paratuberculosis is a fatal, chronic enteritis of ruminants that has detrimental effects on production and health and significantly reduces animal welfare. A recent study found that Johne's positive cows reduced their lying time around peak lactation compared to Johne's negative cows, but step count was not different. This suggests that the cows were standing idle or possibly standing eating. Feeding behaviour and intake were not recorded. Therefore, the objective of this study was to compare feed intake and activity of cows naturally infected with JD (Johne's positive) to Johne's negative cows from early to mid-lactation. Ethical approval was given by The Animals Experimental Committee of Harper Adams University. We hypothesized that due to the impact of JD on the intestinal tract which can affect the absorption of nutrients and proteins that Johne's positive cows will increase their feed intake when production demand is at its highest. The study was conducted using 66 multiparous (3.0±0.17 (Mean±SEM); range: 2-7 lactations) Holstein Friesian cows at Harper Adams University, UK. The cows were fitted with an IceQube® accelerometer (IceRobotics Ltd, Edinburgh, UK) on the back left leg. The sensors recorded lying duration (h/d), frequency of lying bouts (LB/d), average lying bout duration (LBD; min/bout) and steps (S/d). Feed intake was recorded using RIC intake feeders (Marknesse, the Netherlands). Approximately every three months the cows were milk sampled, and subsequently tested for JD using an ELISA. Cows in infection groups JD0-JD2 were classed as low risk and cows in infection groups JD3-JD5 were classed as high risk. JD0 cows (Johne's negative; n=38 (>2 repeat ELISA –ve)) were matched to JD5 cows (Johne's positive; n=28 (>2 repeat ELISA +ve)) based on parity and age. Feed intake and activity data were averaged and analysed weekly. Two sample T-Tests (Genstat, 18th edition, VSN International Ltd, UK) revealed that JD5 cows ate significantly more than JD0 cows during weeks 5, 7, 9 and 11 of lactation (P<0.05). The largest difference was observed during week 11 with average DMI of 3.1 kg/d higher for JD5 cows (P=0.006; 24.3 vs 21.2 kg of DM/d for JD5 vs JD0 cows, respectively). Lying duration was different (P<0.05) with JD5 cows spending less time lying during weeks 5, 7 to 11 and week 13 of lactation. During week 8, JD5 cows spent, on average, 1.4 h/d less lying compared to JD0 cows (P=0.005; 9.5 vs 10.9 h/d, respectively). JD5 cows also had fewer lying bouts/d (P<0.05). During week 11, JD5 cows had three lying bouts less than JD0 cows (P<0.001; 9.2 vs 12.2 LB/d, respectively). Milk yield was higher (P<0.05) for JD5 cows during weeks 13 and 17 (week 17; P=0.017; 40.3 vs 36.9 kg/d for JD5 vs JD0, respectively). The results show that during peak lactation, Johne's positive cows alter their lying behaviour, and increase their food intake over several weeks but only marginally increase their milk yield. The integration of data from commercially available sensors that can monitor these behaviours has the potential to help in the on-farm diagnosis of JD in dairy cows.

## Animal welfare: an integral part of the United Nations Sustainable Development Goals?

*Linda Keeling[1], Håkan Tunon[1], Gabriela Olmos[1], Charlotte Berg[1], Mike Jones[1], Leopoldo Stuardo[2], Janice Swanson[3], Anna Wallenbeck[1], Christoph Winckler[4] and Harry Blokhuis[1]*
*[1]Swedish University of Agricultural Sciences, Almas Allé 8, 750 07 Uppsala, Sweden, [2]World Organisation for Animal Health, 12 rue de Prony, 75017 Paris, France, [3]Michigan State University, E Lansing, MI 48824, USA, [4]University of Natural Resources and Life Sciences, Gregor Mendel Str 33, 1180 Vienna, Austria; linda.keeling@slu.se*

Is animal welfare (AW) affected by the United Nations Sustainable Development Goals (SDGs) and does the improvement of AW have a role in achieving the SDGs? Our aim was to systematically evaluate these questions. The initial analyses were based on a series of discussions and individual scoring among 12 experts from environmental, agricultural and veterinary sciences. The strengths of the links between improving AW and achieving each of the 17 SDGs (and vice versa) were scored on a 7-point scale, ranging from being completely indivisible (score +3) down to where it was considered impossible to reach both the SDG and improved AW at the same time (score -3). The mean of the scores for each of the 17 SDGs was always positive, indicating that working to achieving the SDGs is compatible with working to improve AW and that overall there is a win-win scenario even if AW is never explicitly mentioned in the SDGs. However, the impact of achieving the SDGs on AW was considered to be slightly better (mean score=1.15) than the impact of improving AW on achieving the SDGs (mean score=0.89). The exception to this was SDG 2 (zero hunger), where the impact of improved AW on enabling the SDG was stronger than the effect of achieving the SDG on improving AW (P<0.05, Wilcoxon). The strength of the associations helped to identify three broad categories of SDGs in relation to AW. The first category can be described as mutually reinforcing agendas, the related SDGs mostly dealing with 'social' issues like poverty alleviation, production and consumption. It includes two SDGs for which there was strongest mutual reinforcement (win-win): SDG 12, dealing with responsible production and consumption, and SDG 14 dealing with life below water. This is not surprising considering the long established discussions on sustainable agriculture and fishing practices. A second category included SDG 9, with the weakest link to AW, dealing with industry, innovation and infrastructure. The remaining SDGs in this second category also dealt with 'economical and technical' issues, such as affordable energy, sustainable cities and climate. As for the final category, the effects were seen as win-win but, unlike the other categories, the impacts were not symmetrical. Most of these SDGs were anthropocentric, related to education, equality, peace and justice. Our conclusions relate mainly to the methodological approach within a pilot context. These results need to be confirmed with a larger, more diverse group of stakeholders. Moreover, we foresee that this methodology will highlight different associations under case specific scenarios, thereby targeting areas where progress is most effective. Where asymmetries exist, the methodology identifies which approach may be the better 'driver' towards progress.

## Systems modelling of the UK pig industry: implications for pig health, behaviour and welfare

*Conor Goold[1], Mary Friel[1], Simone Pfuderer[2] and Lisa Collins[1]*
*[1]University of Leeds, Faculty of Biological Sciences, Leeds, LS2 9JT, United Kingdom, [2]University of Reading, School of Agriculture, Policy and Development, Reading, RG6 6AH, United Kingdom; c.goold@leeds.ac.uk*

Pork is the most consumed meat on the planet, with per capita consumption predicted to double by the year 2050. The UK pig industry is known for its high welfare standards, with 40% of the national breeding herd being managed on outdoor units. For the UK pig industry to benefit from the growing global market, it needs to respond to a number of challenges and become resilient to future shocks. These shocks include volatile feed prices, competition from international trade, disease risks, changing consumer attitudes and climate change. Nonetheless, there are no predictions of how resilient the UK pig industry is against these potential shocks. Importantly, the effects of shocks on the health, behaviour and welfare of pigs themselves are unknown. For example, rising ambient temperatures could increase the chances of heat stress in pigs, which would be expected to decrease farrowing rates, decrease food intake, slow growth rates and increase the chances of welfare-compromising behaviours such as tail biting. The PigSustain project is investigating the resilience of the UK pig industry to future shocks and changes using systems modelling techniques. Systems modelling entails taking a holistic approach to simulating the activities of, and interactions between, components of complex biological systems. In particular, PigSustain is applying agent based and system dynamics modelling to simulate the behaviour of agents within the pig supply chain. This approach is informed by both quantitative data collected on the UK pig industry (e.g. industry structure, farm-level production and financial data) as well as qualitative data gathered during stakeholder engagement. In this presentation, we illustrate how we are using systems modelling to examine the relationship between pig health, welfare and behaviour on the resilience of the UK pig industry. The model agents include different types of pig farms (e.g. breeding and finishing sites), pork processing units, and retailers. Agents have their own properties (e.g. breeding farms hold a number of breeding sows, retailers hold kilograms of pork) updated by dynamic submodels (e.g. the breeding cycle of pigs, product pricing mechanisms, consumer demand). The model includes heterogeneity in production parameters (such as mortality rates, sow productivity, and pig growth) as well as variation in production systems (e.g. indoor/outdoor, batch-farrowing systems). Variables such as pig growth, mortality rates, sow productivity and disease dynamics are being encoded as functions of current industry dynamics (e.g. changes in consumer demand) as well as exogenous variables such as changes in disease risk and feed prices. This will allow us to make forecasts of these variables under different future scenarios presenting challenges to the industry, as identified through stakeholder engagement. We conclude that systems modelling is a powerful tool to understand animal behaviour and welfare within complex anthropogenic systems like the UK pig industry.

## Environmental impact of animal welfare improvement measures in dairy farming – model calculations for Austria

*Anna Christina Herzog[1], Stefan Hörtenhuber[1,2], Christoph Winckler[1] and Werner Zollitsch[1]*
*[1]University of Natural Resources and Life Sciences, Vienna, Department of Sustainable Agricultural Systems, Division of Livestock Sciences, Gregor-Mendel-Strasse 33, 1180 Vienna, Austria, [2]Research Institute of Organic Agriculture (FiBL) Austria, Doblhoffgasse 7/10, 1010 Vienna, Austria; a.herzog@students.boku.ac.at*

Dairy farming has a considerable impact on total anthropogenic greenhouse gas emissions (GHGE) and is therefore attributed a significant share in achieving global sustainability goals. Among numerous measures to mitigate emissions, good animal health and welfare are considered essential to keep the sector's environmental impact low. However, little is known about the effect of interventions aimed at improving welfare on the emission level of dairy farms. The aim of the current study was therefore to quantify the impact of selected welfare improvement measures on a production system's potential for global warming ($GWP_{100}$), acidification (AP) and freshwater eutrophication (EP), using a case study for Austria. To this end, three farm models were built, representing typical production conditions in different agricultural areas of the country. After defining region-specific feeding regimes and corresponding yield levels, as well as housing and manure management systems, we calculated total greenhouse gas and ammonia emissions from cradle to farm gate for each production system, based on IPCC and EMEP EEA guidelines. The environmental impact potential was assessed in kg $CO_2$, $SO_2$ and P equivalents per unit of product (ECM) using LCA methodology. In a second step, we assumed that commonly suggested measures to improve dairy cow welfare, notably soft flooring, heat mitigation and increased cleaning frequency, would be implemented. Expected improvements in welfare and performance as well as respective changes in the farms emission potential were modelled based on current literature. Subsequently, we recalculated GWP, AP and EP and compared the results to the reference values of each farm obtained in the first assessment. Preliminary results show that under Austrian production conditions, the implementation of rubberized alley flooring may reduce GWP, AP and EP by up to 2.0, 1.6 and 0.7% per unit of product, respectively, although the improvement measure per se accounts for an increase in the overall emission potentials ranging between 0.1 and 0.4% compared to the reference. Similar results have been calculated regarding the measure heat mitigation. Uncertainty analysis is still to be conducted to evaluate the relevance of these results in contrast to variations in input parameters. Nevertheless, our present findings confirm assumptions in the literature about welfare improvement as a means to promote emission mitigation in dairy farming, since the emission reducing effect of good welfare seems to more than outweigh the emission potential of the welfare improvement measure.

## Automated tracking of individual activity of broiler chickens

*Malou Van Der Sluis[1,2], Britt De Klerk[3], T. Bas Rodenburg[1,4], Yvette De Haas[2], Thijme Hijink[4] and Esther D. Ellen[2]*
*[1]Utrecht University, Animals in Science and Society, P.O. Box 80163, 3508 TD Utrecht, the Netherlands, [2]Wageningen University & Research, Animal Breeding and Genomics, P.O. Box 338, 6700 AH Wageningen, the Netherlands, [3]Cobb Vantress, Koorstraat 2, 5831 GH Boxmeer, the Netherlands, [4]Wageningen University & Research, Adaptation Physiology Group, P.O. Box 338, 6700 AH Wageningen, the Netherlands; malou.vandersluis@wur.nl*

There is a growing interest in quantifying individual behaviour of group-housed animals and its relation to individual performance. Broiler chickens are an example of a livestock species for which individual data can be valuable. Broiler breeding goals focus on efficient growth and reproduction, as well as welfare indicators for balanced genetic improvements. Recording of broiler behaviour can provide insight into welfare indicators, such as activity or general leg health. However, monitoring individual behaviour in group-housed animals is a challenge. Often, video analyses are used, but these are time-consuming and prone to human error. Therefore, automated systems for monitoring individual animals are desired. Here, we studied whether individual broiler activity could be tracked using an ultra-wideband (UWB) system. Birds were fitted with UWB tags that sent out signals to four beacons. The location of the birds was determined using triangulation of the signal, allowing calculation of distances moved over time. Distances moved according to the UWB system were compared to those found on video for twelve birds. A moderately strong correlation between the UWB system and video tracking was found (Repeated measures correlation, $r=0.71$ (95%-CI: 0.64-0.77), df=209, P<0.001). Furthermore, the UWB system was used for assessing individual levels of activity. In total, 137 birds from different genetic crosses were tracked near-continuously for seventeen consecutive days, starting on day 16 of life, and their weight was determined at the start and end of this 17 d period. Data were analysed using an LME-model in R. First analyses showed that activity, measured as the average distance moved, decreased over the seventeen days in all genetic crosses ($F(1,127.00)=301.47$, P<0.001). Furthermore, in all genetic crosses, birds with a lower weight at the start of the trial were on average more active ($F(1,125.14)=9.16$, P<0.01). Overall, the UWB system appears well-suited for activity monitoring in broilers and the longitudinal information on individual differences in activity can potentially be used to monitor health, welfare and performance at the individual level. Unfortunately, the UWB tags are too large and heavy for day-old chicks to wear and can only be implemented later in life. Therefore, current work is focussing on the implementation of a passive radio frequency identification (RFID) system to track individual broiler activity, using smaller, lightweight tags that can be attached to the broilers' legs at day-old. This system may be able to track individual activity of broiler chickens throughout the entire life.

## Assessment of open-source programs for automated tracking of individual pigs within a group

*Kaitlin Wurtz[1], Tomas Norton[2], Janice Siegford[1] and Juan Steibel[1,3]*
*[1]Michigan State University, Department of Animal Science, 474 S. Shaw Ln Room 1290, East Lansing, MI 48824, USA, [2]KU Leuven, Department of Biosystems (BIOSYST): Division Animal and Human Health Engineering, Kasteelpark Arenberg 30, 3001 Leuven, Belgium, [3]Michigan State University, Department of Fisheries and Wildlife, 480 Wilson Rd #13, East Lansing, MI 48824, USA; wurtzkai@msu.edu*

Knowledge of individual pig behavior within groups is necessary to maintain efficient production and positive welfare. As increasing numbers of pigs are being housed in groups worldwide, and as these groups grow in size, there is increasing demand for automated behavioral monitoring. Precision livestock farming tools such as animal-based sensors (e.g. RFID, accelerometers, or GPS) and environment-based sensors (e.g. microphones, digital cameras, or depth cameras) can generate continuous, real-time data for management, research, and selection. The aim of this study was to select and test various freely-available visual detection software programs using video footage of pigs that was previously decoded manually. Understanding the limitations of detection programs can help make systems more robust and user-friendly for future use. Seven open-source tracking software were identified for testing: idTracker, ToxTrac, Tracktor, BioTracker, CADABRA, Ctrax, and SwissTrack. The video consisted of 1 min recordings of the following: (1) top-down, single purebred Yorkshire pig; (2) top-down, pair of Yorkshire pigs; (3) top-down, pair of multi-colored pigs; (4) angled-view, pair of multi-colored pigs; and (5) top-down, pen of 8 purebred Yorkshire pigs. Recordings were strategically selected to address areas where we know software fails: occlusion (1 vs 2), segmentation (2 vs 3), angled camera angle (3 vs 4), and pens of multiple pigs (5). Systematic testing of software was performed by an ethologist with minimal coding knowledge. Systematic records of ease of program installation, system requirements, simplicity of setting tracking parameters, system failures and errors, as well as system outputs were made. At present, preliminary testing has been conducted with idTracker and ToxTrac on video segments 1 and 2. Installation and start-up of both programs were intuitive. After importing video, idTracker required selection of a region of interest, removing the background (done automatically by the program), adjusting the pixel intensity threshold until the pigs were highlighted, and entering the minimum pixel grouping size to be counted as an individual. There was also an option to invert the video contrast to account for light animals on a dark background. ToxTrac required a similar procedure, however this program required more trial and error as a range was required for the color values (pixel intensity) and both a minimum and maximum pixel grouping size were necessary. Both programs were able to track 1 or 2 individuals from 1 min of video in about 3 min. Further testing will be performed on the remaining programs, and a detailed assessment of accuracy and precision will be computed for each using a standardized method. Our results thus far suggest that theoretical applications of these software are promising; however, in practice these technologies may not yet ready to be incorporated off the shelf into real-world applications.

## What do we know about the link between ill-health and tail biting in pigs?

*Anna Valros[1], Andrew M. Janczak[2], T. Bas Rodenburg[3], Janicke Nordgreen[4] and Laura Boyle[5]*
*[1]Faculty of Veterinary Medicine, University of Helsinki, Research Centre for Animal Welfare, P.O. Box 57, 00014 Helsinki, Finland, [2]Faculty of Veterinary Medicine, Norwegian University of Life Sciences, Department of Production Animal Clinical Science, Ullevålsveien 72, 0454 Oslo, Norway, [3]Faculty of Veterinary Medicine, University of Utrecht, Department of Animals in Science and Society, Yalelaan 2, 3584 CM Utrecht, the Netherlands, [4]Faculty of Veterinary Medicine, Norwegian University of Life Sciences, Department of Food Safety and Infection Biology, Ullevålsveien 72, 0454 Oslo, Norway, [5]Teagasc, Animal and Grassland Research and Innovation Centre, Moorepark, P61 C996 Fermoy, Co. Cork, Ireland; janicke.nordgreen@nmbu.no*

Tail biting in pigs is a serious and prevalent damaging behaviour leading to reduced welfare and production. Anecdotal reports suggest that health challenges increase the risk of tail biting in pigs. Further, several risk factors are common to both tail biting and poor health. One aim in COST Action GroupHouseNet is to review studies on the link between health and tail biting and to propose hypothetical causal relationships. Tail biting causes lesions which may become infected and potentially act as a route for the spread of infection to other parts of the body. On-farm studies suggest a link between health problems such as respiratory disease and lameness, on the one hand, and tail biting on the other. Recent experimental studies suggest a possible causal link between an artificially-induced inflammatory response and an increase in tail biting-related behaviour, and a link between cytokines and tail biting-related behaviour. Although the underlying mechanism is unknown, one suggestion is that the sickness response might act to increase tail biting via changes in cytokines and altered neurotransmission. Suboptimal management and housing can increase the risk of both tail biting and health problems. Risk factors common to tail biting and health problems include suboptimal thermal climate, poor hygiene, high stocking density and poor feed quality. Therefore, it is plausible that improvements to management and housing to increase the health of pigs will also reduce the risk of tail biting. Tail biting can also increase the risk of health problems. Firstly there is a direct spread of pathogens between biters and victims, and secondly, pathogens can enter the body via the tail lesion. Once infected, systemic spread of infection may occur mainly via the venous route targeting the lungs, and to a lesser extent via lymphatic spread. Studies of slaughtered pigs report an increase in lung lesions, abscessation and arthritis in carcasses with tail lesions. In summary, there is a clear link between poor health and tail biting in pigs, but there is only preliminary evidence suggesting causal relationships, so caution is warranted.

## Validation of accelerometers to automatically record postures and number of steps in growing lambs

*Niclas Högberg[1], Johan Höglund[1], Annelie Carlsson[2] and Lena Lidfors[2]*
[1]*Swedish University of Agricultural Sciences, Department of Biomedical Sciences and Veterinary Public Health, BVF, Box 7036, 75007 Uppsala, Sweden, [2]Swedish University of Agricultural Sciences, Department of Animal Environment and Health, HMH, Box 234, 53223 Skara, Sweden; niclas.hogberg@slu.se*

We validated the accuracy of two commercially available activity loggers for cattle in determining lying and standing durations, number of lying bouts and number of steps in growing lambs. Ten growing lambs divided into two weight classes were fitted with an IceTag logger (IT) (IceRobotics Ltd) on the right rear leg and an IceQube logger (IQ) on the left rear leg. The outer plastic casing of the IT was removed to minimize weight and size. Loggers were attached above the metartarso-phalangeal joint with veterinary self-adherent bandage and padding to minimize chafing. No logger fell off during the study. The IT logger reports activity per second, whereas the IQ reports activity in 15-min periods. To enable comparison between loggers, IT data were also summarized in 15-min periods. A computed indication for the start of a lying bout of durations >10 sec was also performed. Analyses of the lambs body posture and number of steps per second from video recordings of totally 50 hours were used as a golden standard to determine the accuracy of the two loggers. Statistical analyses were performed in R (3.4.3). Two observers scored the two different groups and inter-observer reliability was consistent for standing, lying and number of lying bouts (Cohen's Kappa=0.99). Even though observers defined step differently, no difference of scoring could be observed (paired t-test, P=0.92). A linear regression analysis showed that the standing and lying time given by both loggers were correlated with the video recordings ($IT_{stand}$: $R^2>0.99$; $IT_{lie}$: $R^2>0.99$; $IQ_{stand}$: $R^2>0.99$; $IQ_{lie}$: $R^2>0.99$). The positive predictive value (PPV), sensitivity and specificity of the IT logger compared to video recordings per second for standing and lying were all >91.5%. The IT logger showed a poor PPV (<44%) and sensitivity (<91%) for lying bouts, whereas the IQ logger showed a better PPV (<92%) and sensitivity (<88%). The performance improved with the computed indication for lying bouts, for IT (PPV: 100% and sensitivity: 89%) and IQ (PPV: 98% and sensitivity: 100%), respectively. However, a difference between the loggers ability to correctly record step count was found. The IT logger corresponded better with the video recordings ($R^2>0.93$ and $R^2>0.91$) whereas the IQ showed a poorer correlation ($R^2>0.79$ and $R^2>0.85$) for number of steps. We conclude that both loggers are able to record standing and lying time accurately. The ability to record number of lying bouts is poor for the IT logger but increases if bouts<10 sec is disregarded, whereas the ability of the IQ logger to record true lying bouts is adequate. Furthermore, the IT logger is able to record step count accurately, while the IQ is less accurate.

## Have the cows hit the wall: validation of contact mats to monitor dairy cow contact with stall partitions

*Caroline Freinberg[1,2], Athena Zambelis[1] and Elsa Vasseur[1]*
*[1]McGill University, Animal Science, 21111 Lakeshore Road, Sainte-Anne-de-Bellevue, QC H9X 3V9, Canada, [2]Columbia University, Environmental Biology, 116th St & Broadway, New York, NY 10027, USA; elsa.vasseur@mcgill.ca*

The shift of animal agriculture to more intensive systems has led to an increase in indoor housing, which restricts animals' movement and may cause potential for discomfort and injury. Repetitive contact with the stall partitions may reflect problems between the housing environment and the cow, by reducing the quality of rest and hindering ease of movement. However, little is known about the actual frequency of contact between the animal and the confines of its housing environment. The objective of this study was to validate the ability of a contact mat (CM) system to monitor cow contact with stall dividers and tie-rail when compared to visual observation. Eleven lactating cows were monitored both visually and with the CM system for 4 h/d for 4 consecutive days (176 h of observation total) in a tie-stall housing system. Individual CM were affixed to the stall dividers and tie-rail to record the frequency of cow contact per second based on a minimum of 11.3 kg of pressure required to produce a digital output. Two trained observers recorded the frequency of cow contact against the stall partitions on a per second basis using 3 criteria: cow contact with the stall dividers or tie-rail regardless of CM contact (Total Rail Contact), cow contact with the CM regardless of the placement or force used (Total CM Contact), and cow contact with the CM through proper placement and sufficient force to effectively create an output signal (Effective CM Contact). Spearman's correlations were run to compare visual observation and CM system agreement on the ranked position of each cow based on the average number of contacts per min with the dividers or tie-rail. The correlation strength between visual observation and CM system varied from moderate to very high depending on the visual observation approach used (Dividers – Total Rail Contact: $r_s$=0.68, Total CM Contact $r_s$=0.90, Effective CM Contact $r_s$=0.90; Tie-Rail – Total Rail Contact: $r_s$=0.71, Total CM Contact $r_s$=0.66, Effective CM Contact $r_s$=0.58). The results suggest that CM can be used to accurately rank dairy cows based on individual frequency of contact with the stall dividers. CM can be used to identify individual cows that may require intervention for stall comfort risk factors, as well as to assess whether different indoor housing environments offer fewer constraints and therefore better movement opportunities for dairy cows. A generalization of this assessment tool could be made to other species of animals kept in confinement.

## Welfare aspects related to virtual fences for cattle

*Silje Eftang[1,2] and Knut Egil Bøe[2]*
*[1]Nofence AS, Evjevegen 8, 6631 Batnfjordsøra, Norway, [2]Norwegian University of Life Sciences, Department of Animal and Aquacultural Sciences, P.O Box 5003, 1432 Ås, Norway; silje.gunhild.eftang@nmbu.no*

Virtual fencing (VF) is new technology facilitating extensive production by giving livestock flexible access to pastures. The Nofence VF system establishes specific pasture borders defined by GPS-coordinates conveyed to animals via signals from a neck collar. When an animal crosses a virtual border, the collar starts playing a warning sound (WS). A mild electric shock (ES) is delivered by the collar if the animal does not respond by turning around upon hearing the sound. If an animal ignores three successive WS and ES, it is registered as escaped and the system is switched off until the animal has returned to the pasture. During autumn 2018, we performed two studies in Norway relating to the welfare of cattle fitted with Nofence collars. These studies received ethics approval from the Norwegian Food Safety Authority. Study 1 aimed to investigate responses to three different strengths of ES, to find a magnitude that was aversive enough to keep all the cattle within the VF while still being acceptable for animal welfare. Norwegian Red (NR) heifers were divided among three pastures, six in each, and a virtual border was established on one side of each pasture. The groups were exposed, respectively, to: A) Weakest strength: 1.5 kV/ 0.1 Joule, B) Double strength: 1.5 kV/ 0.2 Joule, or C) Highest strength: 3 kV/ 0.3 Joule. Heifers in group A received a mean (SD) of 15.6 (±37.5) WS and 1.9 (±2.2) ES/ animal over a period of 11 days (including the first days of learning). Four animals in this group escaped once. Group B heifers received a mean of 7.7 (±9.0) WS and 1.1 ES (±1.3) while in group C, the means were, respectively, 4.4 (±6.9) and 0.9 (±1.7)/ animal. No Group B or C animals escaped. Study 2 aimed to investigate how two different learning methods for adapting to VF, assisted and unassisted, affected the number of ES received. NR heifers were divided among two pastures, six in each, with collars set at 1.5 kV/0.2 Joule based on results of study 1. A virtual border was set on one side of each pasture. One group received an assisted introduction to the location of the border, with humans present outside the border preventing the heifers of running forward and out of the pasture, while the second group was unassisted. Heifers with assisted learning received a mean of 6.6 WS (±6.4) and 1.2 ES (±1.6)/ animal over a period of 5 days. In the unassisted group, heifers received a mean of 6.0 WS (±5.6) and 1.5 ES (±1.9)/ animal. The mean for both groups regarding WS was reduced from 11.4 (±6.4) at Day 1 to 3.6 (±4.5) at Day 5, while the mean number of ES was reduced from 4.3 (±1.4) at Day 1 to 0.6 (±0.8) at Day 5. The weakest electrical signal resulted in the most shocks and was insufficient to prevent escapes, and the assisted method did not seem to improve learning. These results are important for practical implementation of VF while gaining the welfare benefits of access to pasture in areas where physical fencing is unfeasible.

# Evaluating unmanned aerial vehicles for observing cattle in extensive field environments

*John Scott Church, Justin Mufford, Garrett Whitworth and Joanna Urban*
*Thompson Rivers University, Natural Resource Sciences, 805 TRU Way, Kamloops, British Columbia, V2C 0C8, Canada; jchurch@tru.ca*

The aim of this study was to determine the utility and limitations of consumer vs professional level unmanned aerial vehicles (UAVs), commonly known as drones, for use in identifying individual cattle and observing different behavioural states such as grazing in extensive field settings. All drones tested were from one manufacturer (DJI Phantom 4, DJI Inspire 1, DJI M600; SZ DJI Technology Co. Ltd.) due to their reliability, affordability, ease of use, and dominant market share. We attempted to locate individual cattle in three different environments; pasture (cut blocks), a low-density and/or a high-density forested setting during multiple flights at aerial heights between 40 and 100 m. DJI quadcopter drones were equipped with a conventional red, green blue (RGB) camera (DJI Phantom 4K UHD, 20.8 MP) or an infrared thermography (IRT) camera (DJI FLIR XT, 60 Hz frame rate, 336×256 resolution; DJI Inspire 1). A larger hexacopter was used to carry a higher magnification camera (DJI Z30, RGB 30x optical, 6x digital; DJI M600). Cattle were easily located by either RGB or IRT in open pasture land as well as in low density forested sections at distances up to 400 m; however, animals were often concealed by high-density forest canopies. In addition, consumer level RGB and IRT cameras are limited in identifying specific individual animals. The DJI Z30 zoom camera; however, provided exceptional detail of the under-canopy area at an angle of 45 to 60 degrees from the horizontal plane of the UAV, even under high-density conditions. Using this camera, we could explore significant areas of rangeland (~100 ha) and were able to successfully identify cattle underneath the forest canopy at over double the distance (300 vs 750 m) of conventional and IRT camera payloads (P<0.01). In addition, by using the DJI Z30 camera we could easily identify individual animals from conventional dangle ear tags at an impressive height of between 70 and 100 m, whereas we could not identify animals with the conventional camera without disturbing them. Furthermore, from this height we could still identify different behavioural states such as grazing, walking, resting and even ruminating, which was generally not possible with the consumer grade UAV or IRT cameras. In conclusion, we determined that inexpensive DJI drones equipped with consumer grade cameras provide some utility as a tool to study cattle behavior at lower altitudes (i.e. <50 m); but individual identification is challenging, especially under forest canopies and will likely be noticed by some animal subjects. The upgrade to a more professional level camera (DJI Z30) and UAV (DJI M600) such as the one used in this study proved to be very effective in both identifying individual cattle and noninvasively observing their behaviour from impressive aerial heights (~100 m). The emerging use of professional level drone technology is unprecedented as an ethological tool, and will likely revolutionize the study of animal behaviour in a wide range of other settings such as feedlots, zoos, and parks by minimizing adverse observer effects on animals.

## Are feeding behaviors associated with recovery for dairy calves treated for bovine respiratory disease (BRD)?

*Melissa C. Cantor, Megan M. Woodrum and Joao H.C. Costa*
*Dairy Science Program, Department of Animal and Food Sciences, 404 W.P. Garrigus Building, Lexington, KY, 40546-0215, USA; melissa.cantor@uky.edu*

Changes in feeding behavior patterns recorded by an automatic milk feeder have been associated with pre-clinical signs of Bovine Respiratory Disease Complex (BRD), and may also be used as an indicator of recovery. This study aimed to investigate the association of feeding behavior patterns with calves failing to recover from BRD after treatment with antibiotics. This retrospective cohort study identified dairy calves (n=10 recovery, n=10 failure-to-recover) treated with antibiotics for the first time for BRD on one research facility using an automatic milk and solid feed feeder (*Förster-Technik*, Engen, Germany). Calves were weighed at birth and placed in a pen of 6±3 calves (mean ±SD) at 3±2 d of age. Milk replacer was offered at 10 L/d and calf starter was offered *ad libitum* from the automatic milk and solid feed feeders. Behaviors were recorded automatically: milk intake (l/d), average drinking speed (l/min/d), rewarded and unrewarded visits (visits/d), and calf starter consumption (kg/d). Calf symptoms for BRD (Calf Health Sheet, UW-Madison, Wisconsin, USA) were recorded daily between 7:00 to 9:00 hr and included assigning a symptom score of 0 to 3 for nasal discharge, eye discharge, ear position, coughing, and temperature. Then, the nasal score, highest eye or ear score, cough score, and temperature score were summed to assign a daily BRD score. A clinically sick calf was defined as a (>6 BRD score), and resulted in antibiotic treatment. In order to determine recovery status, from d 6 to d 10 after antibiotic treatment, daily BRD scores were averaged across the 4 ds for each calf; an average (<4 BRD score) was identified as recovery, the lack of response (>4 BRD score) was identified as a failure-to-recover based on Calf Health Sheet criteria. The MIXED procedure (SAS 9.4, Cary, NC, USA) was used to evaluate the fixed effect of recovery on feeding behaviors for 3 to 10 d after treatment. Treatment d was the repeated measure and calf was the subject. Calves were matched between the treatment by treatment age as a random effect. Milk intake (LSM ± SEM; recovery 6.15±0.70; failure-to-recover 6.05±0.70 l/d; P=0.92) and drinking speed (recovery 0.74±0.06; failure-to-recover 0.80±0.06 l/min; P=0.49) were not associated with recovery. Rewarded visits (recovery 3.46±0.54; failure-to-recover 3.90±0.54 visits/d; P=0.59) and unrewarded visits, (recovery 3.44±1.13; failure-to-recover 4.52±1.13 visits/d; P=0.52) were also not associated with recovery. Calf starter consumption was associated with recovery (recovery 0.47±0.07; failure-to-recover 0.22±0.07 kg/d; P=0.03.) Using calf starter consumption as a tool to identify calves failing to respond to antibiotic treatment may assist in earlier re-intervention, or identify calves to watch closely. We conclude that grain feeding behavior may be an indicator of recovery from Bovine Respiratory Disease Complex, while milk feeding behaviors may not be associated with a positive or negative response to antibiotic treatment.

## Smart ear tags for measuring pig activity: the first step towards an animal-oriented production environment

*Oleksiy Guzhva[1] and Sangxia Huang[2]*
*[1]Swedish University of Agricultural Sciences, Biosystems and Technology, Sundsvägen 16, Box 103, 230 53, Alnarp, Sweden, [2]Sony Mobile Communications AB, Mobilvägen 4, 221 88, Lund, Sweden; oleksiy.guzhva@slu.se*

Throughout past years, there has been an increasing interest towards the understanding of underlying factors influencing pig welfare, health and natural behaviour in modern production systems. With the vast variety of housing solutions, management routines, different enrichment options as well as local interpretations of welfare legislation, many factors are affecting the initial state of the animal. There is evidence showing that changes in animal activity and behaviour could be used as early warning indicators of ongoing pathophysiological processes (e.g. subclinical disease, stress) as well as possible flaws in the animal housing environment. However, manual observation of pig behaviour/activity is quite a time-consuming task and is impractical in the everyday farm workflow, considering the level of detail needed to be monitored for early intervention and understanding of the underlying factors. Therefore, sensor-based continuous monitoring of subtle changes in pig behaviour and activity could offer insights into animals' inner physiological state and allow more precise animal-based management practices. The aim of this pilot study was to investigate the potential of using smart ear tags for studying and classifying animal activity. The pilot study was performed on a conventional Swedish pig farm where animals were housed in groups (15 individuals per group, deep straw litter, manual feeding 2-times daily) and had smart tags to measure their activity. Custom-made smart ear tags, based on a 3-axillar accelerometer sampling at 10 Hz, were used for continuous (24/7) real-time monitoring of five different behaviours (lying, sitting, standing, walking, exploring) in sows. The behavioural activity data was recorded in three batches resulting in 6 weeks of continuous recordings for 18 individual sows. The performed behaviours were confirmed by experienced observers from video recordings and then converted into two activity states – passive (lying/sitting) and active (all the other behaviours). These two condensed activity states were needed to ensure robust prediction model performance under varying conditions. To build the initial classification model distinguishing between active and passive states, 20 hours of annotated activity data with an observational resolution of 1 second from 8 sows was used (2.5 hours per individual sow). The classification model was based on the Support Vector Machine algorithm, and its performance was validated via a 4-fold cross-validation procedure (folds were split by individuals). The study showed that by observing changes in daily activity (while comparing an individual sow to itself and the rest of the group) the detailed time budget could be formed and visualised, allowing earlier interventions in case of potential disease or changes in group dynamics.

## Sensor technology and its potential for objective welfare monitoring in pasture-based ruminant systems

*Eloise Fogarty[1], David Swain[1], Greg Cronin[2] and Mark Trotter[1]*
*[1]Central Queensland University, Institute of Future Farming Systems, CQIRP Rockhampton, QLD 4701, Australia, [2]The University of Sydney, Faculty of Science, Camden, NSW 2570, Australia; e.fogarty@cqu.edu.au*

Ruminant producers are often subject to multiple demands, usually with conflicting requirements, e.g. expectations to increase food and fibre production, reduce their environmental footprint, and still maintain high animal welfare standards. Digital technologies may provide a method of addressing these demands, though careful consideration of system design based on objective scientific principles is essential. The Five Domains Model (FDM) is a well-known system for welfare assessment, addressing five aspects of welfare that may be compromised or enhanced. The aim of this paper is to explore the fundamental requirements of sensor-based welfare monitoring systems for pasture-based livestock, using FDM as an evaluation tool. Nutrition: At a basic level, the use of sensors to monitor nutritional aspects of welfare will require assessment of food or water intake. This can be addressed through physical movement (e.g. jaw sensor) or aural cues (e.g. acoustic monitors). Supplementary measurements of food quality could also indicate welfare enhancement; e.g. through integration of location sensors (e.g. GPS) with external sensors (e.g. remote sensing of pastures) to relate measures of animal location with potential intake of quality feed. Environment: Measuring animal welfare within an environmental context is often linked to provision of appropriate thermal conditions and adequate space for movement. Given these are mostly extrinsic factors, sensors which monitor use of space (e.g. GPS) will be valuable. Again, integration with external sensors (e.g. climate sensors) could further improve welfare monitoring, especially in extreme conditions such as heavy rainfall or high temperatures which may be experienced in pasture-based systems. Health: Monitoring health could occur via two major sensor types; physiological (e.g. heart rate (HR) monitor) which provide a relatively direct measure and motion sensors (e.g. accelerometer) from which health status might be inferred. In these situations, sensors can indicate the likelihood of 'healthy' or 'unhealthy' status by monitoring deviation from baseline patterns. Behaviour: Sensors can monitor different behavioural states including feeding, sexual behaviour and movement and thus provide great benefit for welfare assessment. However, given behaviour is plastic and often contextual, some level of human operator judgement will be required to determine if the behaviour is related to 'good' or 'bad' welfare. Mental state: Mental state is fundamentally intrinsic and difficult to objectively measure. The exception is monitoring HR responses (e.g. rising HR indicating sympathetic system activation, which can then be related to affective state). This domain represents one of the greatest challenges in terms of sensor-based quantification. Digital technologies provide great potential for developing welfare monitoring systems, provided they are implemented objectively using systems such as FDM. Commercial systems will likely require integration of multiple sensor types for holistic assessment, incorporating information on animal location, activity, and potential affective state.

## Comparison of heat stress behaviour between different Canadian *Bos taurus* cattle breeds using unmanned aerial vehicles

*Justin Mufford, Mark Rackobowchuck, Matthew Reudink, Carmen Bell, Stefany Rasmussen and John Scott Church*
*Thompson Rivers University, Biological Sciences, 805 TRU Way, Kamloops, British Columbia, V2C 0C8, Canada; jt.mufford@gmail.com*

Heat stress is an emerging cause of mortality and production loss in *Bos taurus* beef cattle production in North America. As heat waves and summer temperatures are projected to increase, there is a growing need to investigate heat tolerant breeds, or adapt commonly raised breeds, for beef production. Despite the recent occurrence of extreme heat events in Canadian pastures and feedlots, there is very little heat stress research conducted in Canadian settings. The purpose of this study was to develop a non-invasive method to compare behavioral and physiological indices of heat stress between different Canadian cattle breeds. In August 2018, we compared respiration rates and standing behavior between breeds varying in coat colour while in feedlot pens, including Black Angus, Red Angus, Hereford, Simmental, Charolais, and Canadian Speckle Park. All animals included in the study were steers between 1-2 years in age. The feedlot was comprised of 64 pens that held 100-150 animals; each pen was comprised of different breeds and there was a difference in the amount of variation in breeds between pens. Over a 4-day collection period, we recorded 3-minute videos of cattle using a UAV in the morning (8.30-11.30 hours) and in the afternoon (14.00-17.00 hours). Throughout each time period, we flew the drone to various lots that were randomly selected. The UAV was equipped with a 4K video camera positioned at nadir and we hovered the UAV directly overhead at a height of ~10-15 m to obtain a clear view of their flank movements. We later used Observer XT software to quantify their respiration rate, the time spent lying, standing, and walking. Prior to the data collection periods, cattle were given a week to acclimate to the sound of the UAV hovering above them. None of the cattle included in the study showed any behavioral responses to the UAV. The mean respiration rate in breaths per minute (BPM) for black-coated cattle (97 BPM, SD=20) and red-coated cattle (96 BPM, SD=20) was higher than white-coated cattle during the afternoon (87 BPM, SD=21) (ANOVA, Tukey Test; P=0.006). There was no significant difference in mean respiration rate between cattle during the morning period. As the average temperature in the afternoon (32.6±1.6 °C) was higher than that of the morning (29.1±4.3 °C), it was expected that the differences in respiration due to coat color would be larger in the afternoon. Dark-coated cattle should show stronger physiological responses to hot temperatures since dark coats with a lower albedo absorb more solar radiation than lighter coats. There were no differences in time spent standing between cattle of different coat color, which was unexpected given that standing allows for greater convective heat loss compared to lying. Cattle spent more time standing in the afternoon than in the morning; however, it is uncertain whether this was linked to diurnal variation in activity or if it was linked to heat stress. Taking longer bouts of behavioral video may improve the efficacy of sampling behavioral measures. We conclude from our data that UAVs are potentially an effective, non-invasive tool to study cattle heat stress behavior.

## Voluntary heat stress abatement system for dairy cows: does it mitigate the effects of heat stress on physiology and behavior?

*Lori N. Grinter, Magnus R. Campler and Joao H.C. Costa*
*Dairy Science Program, Department of Animal and Food Sciences, 404 W.P. Garrigus Building, Lexington, KY 40546, USA; costa@uky.edu*

Heat stress in dairy cows is a major concern in the dairy industry around the world. A myriad of heat abatement strategies are used to decrease the negative effects of heat stress. Recently, focus on individualized heat stress tolerance and heat abatement has increased, especially in combination with precision dairy management strategies. Thus, the aim of this study was to investigate the heat abatement effect of a voluntary soaker with or without mandatory use on behavior and physiological measures in dairy cows. Holstein cows (n=15) were enrolled over 8 1-wk periods in a random block design consisting: (1) free access to a voluntary soaker (NMS); or (2) mandatory soaker (MS) with two mandatory soakings/d administered when exiting the milk parlor in addition to free access to the voluntary soaker in the pen. All cows were trained and acclimated to the soakers during a 4-week period prior to the study. Hourly and daily mean temperature humidity index (THI) was calculated using data obtained from a portable weather station on site. A reticulorumen temperature bolus (HerdStrong) was used to record temperature every 10 min. Cows were fitted with a behavior monitoring collar (MooMonitor +) which recorded ruminating time. Feeding intake and behavior was measured with the use of feed intake bins (Insentec). The number of soaker visits per cow was recorded manually from video observations. Mixed linear models were used to determine any treatment differences including fixed effects of treatment, soaker use (no./d) and mean daily THI and its interactions with day as a repeated measure. A linear regression was used to model the relationship of each variable and voluntary soaker use. Cows responded as expected to the increase in heat exposure by going to the voluntary shower more often; whereas THI had a positive relationship with voluntary soaker use (P<0.01). Treatment did not influence soaker use (MS: 12.4, NMS: 14.8, SEM: 1.4; P=0.21), measures of physiology and behavior (P>0.19), nor respiration rate (minimum; P=0.20;, mean; P=0.24); or maximum; P=0.92). Mean, minimum and maximum reticulorumen temperature (RT) was not different between treatments (P≥0.11). Mean (R2=0.14; P<0.01) and maximum RT (R2=0.09; P=0.01) were correlated with increasing voluntary soaker use. Cows did ruminate longer per day in the MS treatment compared to the NMS treatment (558.6 vs 543.3, SEM: 5.2 min/d; P=0.01). No relationship was observed for daily rumination and daily soaker use (P=0.21), daily feeding time and treatment (P=0.75) or daily soaker use and treatment (P=0.57). Overall, limited differences in heat stress alleviation were observed between the two treatments, suggesting that MS cows had the same level of heat alleviation compared to NMS cows despite two forced soakings. Future research could investigate the magnitude of heat alleviation of soakers in dairy cows on dairies with different heat abatement strategies or include a treatment of no voluntary soaker.

## Are Freewalk systems offering more comfortable housing conditions for cattle?

*Isabel Blanco Penedo and Ulf Emanuelson*
*SLU, Department of Clinical Sciences, SLU, Institutionen för kliniska vetenskaper, Box 7054, 750 07, Sweden; isabel.blanco.penedo@slu.se*

The Free Walk systems (FWS), i.e. 'Bedded pack barn' or 'compost barn' and 'Cow garden', are promising housing from the point of view of animal welfare and environment. Our aim was to give some insights about the comfort in FWS compared to cubicle housing systems (C) on a wide range of European production conditions as part of the Freewalk project. A total of 44 dairy and suckler farms (22 FWS vs 22 C) of 6 European countries were visited during Winter and Summer 2017-18 to conduct an adaptation of the Welfare Quality® protocol by the same trained observer. Information from 4,183 animals and 730 behavioral observations were obtained. Data from individuals were expressed as the number of animals affected out of the total number of animals assessed in each farm. Comfort around resting data were means of time to lie down in seconds, scores of ease of rising up and percentage of collisions. Behavioral data were the percentage of cows adopting specific natural lying positions. All statistical analyses were performed using the program STATA Software v. 11 (Stata Corporation, College Station, TX, USA). The statistical significance of the housing system's effect on welfare measures in the studied farms was determined by $Chi^2$ tests, Fisher's test and the Kruskal-Wallis test. Cleanliness of udders did not differ significantly between both systems ($P>0.05$) but a lower prevalence of dirty lower legs ($P=0.004$) and flank/upper legs ($P<0.001$) were observed in FWS vs C. The prevalence of not lame cows was 78.8% in FWS vs 73.5% in C, light lameness was 16.3% in FWS vs 20.3% in C and prevalence for severe lameness was 4.8% in FWS vs 6.1% in C ($P<0.001$). Time spent on lying was shorter in FWS vs C (5.32 vs 6.08 seconds) ($P<0.001$). Percentage of collisions with housing equipment during lying down was 6.7% (FWS) vs 42.2% (C). The number of warnings and alarms by a farm (regarding the time needed to lie down) were 30.2% in FWS vs 30.6% in C that was categorized as a moderate problem. But, it varied significantly when reached the level of a serious problem between systems, 25% in FWS vs 44.7% in C ($P<0.001$). Percentage of animals lying partly or completely outside the supposed lying area reached >5% in 50% of the events what was considered a serious problem ($P<0.001$). Regarding different cow lying postures, cows in FWS more often adopted different comfortable lying positions compared with C. A higher occurrence of long position (28.8 vs 13.9%) ($P=0.010$), wide position (50.8 vs 32.6%) ($P<0.001$) events were observed in FWS vs C. Whilst short position (75.5 vs 66.9%) was slightly higher on FWS and narrow position was a similar event. In addition, a significantly higher number of cows on long and wide positions was also observed in FWS vs C. Rising score varied significantly in both systems 2.32 (FW) vs 2.91 (FS) ($P<0.001$). Animals colliding with housing equipment or cows was significantly lower in FWS (18.1%) vs C (75.8%) ($P<0.001$). The number of cows ruminating was higher in FWS than C ($P=0.021$). FWS seem to be more comfortable for cattle. Further integrative analysis is still needed specifically considering the effects of space allowance and bedding conditions to provide firm conclusions.

## A cow in motion: are we really providing 'exercise' to dairy cow?

*Elise Shepley[1], Joop Lensink[2] and Elsa Vasseur[1]*
*[1]McGill University, Animal Science, 21111 Lakeshore Rd, H9X3V9, Canada, [2]Yncréa Hauts de France, ISA Lille, 48 Boulevard Vauban, 59046, France; elise.shepley@mail.mcgill.ca*

As humans, we recognize the importance of exercise. We go to the gym, walk our dogs, and ride our horses. Even our hamsters get a wheel to run on. Considering this, it is surprising that, when it comes to the production animals that make up the bulk of domestic animals worldwide, the concept of 'exercise' becomes more muddled. There is empirical evidence of improvements to health and behavior that have been linked to what has been perceived in the literature as the provision of exercise to dairy cows as well as evidence of the preference for cows to access environments like pasture often associated with exercise. More unclear, however, is whether these benefits and preferences are a direct result of exercise or due to the increased movement opportunity provided by the housing and management methods that are implemented in these studies. To address this conceptual quandary, it is necessary to turn to the methods by which exercise is measured and applied in research. For quantitative evaluations of exercise, researchers turn to measures of the cow's locomotor activity, looking at the distance, speed, and or duration of time spent moving. These measures use visual observations of time spent moving and estimates of distances traveled, with more recent studies utilizing technologies such as GPS and pedometers to obtain a more accurate, comprehensive analyses of locomotor activity. Equally important to the method of measurement of locomotor activity are the factors that research studies investigate as the cause for the level of locomotor activity recorded: (1) changes to characteristics of the cow's housing environment (e.g. space allowance/cow, walking surface, stall hardware); and (2) changes to the duration of time and or frequency that cows are managed under certain housing options (e.g. provision of outdoor access). These factors also elicit other behavioral activities – from exploration and social interactions with herd mates to displays of estrous behaviors – in which the cow exerts energy and movement that deviates from more quantitative measures of exercise (i.e. locomotor activity). Moreover, providing a cow with an environment in which she can exercise does not guarantee that she will utilize the resources provided to do so. As such, it becomes more apparent that the positive outcomes of these studies cannot be attributed to exercise, per se, but to the level of movement opportunity provided to the cow – movement opportunity dependent on the level of locomotor and other activities an individual cow opts to engage in based on her preferences/need, but that ultimately is either hindered or promoted by the cow's environment and or management. This change in perspective creates a more direct relationship between what we find in research and what could be applied on-farm. Most importantly, it leads to an improved ability to make suitable recommendations to producers on how to keep their cows healthy, fulfill a wider range of behavioral needs, and, of course, to keep their cows in motion.

## Evaluating the effects of mean occupation rate and milk production on two automatic milking systems

*Amanda Lee[1], Peter Krawczel[1], Emma Ternman[2], Liesel Schneider[1], Peter Løvendahl[2] and Lene Munksgaard[2]*
*[1]University of Tennessee, 2506 River Dr Knoxville, 37996 Tennessee, USA, [2]Aarhus University, Blichers Alle 20, 8830 Tjele, Denmark; alee90@vols.utk.edu*

Stocking density recommendations on automatic milking system (AMS) are based on company guidelines rather than empirically-derived quantities. More cows sharing a resource, such as an AMS, can increase competition and decrease milk production. Therefore, determining the effects of cow number on milk production is necessary for maintaining cows' production. The objectives of this study were to determine how: a) number of hourly AMS visits varied with occupation time (total hours available for visiting AMS, deducting time for cleaning, repairs, and training heifers) and b) cows' fat plus protein corrected milk (FPCM) varied with stocking density (daily number of cows milked by an AMS). A retrospective analysis was performed on data collected during 2004 to 2006 and 2012 to 2018 on two AMS (Voluntary Milking System™, DeLaval, Tumba, Sweden) located within the same barn at the Danish Cattle Research Center (Foulum, Denmark). The gap occurred between 2006 and 2012 due to missing cleaning time data, but variation within the model was accounted for with week and year measurements. From 2004 to 2018, other experiments were ongoing, but stocking density within freestalls never exceeded 100%. Hourly AMS visits were assessed via an analysis of variance (GLIMMIX Procedure, SAS 9.4, Cary, NC, USA). Fixed effects included days in lactation, parity (primiparous and multiparous), occupation rate, hour, and occupation rate by hour, with year and AMS by year as random effects. Fat plus protein corrected milk was assessed by linear mixed models using the MIXED procedure. Polynomial effects of stocking density were tested for a quadratic relationship. Fixed effects of parity and lactation week were included in all models. Random effects were defined as AMS, week, and year. Weekly mean stocking densities per AMS were 56±4 and 56±5 cows (range: 38 to 64), respectively. Parity, days in lactation, occupation rate, hour, and occupation rate by hour predicted hourly AMS visits ($P<0.001$). Hourly AMS use was lowest from 3:00 to 4:00. As occupation rate increased, cows were more likely to visit the AMS from 10:00 to 12:00 and 20:00 to 22:00 ($P<0.01$). Automated un-manned cleanings occurred at 4:00 and between 18:00 and 20:00, indicating the cleaning cycle at 20:00 may need to shift based on increased stocking density. Fat plus protein corrected milk was greatest when stocking density was 55 to 64 cows (range 29.3 to 30.6 kg/d). Quadratic relationships ($P<0.05$) indicated FPCM varied with stocking density, but was primarily driven by a linear relationship ($P<0.001$). Study results suggest a stocking density of 55 to 64 cows within an AMS, while considering cleaning times, may help maximize productivity.

## Virtual fences for goats on commercial farms, an animal welfare concern?

*Knut Egil Bøe and Silje Eftang*
*Norwegian University of Life Sciences, Faculty of Biosciences, Department of Animal and Aquacultural Sciences, P.O. Box 5003, 1432 Ås, Norway; knut.boe@nmbu.no*

A virtual fencing system for goats has been developed by the Norwegian company Nofence, and is currently being used on nearly 1000 commercial Norwegian goat farms. The animals get warning signals (sounds) when approaching/getting near to the virtual border, and a small electric shock (0.1 J, 3 kV) if passing the virtual border. This shock is far less than from an ordinary electric fence. The aim of this study was to investigate how goats on commercial farms responded to a virtual fencing system, both goats familiar with the virtual fencing system and goats with no experience with the system. The studies were approved by the Norwegian Food Safety Authority. Study 1 included 10 commercial Norwegian goat flocks (Kashmir and Boer goats) with 4-20 goats per flock, with a total of 92 goats on pasture. These goats had at least 10 days of experience with the virtual fence system. The pasture size varied between 350 and 5,000 $m^2$/animal. The herds were visited by trained observers and data on warning sounds and electric shocks were logged for the next 7 days. Study 2: Six commercial Norwegian goat flocks (Kashmir and Boer goats) with 5-16 goats per flock, a total of 53 goats, with goats that had no experience with virtual fencing systems were visited. The goats had been on the pasture area for at least two days with ordinary fencing. At day 0, parts of or the whole physical fence was removed and replaced by virtual borders. Four of the six flocks got some assistance in what direction to turn when exposed to warning signals. Data on warning sounds and electric shocks for each individual were logged for the next five days. In study 1, 2 animals escaped from the intended pasture area at day 1 in flock 7 and in flock 8, one animal escaped day 1 and 2, and at day 3, 2 animals escaped. Apart from that, no animals escaped. In 6 of the flocks the mean number of warning signals per animal and day was ≤5 and between 28 and 51 in the 4 other flocks. The mean number of electric shocks per animal and day varied between 0.09 and 0.98, and the maximum number of electric shocks received by an animal for the 7 day observation period varied between 1 and 13 in the 10 flocks. 14 animals (of 92 in total) did not receive any electric shocks during this period. Observations on goats that got electric shocks suggests that the behavioural reaction was short-lived. In study 2, both the number of warning sounds and electric shocks clearly declined during the 5 day observation period. The mean number of electrical shocks per animal and day varied between 0.20 and 5.47 and the mean number of warning sounds per animal and day varied between 0.96 and 13.20, showing quite large differences between flocks. We conclude that goats can understand the virtual fencing system and, among goats familiar with the virtual fencing system, the number of electric shocks received was low.

## Introduction of a new welfare labelling scheme in France

*Lucille Bellegarde[1], Mathilde Bibal[2] and Amélie Legrand[1]*
*[1]Compassion In World Farming, River Court, Mill Lane, Godalming, Surrey, GU7 1EZ, United Kingdom, [2]Casino Group, 1 Cours Antoine Guichard, 42008 Saint-Etienne, France; lucille.bellegarde@ciwf.fr*

Over 90% of French consumers want clear honest labelling about the way animals were reared and slaughtered, when buying meat and dairy products. In December 2018, a French retailer (Casino) and three animal welfare organisations (CIWF, OABA, LFDA) have launched the first animal welfare labelling scheme in France, with chicken meat being the first product labelled in stores. The aim of this initiative is to set up a harmonised welfare labelling scheme in France and potentially in Europe, to provide full transparency to consumers and drive demand for higher welfare products. Products are labelled on a scale from D to A, with A being the highest level of welfare and D the lowest. All product ranges can thus be labelled transparently, including products from standard intensive production. The scoring is based on a welfare assessment carried out at all stages of the animals' life, from birth to slaughter, and composed of 225 items, all evaluated once a year by trained third party auditors. Items include outdoor access, stocking density, breed, environmental enrichment (such as perches and pecking substrates), transport time and CCTV in the slaughterhouses. Existing welfare assessment protocols for broilers such as Welfare Quality or RSPCA AssureWel protocol were used to select relevant items and establish the thresholds and scoring system. The scoring system was then validated through on farm pilots. The selected criteria consist of welfare inputs and welfare outcomes encompassing the health, physical and mental wellbeing of the animals as well as the expression of natural behaviours. As a score is attributed at the individual farm and slaughterhouse level, different products sold under the same brand can thus be noted differently. The labelling scheme is now opening up to other companies and expert bodies that will respectively be able to use the labelling and contribute with their expertise to its evolution, including developing auditing and scoring protocols for other species.

## Development of a theoretical model to explain consumers' willingness to purchase animal welfare products in Japan

*Takuya Washio[1], Miki Saijo[1], Hiroyuki Ito[1], Ken-ichi Takeda[2] and Takumi Ohashi[1]*
*[1]Tokyo Institute of Technology, 2-12-1 Ookayama, Meguro-ku, Tokyo, 152-8550, Japan, [2]Shinshu University, 8304 Minamiminowa, Kamiina, Nagano, 399-4598, Japan; washio.t.aa@m.titech.ac.jp*

As the domestic market for dairy farm products shrinks in Japan, the industry is being compelled to undertake challenges in other areas, such as producing dairy products for export. Among the measures being explored, adding value by promoting animal welfare and enhancing the transparency of the production processes of dairy products are anticipated. At the same time, Japanese consumers' awareness of animal welfare (AW) is still low, and high animal welfare products (AWP) produced at sites recognized for high animal welfare, marketed in Japan are still in a developmental stage. Because of this, greater understanding of consumer behavior related to AWP is necessary. Theory of Planned Behavior (TPB) argues that an individual's behavioral intentions and behaviors are shaped by *attitude to behavior*, *subjective norms*, and *perceived behavioral control*. In this study, we aimed to develop a model based on the TPB to explain the factors that influence consumers' willingness to purchase AWP. We began in November 2018 in Tokyo by conducting a questionnaire survey of 353 high school students deemed suitable for an initial exploration of people's reactions by giving the AW-related formal information. The survey had received approval from the high school. Participants were given a 90-minute lecture focused on the problems of communication among people with different backgrounds, the gap in awareness between farmers and consumers, and the Five Freedoms of AW. AW was described as 'an affective state of an animal needs justifying based on the Five Freedoms.' After the lecture, the participants were asked questions about their affinity for dairy products, dairy product consumption frequency, initial awareness of AW, empathy for AW, and willingness to purchase AWP. A five-point response scale and open-format questions allowed the participants to state the reasons for their choices. A total of 293 participants completed the survey (83.0% completion ratio). Spearman's rank correlation test was conducted (SPSS Statistics 25), and concepts were elucidated from the descriptive answers. Among the respondents, only 4.1% answered 'I was aware of the concept of AW before the lecture'. This suggests that awareness of AW is still low among Japanese high school students. We found a positive correlation between the survey participants' *willingness to purchase AWP* and *empathy for AW* ($\rho$=0.441; P<0.001). According to the keywords elucidated from the descriptive answers, factors that appeared to most strongly influence the respondents' willingness to purchase AWP were *empathy for animals*, *empathy for farmers*, and *expectations for added value*. Although there is a limitation that high school students are not primary household purchasers, this suggested that there was a need for our analysis to take into consideration the psychological factors affecting empathy. We therefore extended the TPB model by including empathy as an interim factor influencing consumers' ethical consumption between the two factors of *subjective norms* and behavioral intentions.

## Do human-dog interactions affect oxytocin concentrations in both species?

*Lauren Powell[1], Kate Edwards[2], Paul McGreevy[3], Adrian Bauman[1], Adam Guastella[4], Bradley Drayton[1] and Emmanuel Stamatakis[1]*
*[1]The University of Sydney, Prevention Research Collaboration, School of Public Health, Faculty of Medicine and Health, Orphans School Creek Lane, Camperdown 2050, Australia, [2]The University of Sydney, Faculty of Health Sciences, Orphans School Creek Lane, Camperdown 2050, Australia, [3]The University of Sydney, Sydney School of Veterinary Science, Regimental Dr, Camperdown 2050, Australia, [4]The University of Sydney, Brain and Mind Centre, Sydney Medical School, 94 Mallett St, Camperdown 2050, Australia; lauren.powell@sydney.edu.au*

Oxytocin is regularly proposed as the primary hormone mediating human-dog bonding, with a limited body of evidence showing human and canine oxytocin concentrations increase after human-dog interactions. However, the possible joint influence of human-dog interaction and exercise is unexplored. The primary aims of this research were to *a)* examine the effect of dog-walking and affiliative human-dog interactions on human and canine oxytocin concentrations, and *b)* investigate any putative moderating role for the strength of human-dog attachment on such responses. 29 owners (83% female) and their privately owned male dogs completed a four-condition random order cross-over trial: dog-walking (DW); walking without the dog (W); affiliative human-dog interaction (H-DI); and the control, resting quietly (C). Each condition was performed in the participant's home or surrounding area (DW and W) for approximately 15 minutes. Human saliva and canine urine samples (DW and H-DI only) were collected before and after each condition. Oxytocin concentrations were quantified using an ELISA. The Monash Dog Owners Relationship Scale was applied to measure the human-dog relationship. Linear mixed models were used to analyse data with the participant considered a random effect; condition, order of conditions, condition duration and latency from initiation of condition to urine sample collection (where applicable) as fixed effects. Human oxytocin concentrations were not significantly different from the control following DW (mean change 0.39 pg/ml, 95% CI -3.29, 4.07), W (-0.29 pg/ml, 95% CI -3.91, 3.34) or H-DI (-2.89 pg/ml, 95% CI -6.45, 0.67). Similarly, canine urinary oxytocin concentrations were not significantly different following DW (mean change -14.66 pg/mg Cr, 95% CI -47.22, 17.90) or H-DI (6.94 pg/mg Cr, 95% CI -26.99, 40.87). Considering moderation by the strength of the human-dog bond, pairwise comparisons between conditions revealed that oxytocin concentrations increased among owners with lower levels of attachment after DW compared to W (mean change 9.32 pg/ml, P=0.003) and after H-DI compared to C (3.90 pg/ml, P=0.02). In owners with high levels of attachment, oxytocin concentrations were not significantly different between DW and W (P=0.97), or H-DI and C (P=0.32). The strength of the human-dog bond did not moderate the canine oxytocin response to either condition. Overall, we did not find a consistent pattern for positive human or canine oxytocin responses to human-dog interactions. The strength of owner-dog attachment had a moderating effect on human oxytocin responses, suggesting that human-dog interactions may elicit greater physiological responses in low attachment individuals.

## Prioritising cat welfare issues using a Delphi method

*Fiona C. Rioja-Lang, Heather Bacon, Melanie Connor and Cathy M. Dwyer*
*University of Edinburgh, Jeanne Marchig International Centre for Animal Welfare Education,*
*Royal (Dick) School of Veterinary Studies, Easter Bush Campus, Midlothian, EH25 9RG, United*
*Kingdom; cathy.dwyer@sruc.ac.uk*

In order to determine where limited funding resources should be directed, or to raise awareness of best practice, it is sometimes necessary to prioritise particular welfare issues to identify those needing special consideration. A modified Delphi method was used to generate expert consensus on priority welfare issues for cats in the UK, as part of a wider, multi-species study. The study involved 14 cat welfare experts, recruited from a range of disciplines including veterinarians, academics, charity sector employees, and industry representatives. An expert was defined as someone who had worked in their field of expertise for more than three years and all experts were based in the UK. Experts initially participated in an anonymous, online discussion board, and they generated a list of 118 welfare issues for cats. Two rounds of surveys were then conducted online using the Online Surveys tool (formerly BOS), to rank these issues separately for the severity, duration and perceived prevalence of each welfare issue on a 6-point Likert scale, where 1 = never/none, and 6 = always/high. In round 2, all welfare issues which had a median score of 3 or above (n=30, 43 and 38 for prevalence, severity and duration, respectively) were included. Participants were asked whether they agreed or disagreed to the rankings generated from round 1. The response rate of cat experts for the first survey was 86%, and 79% for the second round. The mean age of the experts in the cat group was 42 years old (min. 27; max 58), 62% of experts were female. All research generated from this study was approved by the University of Edinburgh's Human Ethics Review Committee. The final stage of the study was a 2-day workshop with a subsection of experts (n=21; n=2 cat experts) which consisted of a combination of small group and large group exercises and discussions in order to finalise the priority welfare lists for each species group and to rank them. The most severe or prolonged welfare issues from the workshop were considered to be social behaviour issues, diseases of old age, obesity, owners not seeking veterinary care, poor pain management, issues with being kept in shelters, issues for unowned cats, delayed euthanasia, neglect or hoarding and inherited conformational defects or diseases. For the UK cat population, the most prevalent welfare issues were neglect or hoarding, delayed euthanasia, inherited conformational defects or diseases, social or environmental restriction, poor pain management, diseases of old age, owners not seeking veterinary care, obesity, issues with being kept in shelters, and inadequate stray cat management. The Delphi process resulted in consensus on the most significant welfare challenges faced by cats in the UK and could be used to guide future research, policy and education priorities.

# Dogs with canine atopic dermatitis exhibit differing behavioural characteristics compared to healthy controls

*Naomi D. Harvey, Peter J. Craigon, Steve C. Shaw, Sarah C. Blott and Gary C.W. England*
*The University of Nottingham, School of Veterinary Medicine & Science, College Road, Sutton Bonington, Leicestershire, LE12 5RD, United Kingdom; naomi.harvey@nottingham.ac.uk*

Canine atopic dermatitis (cAD) is a common, chronic allergic skin condition and cause of pruritus (itching). Quality of life has been shown to be reduced in dogs with cAD, but behavioural differences between affected and unaffected dogs have not been characterised. Owners of Labrador and Golden retrievers that reported their dog's dermatological medical history for the Itchy Dog Project were invited to complete the canine behaviour and research questionnaire (C-BARQ). C-BARQ data was gathered for 343 dogs with a diagnosis of cAD (cases), and 552 controls with no history of skin problems. We tested the hypothesis that pruritus would be associated with cases displaying more problem behaviour and lower scores for Trainability (a score that requires sustained attention) but not differ elsewhere from non-pruritic controls. Zero-inflated C-BARQ data were transformed to binary scales using the median. Multivariate regressions were used with each C-BARQ score acting as the dependent variable in its own model. Case/control was included as the main fixed effect to evaluate any possible relationship between cAD diagnosis and behaviour, with potential confounding variables (age, sex, breed, neuter status or other health problem) controlled for as fixed effects. Linear, logistic or Poisson models were used dependent on the C-BARQ score distribution (normal/binary/right-skewed). Additionally, each model was re-run with scores on the Edinburgh Pruritus Scale (a six-point Likert scale) as the main effect in place of case/control to elucidate what role pruritus alone had on the dogs' behaviour, accounting for dogs with differing pruritus severity. Q-values based on false discovery rates were calculated in R to control for multiple testing. Dogs with cAD displayed significant differences from healthy controls; being more likely to display problem behaviour that could be indicative of emotional stress (chewing Q=0.003, OR 1.40; mounting Q<0.001, OR 2.14; coprophagia Q<0.001, 1.16; hyperactivity/restlessness Q<0.001, OR 1.64; leash pulling Q<0.001, OR 1.29), comfort-seeking (begging for food Q<0.001, OR 1.29; attachment-attention seeking Q<0.001 T=3.70) and grooming related behaviour (self-grooming Q<0.001, OR 6.58; allo-grooming Q<0.001, OR 1.88; sensitivity to touch Q=0.001 OR 1.49), and were less trainable than controls (Q<0.001 T=-2.92). Whilst causation cannot be established, each of these findings (and non-significant findings) were replicated in the models where pruritus scores were the dependent instead of cAD status, pointing to pruritus as being the main factor involved in these behavioural differences. The results of this analysis are the first of their kind to evaluate behavioural differences associated with pruritus. Not only is this a welfare concern for these dogs in terms of their quality of life, but many of the behaviours exhibited by the pruritic dogs are considered to be 'problem' or 'nuisance' behaviour by owners, which could contribute to a break-down in the dog-owner relationship, potentially increasing the risk of relinquishment for dogs with skin allergies.

## Hinged farrowing crates promote an increase in the behavioral repertoire of lactating sows

*Maria C. Ceballos, Karen C.R. Gois, Matthew Herber and Thomas D. Parsons*
*Swine Teaching and Research Center, Department of Clinical Studies, NBC, School of Veterinary Medicine, University of Pennsylvania, 382 West Street Road, 19348, USA; mceballos30@gmail.com*

Concerns about the welfare of lactating sows housed in farrowing crates have emerged in part because crates physically restrict the sows and possibly compromise their behavior and comfort. The aim of this study was to identify how the opening of a hinged farrowing crate on different days (4 and 7) post-farrowing impacts sow behavior. The study was conducted at the University of Pennsylvania's Swine Teaching and Research Center, located in Kennett Square, Pennsylvania, USA. A total of 36 sows (DNA Genetics, Line 241) were studied. The sows were randomly allocated to 1 of 3 treatment groups: TC- crate remained closed until weaning (n=13), T4- crate was opened on day 4 post farrowing (n=12), and T7- crate opened on day 7 post farrowing (n=11). On days 3-8 post farrowing, sow behavior was captured by continuous video recording of each individual sow. Sow behavior was observed daily from 6 am to 6 pm using instantaneous recording (with a 2-minute fixed sampling interval) and focal animal as sampling route. The following behavioral categories were measured: Resting/sleeping-(RES), Investigating piglet-(IP), Exploring environment-(EE), and Standing-(ST). In addition, the location of the sow was recorded as one of 30 possible positions within the crate or pen and then the number of unique positions used by the sow totaled (NP). We analyzed the differences in sow behavior over the study days (3 to 8 post farrowing) and treatments (TC, T4 or T7) using generalized linear mixed models by PROC GLIMMIX in SAS. All models included treatment (TC, T4 or T7) nested with day of assessment (d3 to d8) and parity as fixed effects. We tested individually the size of the litter, room temperature, starting of shoulder lesion, and if the sow was treated with vitamin B12, or not treated and, when significant, we include in the model as random effects. There was a difference between treatments ($P<0.05$), with the average proportion of time spent in IP, EE, ST, but not RES, increasing after opening the crate for both T4 and T7 treatments, compared with TC. Experimental means ± SD for d4 and d7 (opening crate day) for both treatments T4 and T7 vs TC in both days were: ST (10.4±6.6 and 11.5±4.1 vs 6.8±2.5 and 8.1±3.8), RES (65.1±7.4 and 63.8±9.6 vs 72.3±4.6 and 69.6±4.3), EE (3.42±3.0 and 3.96±2.4 vs 1.38±1.3 and 0.96±1.1) and IP (1.36±1.3 and 0.69±0.4 vs 0.47±0.5 and 0.41±0.7). The number of positions used by the sows also dramatically increased following crate opening (T4: 14.3±3.9 and T7: 15.2±2.3 vs TC: 3.0±0). The use of hinged farrowing crates allows the sow to both increase her expression of many motivated behaviors including IP, EE, ST as well as increase the variety of behaviors such as her repertoire of different orientations in the pen. Taken together, these behavioral observations are consistent with increased sow welfare following the opening of a hinged farrowing crate during lactation.

## Relative preference for wooden nests affects nesting behaviour in broiler breeders

*Anna C.M. Van Den Oever[1,2], T. Bas Rodenburg[1,3], J. Elizabeth Bolhuis[1], Lotte J.F. Van De Ven[2] and Bas Kemp[1]*
*[1]Wageningen University, Adaptation Physiology Group, P.O. Box 338, 6700 AH Wageningen, the Netherlands, [2]Vencomatic Group, P.O. Box 160, 5520 AD Eersel, the Netherlands, [3]Utrecht University, Department of Animals in Science and Society, P.O. Box 80166, 3508 TD Utrecht, the Netherlands; anna.vandenoever@wur.nl*

Optimising nest design for broiler breeders has benefits for both the animals and producer. The welfare of the hens will increase by providing preferred housing, while also reducing eggs laid outside the nests. These floor eggs cause economic losses by compromised automatic egg collection and reduced saleability and hatchability. Attractiveness of nests can involve factors as seclusion, material and nest climate. In this study, four nest box designs are offered in a relative preference test: a plastic control nest, a plastic nest with a partition to divide the nest in two areas, a plastic nest with a ventilator underneath to create air flow inside the nest and a wooden nest. Six groups of 100 hens and 9 roosters had access to these four nests in a randomised location during the ages of 20 to 34 weeks. Nest and floor eggs were collected five days a week. Camera images from inside the nests made during the ages 25-26 wk and 27-28 wk were analysed on behaviour. This included general activity, nest inspections, nest visits and social interactions. At 32 wk of age the wooden nests were closed, and the subsequent response of the hens was monitored in terms of number of eggs. We found a clear preference for the wooden nest in number of eggs (69.3±1.0%) compared to the control nest (15.1±0.8%), partition nest (10.2±0.5%) and the ventilator nest (5.4±0.4%; P<0.0001). This preference was also reflected in increased time spent sitting, together with fewer nest inspections and visits per egg laid in the nest. The preference for the wooden nest led to crowding, which caused an increased amount of piling, nest displacement, aggression and head shaking. After the wooden nests were closed, the hens still had preference for nest design, although this was strongly influenced by the location of the nest. We conclude that the broiler breeder hens in this study had a strong preference for the wooden nests and the fact that they were willing to accept the crowded circumstances in these nests, shows the strength of this preference. When denied access to their preferred nest, the hens chose a new nesting location based on nest design depending on proximity to their original nesting location. This study shows how the material used for nests is an important factor in suitability and should therefore be taken into account when designing nests. In future experiments we will investigate gregariousness nesting further in addition to studying the influence of genetics and mobility on nesting behaviour.

## Associations between qualitative behaviour assessments and measures of leg health, fear and mortality in Norwegian broilers

*Guro Vasdal[1], Karianne Muri[2], Solveig Marie Stubsjøen[3], Randi Oppermann Moe[2] and Erik Georg Granquist[2]*
*[1]Animalia AS, Pb 396, Loren, 0513, Norway, [2]Norwegian University of Life Sciences, Faculty of Veterinary Medicine, P.O. Box 8146 dep, 0033 Oslo, Norway, [3]Norwegian Veterinary Institute, Department of Animal Health and Food Safety, Section for Animal Health, Wildlife and Welfare, P.O. Box 750, Sentrum, 0106 Oslo, Norway; guro.vasdal@animalia.no*

Qualitative behavioural assessment (QBA) is an animal-based welfare measure that has been included in several on-farm welfare assessment protocols, including the Welfare Quality® (WQ) protocol for poultry. However, the method has not been validated for broiler chickens, and there is a scarcity of information about how it relates to other animal-based welfare indicators. The aims of this study were therefore to investigate the associations between QBA and selected animal-based welfare indicators commonly used for the assessment of broiler chicken welfare, i.e. lameness, foot pad dermatitis (FPD), fear of humans (touch test), and mortality. A total of 50 commercial broiler chicken flocks were visited by one observer who conducted on-farm welfare assessments using the WQ protocol, close to the time of slaughter (between day 27 and 34). QBA was analysed using principal component analysis (PCA), revealing two main components, labelled arousal (PC1) and mood (PC2). The scores for the other welfare indicators were categorised into dichotomous (touch test) or ordinal scales (gait score, footpad dermatitis score and mortality) to deal with skewed distributions caused by homogenous data. To investigate the associations between QBA and the other welfare indicators, we ran logistic and ordinal logistic regression models with these welfare measures as outcomes, and the two components of QBA as the predictors. Significant negative associations were found between both components of QBA and the chickens' fear of humans, as measured using the touch test. In other words, flocks with higher scores on both arousal and mood were less likely to have any chickens that were possible to touch by the assessor. A possible interpretation of these associations is that both QBA components may indicate greater liveliness and better walking ability in birds that did not accept being touched by the observer. Flocks with a higher arousal score, as measured by the first component of QBA (PC1), were also less likely to be in a higher mortality category. For the other selected animal-based measures, there were no associations with QBA. We conclude that QBA needs further validation for the routine use in the assessment of broiler chicken welfare, but that the method may provide useful supplementary information in overall welfare assessments. This information may be particularly valuable in a production system, like the broiler industry, where management is highly standardised, sometimes resulting in little between-flock variation in other welfare measures.

## Individual laying hen mobility in aviary systems is linked with keel bone fracture severity

*Christina B. Rufener[1], Yandy Abreu Jorge[2], Lucy Asher[3], John A. Berezowski[4], Filipe Maximiano Sousa[4], Ariane Stratmann[1] and Michael J. Toscano[1]*
*[1]Center for Proper Housing: Poultry and Rabbits, Animal Welfare Division, University of Bern, Burgerweg 22, 3052 Zollikofen, Switzerland, [2]National Centre for Animal and Plant Health, San José de las Lajas, Havana, Cuba, [3]IoN, Centre for Behaviour and Evolution, Newcastle University, Framlington Place, Newcastle NE2 4HH, United Kingdom, [4]Veterinary Public Health Institute, University of Bern, Schwarzenburgstrasse 155, 3097 Liebefeld, Switzerland; cbrufener@ucdavis.edu*

Keel bone fractures (KBF) represent one of the greatest welfare problems in commercial laying hens. The aim of this study was to investigate how KBF affect the mobility of individual hens housed in complex systems such as aviaries. Focal hens (60 Lohmann Brown (LB) and 60 Lohmann Selected Leghorn (LSL)) were kept in six identical pens equipped with a commercial aviary system. Strains were mixed (20 focal LSL + 205 non-focal LB or 20 focal LB + 205 non-focal LSL per pen) to maintain comparability with previous studies. A custom-made infrared tracking system was used to collect data at 11 time points throughout the laying period (21, 24, 27, 31, 35, 39, 44, 48, 52, 57 and 61 weeks of age). Infrared receivers were attached to the legs of focal hens and recorded zone-specific codes between five zones (litter, lower tier, nest boxes, top tier, and wintergarden) with a frequency of 1 Hz at six consecutive days per time point. At the end of each data collection period, hens were radiographed to assess KBF severity. Data were analyzed for LB and LSL separately using (generalized) linear mixed effect models. With increasing KBF severity, LB hens spent more time on the top tier (P=0.005) and less time in the litter zone (P<0.0001) and in the lower tier (P=0.001). Duration of stay in the nest box was not linked to age or KBF. The likelihood to cross more than one zone within a movement (e.g. jumping from the nest box to the litter directly instead of moving from the nest box to the lower tier to the litter) increased with increasing KBF severity (P=0.036). In contrast to LB hens, LSL hens spent most time in the nest box zone and top tier and only showed few transitions between zones. With increasing age, LSL hens spent less time in the nest box zone (P=0.018) and more time in the top tier (P=0.006). Irrespective of strain, hens crossed fewer zones with increasing age (LB: P=0.009, LSL: P=0.002) but the total number of transitions per day was not linked to KBF severity. Our findings indicate that hens having KBF prioritized paths among the upper tiers (i.e. between nest box zone and top tier) over paths among the mid and lower tiers (i.e. between litter, lower tier and nest box zones). Behavioral adaptation to pain, social factors as well as the spatial distribution of resources might be important mechanisms driving individual mobility in response to KBF. In complex housing systems such as aviaries where not all resources are provided on all tiers, the accessibility of resources is crucial regarding the ability of hens to cover both their nutritional and behavioral demands especially when coping with physiological challenges.

## Vocal changes as indicators of pain in harbour seal pups (*Phoca vitulina*)

*Amelia Mari MacRae, Inez Joanna Makowska and David Fraser*
*University of British Columbia, Animal Welfare Program, 2357 Main Mall, Vancouver, BC, V6T 1Z4, Canada; amarimacrae@gmail.com*

Vocalizations are potential indicators of pain in animals. We tested whether the vocalizations of captive harbour seal pups (*Phoca vitulina*) changed in response to potentially painful routine procedures of flipper tagging and microchipping. For 31 pups (healthy, >60 d old) at a rehabilitation facility, vocalizations were recorded during these procedures, which are normally done without analgesia. Twenty-one pups were used to compare vocalizations before and after the procedures (Experiment 1), and 10 pups in a cross-over experiment comparing vocal responses to real and sham procedures (Experiment 2). Vocalizations for each seal were analyzed spectrographically. In Experiment 1, the number of vocalizations increased from $5\pm1$ in the 30 s before tagging to $9\pm1$ in the 30 s after (mean ± SEM; $P<0.001$) and the peak frequency increased from $837.1\pm75$ Hz before to $1,041\pm75$ Hz after (mean ± SEM; $P<0.01$). Similarly, for chipping, there were more vocalizations in the 10 s after the procedure ($4.3\pm0.4$) than before ($2\pm0.4$; $P<0.001$) and peak frequencies increased from $848.8\pm79$ Hz before to $1,111.2\pm79$ Hz after (mean ± SEM; $P<0.05$). In Experiment 2, seals also produced more vocalizations in the 30 s after tagging ($8.0\pm1.5$) than before ($3.6\pm1.5$) but there was no similar change after sham tagging ($5.9\pm1.5$ before compared to $3.3\pm1.5$ after), reflected by the highly significant interaction between 'treatment' (actual vs sham) and 'phase' (before vs after) ($P<0.001$; $n=10$). The peak frequency also increased from $872.8\pm155$ Hz before tagging to $1,062.3\pm155$ Hz after (mean ± SEM; $P<0.001$) but there was no difference in average peak frequency before and after sham tagging ($P=0.4$). Similarly, for chipping, the number of vocalizations increased from $2\pm0.5$ before to $4\pm0.3$ after ($P<0.001$) but not after sham chipping ($P=0.4$). In summary, the number and peak frequency of seals' vocalizations consistently increased after both tagging and chipping but not after the sham procedures suggesting that these vocal changes are due to the immediate pain caused by tagging and chipping as opposed to the associated handling and restraint. These results show promise for increases in the number and pitch of vocalizations to be used to assess potentially painful procedures in seal pups.

## Effect of hot-iron disbudding on rest and rumination in dairy calves

*Sarah J.J. Adcock, Blair C. Downey, Chela Owens and Cassandra B. Tucker*
*University of California, Center for Animal Welfare, Department of Animal Science, 1 Shields Ave, Davis, CA 95616, USA; cbtucker@ucdavis.edu*

Hot-iron disbudding, a husbandry procedure that prevents horn bud growth through tissue cauterization, is painful for calves. The resulting burns remain sensitive to mechanical stimulation for weeks, but the procedure's long-term effect on daily maintenance behaviors, such as lying and ruminating, is unknown. We assigned female Holstein calves to 1 of 2 treatments: disbudded with a heated iron at 4 to 10 days of age (n=11) or not disbudded (n=11). Disbudded calves received a lidocaine cornual nerve block and oral meloxicam at the time of the procedure. All methods were approved by the UC Davis Institutional Animal Care and Use Committee. We recorded lying and ruminating behavior using 5-second scans taken every 5 minutes for 24 hours once a week when calves were 10 to 31 days of age. Mixed beta regressions were used to test the effect of treatment and its interaction with age on the daily proportions of time the calf spent lying and ruminating. Calf was fitted as a random effect in the models. Compared to controls, disbudded calves ruminated less in the first 2 weeks after disbudding (mean ± SE: 10±1% vs 18±2% of total time; P=0.003), tended to lie more in the third week after disbudding (73±1% vs 70±1%; P=0.096), and were more likely to lie with their head down and still across all weeks (31±1% vs 26±1% of total lying time; P=0.012). A decrease in ruminating and increase in lying with the head still may reflect an avoidance of head-related movement that could aggravate the calf's disbudding wounds. We conclude that disbudding, in addition to resulting in prolonged sensitivity of the wounds, is severe enough to alter daily behavior patterns for at least 3 weeks. These result raise additional welfare concerns about how long pain lasts after this procedure and our ability to mitigate it.

## Effects of disbudding on use of a shelter and activity in group-housed dairy calves
*Kaitlin Gingerich and Emily K. Miller-Cushon*
*University of Florida, Department of Animal Sciences, 2250 Shealy Dr., Gainesville, FL 32601, USA; emillerc@ufl.edu*

Self-isolation has been described previously in animals experiencing disease, and changes in social behavior may be predictive of animal welfare. In dairy calves, hot iron disbudding is a common practice that results in behavioural changes associated with pain. Our objective was to evaluate effects of hot iron disbudding on use of a shelter in the pen and activity of group-housed dairy calves. We hypothesized that calves may increase use of the shelter to isolate themselves when experiencing pain or distress following disbudding. Holstein bull and heifer calves (n=32; 4 calves/pen) were group-housed (7.4 m$^2$/calf) at 16±2 d of age (age range <7 d), and provided milk replacer *ad libitum* via an automated milk feeder for 6 weeks. Pens were sand bedded and located under an open-sided barn. All pens contained an open-top 3-sided shelter (1.2×1.2 m), built of corrugated plastic allowing for visual seclusion from the rest of the pen. Within pen, calves were randomly assigned to receive either disbudding or handling only at 5 week of age, with the opposite treatment applied 1 week later. Calves received both local anesthetic and analgesic prior to disbudding. Behavior was recorded continuously from video in both observation weeks, for 12 h following either disbudding or handling only, to characterize time spent in the shelter (defined as half or more of the calf's body inside) and lying time, with individual calves identified cased on unique coat markings. Data were analyzed in a general linear mixed model, including fixed effects of treatment (dehorning or handling only), week of observation, and order of treatment. Calves appeared motivated to utilize the shelter, with all calves entering the shelter at least once during the experiment, and there was considerable individual variability in visit frequency (7.8±4.9 visits/12 h; mean ± SD; min=1, max=23) and duration of use (112.5±130.6 min/12 h observation period; mean ± SD; min=27.0 s, max=9.6 h). Following disbudding, calves entered the shelter more frequently (9.1 vs 6.4 visits/12 h; SE=0.81; P=0.022) with similar visit duration (P=0.44). Calves frequently used the shelter together, and time spent sharing the shelter was not affected by disbudding (48.2 vs 41.0% of time in shelter; disbudded vs control; SE=6.9; P=0.46). However, calves entered the shelter when it was unoccupied more frequently after disbudding (7.3 vs 4.0 visits/12 h; SE=1.0; P=0.002), whereas frequency of joining another calf in the shelter was similar between disbudded and control calves (P=0.73). Use of the shelter increased overall between observation periods (P=0.03) but the percentage of time spent using the shelter socially was stable (P=0.51). The frequency (12.7 vs 11.9 bouts/12 h; disbudded vs control; SE=0.68; P=0.39) and duration of lying time (7.9 vs 7.8 h/12 h; disbudded vs control; SE=0.29; P=0.65) was not significantly affected by disbudding. These results suggest that calves use and may benefit from added environmental complexity that provides some visual barriers, and further research is needed to understand how use of these spaces may reflect individual welfare.

## The effect of floor type on lying preference, cleanliness and locomotory disorders in veal calves

*Kees Van Reenen and Maaike Wolthuis*
*Wageningen University and Research, Livestock Research, De Elst 1, 6708 WD Wageningen, the Netherlands; kees.vanreenen@wur.nl*

Veal calves are currently predominantly housed on wooden slatted floors. Floors of this type may be associated with animal welfare problems including an increased risk of slipping and falling, and locomotory problems such as carpal bursitis ('thick knee'). Softer floors may alleviate such problems. Thus, the current study examined the effects of two alternative floor types, i.e. Easyfix Slat Rubber Mats™ (ESRM) and Comfort Slat Mat™ (CSM), respectively, on lying preference, cleanliness, and locomotory disorders in white veal calves in the Netherlands. Alternative floor types consisted of either soft rubber (ESRM) or synthetic air-cushioned mats (CSM) that were attached to the existing wooden slats, and were compared with the normal wooden slatted floors (reference). In the first part of the study, a choice experiment, modified group pens were used, 10 pens per floor type. Half of the floor of each pen was equipped with the reference floor type, and the other half was equipped with an alternative floor type, situated either in the back or the front of the pen. Each pen contained only 3 rather than the usual 5 calves. At the ages of 10 and 25 weeks (2 weeks prior to slaughter) the time spent lying during 24-hours on each floor type (reference vs alternative) was recorded. Lying time on both ESRM and CSM, expressed as % of total lying time, was significantly different from 50% (P<0.01, Wilcoxon Signed Rank Test; lying times on alternative floor types were between 80 and 95% at all ages). The second part of the study took place on 8 veal farms in practice. Within each farm, reference, ESRM or CSM floors were randomly allocated to three physically separated barns or compartments within a barn, each containing between 40 and 200 group pens, with 5-8 calves per pen. Subsequently all pens within a barn or compartment were equipped with the same floor type. Observations were performed during four successive fattening cycles. At individual calf level, cleanliness of the belly, lameness, and carpal bursitis were categorically recorded (0/1) in 10 randomly selected pens in each barn at 10 and 25 weeks of age. All measures were expressed as prevalences within barn and fattening cycle. Lameness was hardly observed during the entire experiment (prevalence <0.1% on all floors). At 10 weeks of age prevalences of calves with dirty bellies, or carpal bursitis were <1% on all floors. At 25 weeks, the prevalence of calves with dirty bellies was lower on the reference floor (2%) compared to ESRM (6%) and CSM (5%), and the prevalence of calves with carpal bursitis was lower on ESRM and CSM (1%) compared to the reference floor (6%) (P<0.01, REML Generalized Linear Mixed Model Analysis). Our findings with regard to the preference study and the assessment of carpal bursitis suggest that calves are more comfortable on the alternative floor types than on the reference floors.

## Pain response to injections in dairy heifers disbudded at 3 different ages

*Sarah J.J. Adcock and Cassandra B. Tucker*
*University of California, Davis, Center for Animal Welfare, One Sheilds Avenue, Davis, CA 95616, USA; sadcock@ucdavis.edu*

Painful experiences near birth can cause lasting changes in pain perception. We assessed whether cautery disbudding with local anesthesia and an NSAID at 3 (n=12), 35 (n=9), or 56 (n=20) d of age affected heifers' pain responses later in life. At 11 months of age, heifers were observed in three 10-min trials while restrained in a stanchion: control (C-1), injection the next day (I0), and a control 6 d later (C6). The vaccine procedure consisted of 2 injections (2 ml each) with an 18-gauge 1.9 cm needle on each side of the neck <30 s apart. In C-1 and C6, an empty, needleless syringe was depressed against each side of the neck. We measured maximum eye temperature every 30 s and continuous R-R intervals beginning 5 min before and ending 5 min after the first injection, as well as frequency of struggling behavior in the 60 s after the first injection. We used mixed models to assess the effects of trial (C-1, I0, C6) and disbudding age (3, 35, 56) on struggling, change in eye temperature, and heart rate (HR) and HR variability (HRV) indices (RMSSD, HF, LF/HF, SD2/SD1, sample entropy). Heifers struggled more when vaccinated (I0) than in needleless control trials (C-1 and C6), and more struggling occurred in C6 compared to C-1 (mean struggling counts/min ± SE: C-1: 20±4; I0: 88±22; C6: 38±11; $P<0.001$). Eye temperature decreased 0.3 °C in the 1 min following the first injection in I0, but not in control trials ($P<0.001$). Eye temperature then increased above baseline between 2-5 min after the first injection/syringe in all trials, but this increase was greater in I0 than C-1 (C-1: +0.3±0.1; I0: +0.6±0.1; C6: +0.4±0.1; $P<0.001$). Others have also observed a short-term drop and subsequent increase in eye temperature after painful procedures. HR increased from baseline in the 30 s following the first injection/syringe in I0 and C6, but not in C-1 (change in bpm: C-1: +4±6; I0: +38±6; C6: +28±5; $P<0.001$). LF/HF and sample entropy were lower in the 5 min after the injections/syringe in I0 and C6 compared to C-1, indicating increased parasympathetic activity and decreased HRV complexity, respectively ($P\leq0.011$). Taken together, these results show that heifers responded to vaccination, compared to the day before, and still had a heightened response 6 d later. We did not observe an effect of disbudding age on struggling or changes in eye temperature. However, some HRV indices in the 5 min before and after the vaccinations suggest increased sympathetic activity in calves disbudded at 35 d, compared to 3 or 56 d. We also found that heifers disbudded at 3 or 35 d had a higher mean HR in the 5 min after the injections in I0 compared to those disbudded at 56 d (3: 97±3; 35: 100±3; 56: 89±2; $P\leq0.065$). We conclude that: (1) injections are sufficiently painful that heifers remember the experience for at least 6 d after only one exposure; and (2) there is some evidence that disbudding at a younger age influences autonomic activity later in life.

## Behavioural response of dairy calves to temperature and wind

*David Bell[1,2], Marie Haskell[2], Alastair Macrae[1], Malcolm Mitchell[2], Amy Jennings[1] and Colin Mason[2]*
*[1]University of Edinburgh, R(D)SVS, Easter Bush, Midlothian, Edinburgh EH25 9RG, United Kingdom, [2]SRUC, West Mains Road, Edinburgh EH9 3JG, United Kingdom; david.bell@sruc.ac.uk*

One of the main objectives of housing dairy calves is protection from adverse climatic conditions such as low temperatures and high winds, with the overall aim of promoting thermal comfort. Thermal comfort can be described as the temperature range in which the calf uses no additional energy to maintain its core body temperature. The lower end of this comfort zone is referred to as the lower critical temperature (LCT) and can be altered by humidity levels and wind. The behaviour that the calf expresses can be used to determine how it perceives its environment and how it adapts its behaviour to compensate for adverse environmental conditions. This experiment examined the response of dairy calves to different temperatures and wind speeds. Eighteen pre-weaned Holstein heifer calves commenced the study at 7 days of age until 25 days and were group housed when not being tested. They were tested at a combination of each of three temperatures (5, 10, 15±2 °C) and three wind speeds (0, 1, 3.3 m/s). All the temperatures were naturally occurring, and the wind speeds created using pedestal fans. Every calf was tested in a random order of temperature/wind speed combinations. The test arena consisted of a pen with two sections. The calf started in the 'test' section, where it was exposed to the wind speed dictated by the test schedule. The adjacent 'shelter' section allowed the calf to retreat from the wind. Two training sessions in the test arena were given to each calf so that this area was familiar to the calf before it was tested. Each test was twenty minutes in duration and video recorded. Observations from each video recording were made for latencies as well as behavioural responses that have shown to indicate irritation or mild aversion such as tail and ear flicks and self-grooming. Other data collected included live weight and milk intake information (time of last feed, volume consumed in previous 24 hours, milk replacer consumed in previous 24 hours). Body surface area for the calf was calculated from live weight. Data was analysed using R by applying mixed effect models. Wind speed had a significant effect on the proportion of time the calves spent in the test pen. Calves spent 0.67±0.046 (mean ± s.e.) (67%) of time in the test pen when the wind speed was 0 m/s, 0.57±0.049 (57%) at 1 m/s and 0.44±0.048 (48%) at 3.3 m/s (P<0.05). Body surface area also had a significant effect on the proportion of time the calf spent in the test pen (0.49±0.050, 0.44±0.080, 0.63±0.048, 0.70±0.046 (mean ± s.e.), P<0.01 for body surface area classes). Wind speed had a significant effect on the latency of 1st movement from test to shelter section (540.2±70.88 s, 462.4±64.97 s, 237.0±53.18 s (mean ± s.e.), P<0.001 for 0, 1, 3.3 m/s respectively) as well as latency to 1st behavioural reaction (205.7 s ±49.91, 90.9 s ±40.68, 2.7 s ±0.13 (mean ±s.e.), P<0.001 for 0, 1, 3.3 m/s respectively). The results show that calves show an aversion to wind which intensifies as wind speed increases suggesting that they should be protected from draughts.

# Effect of environmental temperature on rectal temperature and body-surface temperature in fattening cattle

*Jung Hwan Jeon[1], Kyeong Seok Kwon[1], Saem Ee Woo[1], Dong Hyeon Lee[1], Jun Yeob Lee[1], Hee Chul Choi[1], Jong Bok Kim[1], Ga Yeong Yang[1] and Sang-Ryong Lee[2]*
*[1]National Institute of Animal Science, RDA, Iseo-myeon Wanju-gun Jeollabuk-do, 55365, Korea, South, [2]Jeonju University, Agricultural Convergence, Wansan-gu Jeonju Jeollabuk-do, 55069, Korea, South; jeon75@korea.kr*

Recently, Korea has been affected by impacts of extreme weather events including extended summer and increased temperature caused by global warming and climate change. Especially, the environmental temperature is important in the livestock industry because it is closely related to livestock productivity. The objective of this study was to investigate the influence of different environmental temperature on rectal temperature and body-surface temperature in fattening cattle. These experiments were implemented under the approval of the Institutional Animal Care and Use Committee of National Institute of Animal Science in Korea. Eight fattening cattle were housed in experimental chamber (temperature-humidity controlled room) and divided groups (n=4 per group): 18 °C group (n=4) and 23 °C group (n=4). Being continuously exposed to the environmental treatment of temperatures for 3 days, the relative-humidity was within the range of 35 to 50% in the experimental chamber. Feed and water intake, rectal temperature, and body-surface (neck) temperature were measured during the experiment period. All data were statistically analyzed using the GLM procedure of SAS for a completely randomized design and significantly level was set at 0.05. No differences were found in the feed intake, rectal temperature, and body-surface temperature between 18 °C group and 23 °C group during the experiment. However, water intake in the 23 °C group ($3.22\pm0.05$ kg/d) was higher than in the 18 °C group ($2.59\pm0.13$ kg/d)($P<0.05$). We assumed that the thermal variation in thermal neutral zone had no influence on the rectal temperature and body-surface temperature but contributed on water consumption. The increased behaviour activity of water intake go with the frequency of water intake event, which resulted with the close relationship between the environmental temperature and behaviour activity of water intake. This study focused on identifying the environmental temperature effect on cattle's recital and body-surface temperature. However, further study on the variation of ingestive behaviour on environmental temperature would be useful.

## REM sleep time in dairy cows changes during the lactation cycle

*Emma Ternman[1], Emma Nilsson[2], Per P. Nielsen[3], Matti Pastell[4], Laura Hänninen[5] and Sigrid Agenäs[2]*
*[1]Aarhus University, Animal Science, Box 50, 8830 Tjele, Denmark, [2]Swedish University of Agricultural Sciences, Animal Nutrition and Management, Box 7024, 75007 Uppsala, Sweden, [3]University of Copenhagen, Department of Large Animal Sciences, Grønnegårdsvej 8, 1870 Fredriksbjerg C, Denmark, [4]Natural Resources Institute Finland (Luke), Production Systems, Latokartanonkaari 9, 00790 Helsinki, Finland, [5]University of Helsinki, Production Animal Medicine; Research Centre for Animal Welfare, Koetilantie 2, 00014 Helsinki, Finland; emma.ternman@anis.au.dk*

There are numerous physiological changes a dairy cow undergoes throughout the lactation cycle. Changes in lying behaviours suggest she may need to alter her rest to adapt to these demands, but changes in sleep have yet to be quantified. Using electrophysiological recordings during seven 24 h recording sessions, we investigated changes in rapid eye movement (REM) sleep, non-rapid eye movement (NREM) sleep, drowsing, awake and rumination in 19 dairy cows of the Swedish Red breed kept in single pens with ad libitum access to feed and water. The non-invasive method allows us to accurately record rapid eye movement (REM) sleep, the only vigilance state in which the cow cannot ruminate, and rumination. When the cow is not ruminating, awake, drowsing, and non-rapid eye movement (NREM) sleep can also be distinguished. The PROC MIXED procedure in SAS was used to test for significant differences between the stages of lactation cycle. Stage of lactation and dry period (-2, 2, 7, 13, 22, 37 and 45 weeks relative to calving), time of day (daytime 05-21 and night time 21-05), farm (Farm 1 and Farm 2) and a two-way interaction between stage of the lactation cycle and time of day were included as fixed effects, and cow as the subject of the repeated measure in the model. The recordings on week -2 and 45 were conducted during the dry period, all others during lactation. Total time spent in REM sleep was shorter for cows in week 2 relative to calving ($34.2\pm4.38$ min) compared to week -2 ($50.3\pm4.38$; $P<0.01$; $F6,107=3.49$), and the number of REM sleep bouts were fewer in week 2 ($7.5\pm0.90$) compared to week -2 ($12.4\pm0.90$; $P<0.001$; $F6,107=5.17$). REM sleep was recorded during both day and night, but predominantly performed at night, $5.3\pm1.19\%$ of total night time, compared to $2.1\pm1.19\%$ of total daytime ($P<0.05$; $F6,101=207.34$), and the bout duration was longer during night time ($4.6\pm0.09$ min) compared to daytime ($4.1\pm0.09$ min; $P<0.001$; $F1, 18=17.32$). There was a tendency for time spent in NREM sleep to be shorter in week 2 relative to calving compared to week -2 ($54.1\pm9.55$ and $86.3\pm9.55$ min respectively; $P=0.06$; $F6,107=2.08$). The time spent drowsing was shorter for cows in week 2 ($37.6\pm5.48$ min) and week 13 relative to calving ($41.1\pm5.48$ min) compared to week -2 ($65.2\pm5.48$ min; $P<0.05$; $F6,107=4.39$). In conclusion, REM sleep duration in dairy cows changes over the course of the lactation and dry period. The shortest REM sleep duration was found for cows two weeks after calving and longest two weeks before calving, and the difference was due a higher number of REM sleep bouts in the recording two weeks before calving. REM sleep and rumination predominantly occurred at night but were recorded during both day and night.

## Development of a dairy cattle welfare assessment scale and evaluation of cattle welfare at commercial dairy farms in India

*Madan Lal Kamboj*

*National Dairy Research Institute, Livestock Production Management, National Dairy Research institute, Karnal, Haryana, 132001 Karnal, India; kamboj66@rediffmail.com*

There is a paradigm shift in dairy cattle production system from small scale traditional to large scale commercial production system in India. Commercialization with resultant intensification of production practices is well established to impact welfare of animals. The assessment of level of welfare and identification of factors affecting welfare is a prerequisite for initiating any action for improvement of welfare. The aim of present investigation was to develop a methodology for the assessment of dairy cattle welfare and to assess the level of welfare of different sizes of commercial dairy farms in India. The methodology suggested by Calamari and Bertoni for assessment of dairy cattle welfare was modified to suit Indian farming conditions. The scale comprised of 3 major components viz. housing and other facilities (Component A) feeds and feeding practices (Component B), and animal health, physiology and behaviour (Component C) which were assigned welfare score of 30, 30 and 40 respectively. From these 3 components, 20 welfare indicators (6, 4 and 10 in each component, respectively) were identified; their patterns defined and were assigned a welfare score. The scale was tested for its reliability using Cronbach's alpha and for validity by expert opinion. For welfare assessment, 50 commercial dairy farms from Haryana and 60 from Punjab were selected and were categorized into small (10-20 cows; n=20), medium (21-50 cows; n=20) and large (>50 cows; n=10 in Haryana and 20 in Punjab). The significance of mean welfare scores of different components was tested using one way ANOVA and Duncan's Multiple Range Test in SPSS version 22. The mean values of welfare scores in Component A and B were significantly ($P<0.05$) higher in large farms both in Haryana and Punjab than in other two categories whereas mean scores of component C were similar in all categories of farms in Haryana but higher ($P<0.05$) in large dairy farms in Punjab. The mean total welfare scores of 3 components were $60.5\pm2.74$, $59.35\pm2.17$ and $68.1\pm1.18$ in small, medium and large farms respectively with an overall mean of $62.65\pm1.78$ in Haryana and $60.80\pm2.77$, $68.40\pm2.27$ and $74.60\pm1.70$ respectively with an overall mean of $67.93\pm1.49$ in Punjab. Overall 54.0% farms in Haryana and 76.6% farms in Punjab had an acceptable level of welfare (welfare score >60). Principal Component (PC) analysis performed extracted 6 PC's using Kaiser Rule criterion (retaining components that have Eigen value >1) explained cumulative variance of 83.3% in welfare indicators in Haryana and extracted 7 PC's explaining a cumulative variance of 70.85% in Punjab. In Haryana feeding space, feeding and watering system, mastitis incidence, cow comfort and reproductive efficiency and human-animal relationship were the most compromised welfare indicators. In Punjab floor space availability, floor type, microclimate protection, milking facilities, cow cleanliness, average productivity, reproductive efficiency and body condition were the most compromised welfare indicators. We concluded that dairy cattle welfare was better at large dairy farms in Haryana and both at large and medium dairy farms in Punjab as compared to small farms.

## Effect of the SF6 equipment to estimate methane emissions on dairy cows' behavior

*Daniel Enriquez-Hidalgo[1], Fabiellen Cristina Pereira[1,2], Ana Beatriz Almeida Torres[1], Macarena Fernandez Donoso[1], Dayane Lemos Teixeira[1], Laura Boyle[3], Luiz Carlos Pinheiro Machado Filho[2] and Richard Williams[4]*
*[1]Facultad de Agronomía e Ingeniería Forestal, Pontificia Universidad Católica de Chile, 7820436, Chile, [2]Laboratorio de Etología Aplicada, Universidade Federal de Santa Catarina, 88034-000, Brazil, [3]Animal & Grassland Research and Innovation Centre, Teagasc, P61, Ireland, [4]Agriculture Victoria, Department of Jobs, Precincts and Regions, 3821, Australia; daniel.enriquez@uc.cl*

There has been an increase in the use of precision technologies as research tools in animal science. In some cases this needs equipment being worn by animals, which could be invasive. The objective of this study was to evaluate the effect of the sulphur hexafluoride (SF6) tracer technique equipment to estimate enteric methane emissions (SF6 equipment) on dairy cow behavior. Dairy cows (n=24) kept in individual pens (6×3.5 m) over 6 weeks were allocated to two groups in two consecutive batches and equipped with the SF6 equipment for 7 days. The SF6 equipment consisted of a head halter and two PVC canisters mounted on a padded saddle fitted with a foam horse girth strap and a plastic strap placed around the cows' hindquarters. Cow behavior was assessed through video recordings between milking times (10-14 h and 16-20 h) by scan sampling every 10 min in 4 stages: (1) before cows were equipped with the SF6 (PRE, 3 d); (2) first 2 days of cows being fitted with the SF6 (EARLY) equipment: the 1st day the saddle was fitted and in the 2nd the halter and the canisters; (3) during methane measurements (LATE, 3 d); (4) after the SF6 equipment removal (POST, 2 d). Behaviors: eating, ruminating or idling, lying (with head up or down), walking and others (grooming, drinking, etc.) were registered. The occurrence of social interactions (affiliative or agonistic) with neighbouring cows and discomfort behaviors (scratching or pushing the equipment) were registered during 60 s every 10 min. Individual (n=16) lying time was recorded over 14 days using dataloggers fitted to the cow's hind leg. Milk yield was recorded daily and compared among stages. Data were analyzed using generalized linear mixed models, considering the stage as fixed effect and cow within group as random effect. The day was used as a repeated measure. Cows spend similar times eating (36.0±1.33%) and in other behaviors (6.2±0.71%), ruminated more (25.2 vs 17.8±1.02%; $P<0.05$) and were lying down with their head up more (46.9 vs 39.4±1.7%; $P<0.05$) at LATE stage than at any other stages, but were lying down, with head down longer at POST (9.1 vs 5.2±0.76%; $P<0.05$) than in any other stage. Affiliative behaviors occurred more often during EARLY than in any other stage (2.8 vs 1.4±0.5%; $P<0.05$) and agonistic interactions were not different between stages (0.7±0.77%). Cows tended to lie down less in EARLY compared to LATE (11.8 vs 12.3±0.34 h/d; $P=0.09$), but no other differences were detected. Cows had a similar number of lying bouts (7.1±0.31/d) with similar lying bout duration (1.75±0.08 h). Cows interacted more (0.3 vs 0.1±0.09%) with the equipment in the LATE than in the EARLY stage. The SF6 equipment did not affect milk yield (32±2.2 kg/d). We conclude that there was a minor change in cow behavior regarding the use of SF6 equipment, which apparently did not affect their comfort.

# Lying behaviour in dairy cows is associated with body size in relation to cubicle dimensions

*Neele Dirksen[1,2], Lorenz Gygax[2,3], Imke Traulsen[1], Beat Wechsler[2] and Joan-Bryce Burla[2]*
*[1]Livestock Systems, Department of Animal Science, Georg-August-Universität Göttingen, Albrecht-Thaer-Weg 3, 37075 Göttingen, Germany, [2]Centre for Proper Housing of Ruminants and Pigs, Federal Food Safety and Veterinary Office FSVO, Tänikon, 8356 Ettenhausen, Sweden, [3]Animal Husbandry & Ethology, Faculty of Life Sciences, Humboldt-Universität zu Berlin, Unter den Linden 6, 10099 Berlin, Germany; joan-bryce.burla@agroscope.admin.ch*

In the last two decades, cow body size has increased considerably due to selection for higher milk yield. However, cubicle dimensions of housing systems built many years ago have not been adjusted on most farms. Therefore, cows may be impaired in their lying behaviour. We investigated the lying behaviour of 144 cows with withers heights from 140-163 cm ($\bar{x} \pm$ SD: 149.3±5.2) on eight Swiss dairy farms with cubicle bed lengths of 187-200 cm and lunge space lengths of 47-202 cm. On each farm, lying down and standing up movements, as well as lying positions were observed on three days during 4.5-6 h/d, and the relative proportions of the specific behaviours were calculated in relation to the total number of observations per cow. In total, 703 lying down movements ($\bar{x} \pm$ SD per cow: 5.06±1.93), 655 standing up movements (4.71±1.97), and 3,161 lying positions (21.95±9.8) were observed. In addition, the presence of carpal and tarsal joint lesions were recorded; overall prevalences were 59.7 and 54.2%, respectively. Data of the individual cows (with a specific withers height) were analysed in relation to the given cubicle dimensions on the farms using linear mixed-effects models. An increase in the ratio between bed length and withers height (from 1.2 to 1.4) was associated with reduced proportions of lying down movements with repeated head pendulum movements (-35.6%, P≤0.001), repeated stepping with front legs (-21.0%, P=0.033), and hitting against cubicle elements (-30.3%, P=0.003). Similarly, an increase in the bed length/withers height ratio was associated with reduced proportions of standing up movements with shifting backwards (-32.3%, P=0.023), hesitant head lunge movements (-33.3%, P=0.001), and hitting against cubicle elements (-32.1%, P=0.002). Further, an increase in the bed length/withers height ratio was associated with reduced proportions of lying positions with physical contact with the curb board (-41.1%, P=0.003) or the partitions (-12.8%, P=0.003), and finally, with a reduced proportion of cows with tarsal joint lesions (-40.8%, P=0.02), but not with carpal joint lesions (P=0.165). Conversely, an increase in the ratio between lunge space length and withers height (from 0.3 to 1.4) was associated with a reduced proportion of standing up movements with sideways directed head lunge movements (-54.8%, P=0.001). In summary, cow body size in relation to cubicle bed length strongly affected the cows' lying down and standing up movements, their lying position, and the presence of tarsal joint lesions. Consequently, large-framed cows were restricted to a greater extent than smaller cows under the given cubicle dimensions. Based on the consistency of the obtained results, adjusting cubicles dimensions to the increase in cow body size is recommended.

## The effects of mesh fly leggings on number of flies and fly-avoidance behaviors of dairy cows housed on pasture

*Rielle Perttu, Brad Heins, Hannah Phillips and Marcia Endres*
*University of Minnesota, Animal Science, 1364 Eckles Avenue, St. Paul, MN 55108, USA;*
*pertt009@umn.edu*

Face (*Musca autumnalis* De Geer), horn (*Haematobia irritans* L.), and stable (*Stomoxys calcitrans* L.) flies are ectoparasites that negatively affect grazing distribution, milk production, and health of cows. Flies are a major welfare concern to cows on pasture causing avoidance behaviors. Mesh fly leggings – commonly and successfully used by horses – may be effective and offer relief from flies on pastured dairy cows. The objective of this study was to evaluate the effects of Shoofly Leggins (Stone Manufacturing & Supply, Kansas City, MO) on fly-avoidance cow behaviors. The study was conducted at the University of Minnesota West Central Research and Outreach Center (Morris, MN, USA) from June to July 2017. In this replicated crossover design study, lactating dairy cows housed on pasture (n=80) were randomly assigned to 1 of 2 treatment groups: (1) leggings (Shoofly Leggins on all legs); and (2) control (no Shoofly Leggins). Cows were exposed to their treatment for a two-week period, then switched treatments every period for a total of 4 periods (2 replicates per treatment). Counts for face, horn, and stable flies were recorded on all cows twice daily (morning: 09.30 to 12.30 and afternoon: 13.30 to 16.30 hr), 3 times per week. A random subset of 40 focal cows was observed in 5-minute intervals for frequency of leg stomps, head tosses, skin twitches, and tail swishes. A trained observer counted flies using a method adapted from Dougherty *et al*. Period means were used for the analysis using PROC GLIMMIX of SAS. Poisson models were built for fly count data with fixed effects of treatment, time of day, treatment and time of day interaction, period within replicate, replicate, and order of treatment, and a random effect of cow. For behavior, 3 covariates of each fly species were included, and a negative binomial model was built if the overdispersion term was greater than 1.5. Head toss, skin twitch, and tail swish behaviors were similar between treatment groups and time of day. Leg stomps were greater ($P<0.001$) for the control treatment group than the leggings treatment group (mean ± SE; 2.76±0.34 and 2.09±0.25 per 5-min observation, respectively), and leg stomps were greater ($P<0.001$) in the afternoon than in the morning (2.75±0.33 and 2.10±0.26 per 5-min observation, respectively). If we were to extrapolate this 0.7 extra leg stomps per 5-min observation to a period of 12 hours (day time), the difference between treatments would be approximately 101 stomps. The number of stable flies was a predictor ($P<0.0001$) of all observed behaviors and the number of horn flies was a predictor ($P<0.05$) of head toss, skin twitch, and tail swish behaviors. The number of stable flies on cows was greater ($P<0.001$) in the afternoon compared to the morning (20.6±0.8 and 15.0±0.6 per cow, respectively). The results of this study indicate that flies cause fly avoidance behaviors in cows regardless of the use of leggings. However, leggings effectively reduced leg stomps and may offer some relief from stable flies to dairy cows on pasture.

## Transport of cull animals – are they fit for it?

*Mette S. Herskin[1], Karen Thodberg[1], Kirstin Dahl-Pedersen[1], Katrine Kop Fogsgaard[1], Hans Houe[2] and Peter T. Thomsen[1]*

*[1]Aarhus University, Foulum, Department of Animal Science, P.O. Box 50, 8830 Tjele, Denmark, [2]University of Copenhagen, UCPH, Department of Veterinary and Animal Sciences, Grønnegårdsvej 8, 1870 Frederiksberg C, Denmark; mettes.herskin@anis.au.dk*

Modern pig and dairy production are characterized by high culling rates – 20-40% of cows and 50% of sows are culled/year. Culling reasons are diverse, but mainly related to productivity or health. The majority of the cull animals are transported to slaughter by road. EU Regulation clearly states that animals must be fit for transport. In addition, it is specified that 'all animals shall be transported in conditions guaranteed not to cause them injury'. Based on results already published as part of a recent research project on transport of cull dairy cows and sows, we discuss consequences of current transport practices on the welfare and fitness for transport of cull animals. The discussion integrates results from two observational studies, where sows and cows from private herds were examined clinically before and after transport to slaughter. The 47 and 49 journeys involved in the sow and cow study lasted 232 min (range 46-469) and 187 min (range 32-510), respectively. For both sows and cows, the post-transport clinical examinations showed signs that the animals' condition had deteriorated. For the sows, there was an increased occurrence of wounds, vulva lesions and torn hooves). For the cows the deterioration was seen as increased proportion of lame cows, increased occurrence of milk leakage and an increased proportion of cows with wounds. Results concerning potential risk factors for the deterioration differed considerably between the species. For the sows, risk of deterioration was related to factors associated with the journeys, such as temperature and duration; often in interaction. For the dairy cows, deterioration was related to factors such as lactation stage, body condition score and pelvic asymmetry i.e. pre-transport characteristics of the cows. Depending on the definition of – and threshold of – injury (which is not specified in the EU Regulation), these results may be interpreted as the animals 'being injured'. It is, however, stated in the Regulation, that 'slightly injured animals may be considered fit for transport if transport is not causing additional suffering'. Thus, it can be discussed whether the cull animals actually were fit for transport. The project was conducted in Denmark, where cull animals cannot be transported for more than 8 hours, and only involved direct journeys from farm to slaughter. With this in mind, we argue that the observed deterioration in the clinical condition of the cull animals legally considered fit for transport calls for further research and development into the concept of fitness for transport, as well as a consideration of the implications for animal welfare and strategies to optimize transport of this group of animals.

## A first look at the relationship between skin lesions and cortisol levels in stable groups of pregnant sows

*Martyna Lagoda[1,2], Keelin O'Driscoll[2], Joanna Marchewka[1], Simone Foister[3], Simon Turner[3] and Laura Boyle[2]*
*[1]Institute of Genetics and Animal Breeding of the Polish Academy of Sciences, Department of Animal Behaviour, ul. Postępu 36A, Jastrzębiec 05-552 Magdalenka, Poland, [2]Teagasc, Animal & Grassland Research & Innovation Centre, Pig Development Department, Moorepark, Fermoy, Co Cork, Ireland, [3]SRUC (Scotland's Rural College), Animal and Veterinary Sciences Group, Kings Buildings, West Mains Road, Edinburgh, EH9 3JG, United Kingdom; martyna.lagoda@teagasc.ie*

Group housed sows experience acute and severe stress due to aggression at mixing which is reflected in high levels of skin damage immediately after mixing and high concentrations of cortisol. The aim of this study was to determine whether skin lesion counts in a stable group are related to concentrations of cortisol in hair prior to farrowing, and to determine whether different aggressive strategies employed by sows in acute and/or chronic social situations are reflected in hair cortisol concentrations. 263 sows in 11 groups of 24 with full-length free access stalls on a commercial farrow-to-finish farm were used in the study. Skin lesions were counted on the rear, middle and anterior regions 24 h and 3 weeks post-mixing (c. 28 days post service). Hair samples and back fat measurements were collected 1 week prior to farrowing. Following extraction of cortisol from the hair, cortisol concentrations were determined using EIA (Salimetrics). Total lesion count data (24 h and 3 weeks post-mixing) were split into high and low lesion counts using the median as cut off, and sows with 4 combinations of lesion counts were identified: low-low (n=32); low-high (n=29); high-low (n=28) and high-high (n=31) for 24 h and 3 week lesion counts respectively. Lesion counts, cortisol and back fat data were available for 125 sows and were correlated using Spearman's rank correlation. A one way ANOVA was used to test for differences in hair cortisol concentrations and back fat depths between the 4 aggression strategy (lesion combination) categories, using R. No significant correlations were found, although there was a tendency for high 3 week anterior lesion counts to be negatively correlated with cortisol level ($r_s$=-0.17; P=0.06). High anterior lesions at 3 weeks post-mixing represent sows which were initiating aggressive attacks long after the dominance hierarchy was established. It appears that this strategy may be linked to lower levels of chronic stress. There were no differences in mean cortisol levels (F(3,116)=0.52; P=0.67), or in mean back fat depths (F(3,116)=0.91; P=0.44) for sows in each lesion combination category, indicating that under the conditions of this study, the aggressive strategy employed by sows when considered across these social contexts did not influence levels of chronic stress prior to farrowing. Further work is required to validate skin lesion counts as a proxy for chronic stress in group housed sows.

## An examination of the behaviour and clinical condition of cull sows transported to slaughter

*Louisa Meira Gould, Karen Thodberg and Mette Herskin*
*Aarhus University, Animal Science, Blichers Alle, 8830 Tjele, Denmark; louisa.gould@anis.au.dk*

Despite a worldwide rate of approximately 50% of sows transported to slaughter or killed on farm per year, almost no behavioural data exist on the pre-slaughter period for cull sows. A recent study showed that the clinical condition of sows can deteriorate following transport, identifying ambient temperature, journey duration, duration of stops during the journey and waiting time before unloading as risk factors. A new study is now examining the cumulative effects of potential pre-slaughter stressors for cull sows under commercial conditions in a randomised, controlled design. This presentation introduces the new study and the results of work done to refine and optimize the protocol used for clinical examination of the sows before and after transport to slaughter. The study includes different interventions running in two different study phases, focusing on how transport duration, the presence or absence of a break in the journey, the provision of water during transport, group size within the truck compartments, and thermal environment affect the sows on the day of transport to slaughter. In the current study period we will use blocks of three different journey durations (planned as 3, 5.5, and 8 hours) with each journey duration repeated with or without the currently required 45 minute break by Danish law after the driver has been driving for over 4.5 hours. G-force inside the truck, $CO_2$, temperature and humidity will be recorded during driving and breaks. Cull sows selected by farmers from commercial Danish herds will be clinically examined before and after transport, and their behaviour will be video recorded inside the truck. The study includes clinical measures expected to change such as skin lesions, skin elasticity, hoof condition, and gait score. Other clinical measures will be included to describe the overall condition of the sows before the journey, including lactation status and body condition score. The protocol was subject to several practice trials and inter-observer reliability testing. The six observers involved in the study data collection participated. This allowed for further refinement of measures and to optimise the protocol for many observers. For example, agreement on body condition score and gait score were determined using Kendall's coefficient of concordance. For body condition score scored using three categories, $W=0.83$, with observed scores ranging from 1 to 2 (median=2). For gait score scored using four categories, $W=0.73$, with observed scores ranging from 0 to 2 (median=1). This shows a strong level of agreement between the six observers on these measures. Being able to describe, but also to identify behavioural and clinical patterns among cull sows during the pre-slaughter period will provide useful information about this group of animals, as sows appear to be a more vulnerable group than younger finishing pigs. Knowledge provided from this study will systematically document for the first time how transport to slaughter is affecting the behaviour and welfare of sows.

# Behaviour of piglets before and after tooth clipping, grinding or sham-grinding in the absence of social influences

*Anna R.L. Sinclair[1,2,3], Céline Tallet[1], Aubérie Renouard[1], Paula J. Brunton[2], Richard B. D'Eath[3], Dale A. Sandercock[3] and Armelle Prunier[1]*

*[1]PEGASE, INRA, Agrocampus Ouest, 16 Le Clos, 35590 Saint-Gilles, France, [2]University of Edinburgh, Edinburgh, EH8 9XD, United Kingdom, [3]SRUC, Easter Bush, EH25 9RG, United Kingdom; armelle.prunier@inra.fr*

Needle teeth are routinely resected on pig farms to limit lesions that intact teeth may inflict to other piglets and the sow's udder. Two techniques exist: clipping with pliers and grinding with a rotating grindstone. Both techniques are potential sources of pain and stress. The present study aims to objectify their existence by analysing piglet behaviour just before and after tooth resection in the absence of social influences. In total, 120 piglets from 20 litters were allocated to tooth clipping, tooth grinding, or sham grinding (2 piglets/treatment/litter). In resected groups, the tip of each needle tooth was removed using sterilised pliers or a hand-held rotary grindstone. Piglets assigned to sham grinding were handled and treated as those in the Grind group with a protective covering on the grinder head to prevent tissue damage. One day after birth, each litter was separated from the dam and placed in a heated holding trolley. Selected piglets were taken individually to a separate room to undergo tooth treatment and behavioural observations. Each piglet was placed in an observation box for 1 min to be filmed. Thereafter, tooth treatment was applied by a trained handler and the needle teeth were measured and checked for minor bleeding. The piglet was returned to the observation box and filmed again for 1 min. Once all selected piglets had been treated, the litter was returned to the dam. Behavioural observations focussed on locomotion, oral behaviours, ear position and movements, and vocalizations. For behaviours that were not always visible (e.g. ear position), the percentage of time spent exhibiting the behaviour was calculated after excluding the non-visible period. Rare behaviours were transformed into binary variables. Mean treatment duration was 53, 48 and 46 s in the Sham, Grind and Clip groups, respectively. Minor bleeding was observed directly after resection in Grind and Clip piglets (22.5 vs 97.5%). For quantitative variables, the time × treatment interaction was never significant except for exploring wood shavings. Indeed, its duration increased after treatment only in the Sham group (Pre: 18.5% vs Post: 33.7%, P<0.05) and, in the Post period, it was significantly higher in Sham than Clip piglets (33.7% vs 18.8%, P<0.05) with Grind piglets being intermediate. Regarding binary variables, champing was never observed during the Pre period but, in the Post period, it differed between Sham and Clip pigs (45 vs 80% of pigs with champing, P<0.05) with Grind pigs being intermediate (60%). Numerous variables differed significantly between the Pre and Post periods: walking, exploring walls, and ears back decreased whereas being immobile, not exploring, ears in a plane or front position, and head flick increased regardless of the treatment group (P<0.05). This experiment demonstrates a marked influence of handling on piglet behaviour regardless of the tooth treatment as well as some differences between groups after treatment showing signs of pain in clipped piglets whereas signs of pain are less clear in piglets submitted to grinding.

## Associations between sheep welfare and housing systems in Norway

*Solveig Marie Stubsjøen[1], Randi Oppermann Moe[2], Cecilie Marie Mejdell[1], Clare Phythian[2] and Karianne Muri[2]*

*[1]Norwegian Veterinary Institute, Department of Animal Health and Food Safety, P.O. Box 750 Sentrum, 0106 Oslo, Norway, [2]Norwegian University of Life Sciences, Faculty of Veterinary Medicine, Department of Production Animal Clinical Sciences, P.O. Box 8146 dep., 0033 Oslo, Norway; solveig-marie.stubsjoen@vetinst.no*

Norwegian sheep are commonly housed indoors during winter time, typically in insulated buildings in pens with expanded metal floors. However, various designs of simple, non-insulated buildings are increasingly being used, often containing larger pens with deep litter bedding. This study aimed to explore how different housing systems are related to welfare outcomes in sheep. An electronic questionnaire was distributed to 3,764 Norwegian sheep farmers (response rate 32%, n=1,206). Data on sheep housing conditions, such as insulated vs non-insulated, feeding routines and management practices were collected, and 64 respondents were subsequently contacted and included in this on-farm study. Prior to the lambing season, five trained veterinary assessors performed welfare assessments in 64 farms (35 insulated and 29 uninsulated/open housing designs). A refined version of a welfare assessment protocol previously developed for housed sheep were used, including Qualitative Behaviour Assessment (QBA) and assessment of the human-animal relationship. Signs of clinical disease including lameness, coughing and pruritus were assessed by group observation. This was followed by clinical examination of individual sheep (1,759 ewes in total), sampled according to flock size. Resource-based measures (e.g. space allowance and temperature) were also systematically recorded on all farms. Multivariable regression models were used to investigate possible associations between different housing systems and animal-based measures of sheep welfare. The most prevalent physical conditions observed were callus on carpus (27.5%), dirtiness of the abdomen (18.8%), considerably overgrown claws (18.1%) and wool loss (16.0%). Preliminary results show that the first component of QBA (mood, explaining 35% of the variance) was not associated with the main housing variables. Claw overgrowth was higher in systems where the sheep were kept on deep litter compared to expanded metal floors (OR 7.1, 95% CI 2.6-18.9). However, a lower risk of knee calluses (OR 0.1, 95% CI 0.03-0.4) and dirty hind limbs (OR 0.1, 95% CI 0.02-0.6) were associated with deep litter bedding. Less wool loss was observed in non-insulated (OR 0.5, 95% CI 0.3-1.0) and open building designs (OR 0.2, 95% CI 0.1-0.6) compared to the traditional insulated buildings. Severe skin lesions were higher in flocks provided concentrates three times daily (OR 11.1, 95% CI 3.2-38.4), compared to twice (OR 1.4, 95% CI 0.7-3.2) or once daily. The odds of ewes having improved body condition (scores≥3) were three times higher in those fed concentrates twice daily (95% CI 1.5-7.3) than those fed once daily. In conclusion, this study provided information about physical health and welfare status in different housing systems. Findings of excessive claw overgrowth, dirtiness and skin lesions are most likely to be associated with flooring type, space allowance, pen facilities and feeding regimes and are areas that may warrant further attention.

## First steps for the development of behavioural indicators in dairy sheep: ethogram and functional categorization

*Alejandra Feld, Débora S. Racciatti and Héctor R. Ferrari*
*Facultad de Ciencias Veterinarias, Universidad de Buenos Aires, Animal Welfare, 280 Chorroarin Ave., C1427CWO, Buenos Aires City, Argentina; afeld@fvet.uba.ar*

Animal welfare (AW) can be assessed in a scientific manner using indicators. Direct indicators are relevant because they evince the results of inputs and investments on AW. Behavioural ones are considered early indicators of AW problems. Their assessment can be carried out without restriction, invasion or perturbation. For animals to have lives worth living it is necessary to minimise their negative experiences and to provide them with opportunities to have positive experiences. Information about social aspects, affective states and activity budgets is relevant for these purposes. An ethogram is a list or catalog of an animal's or species' behaviours and their definitions. It´s a quantitative description in particular contexts. Our goal was to make an ethogram of a dairy sheep herd and group the observed behaviours into functional categories, as a step towards developing behavioural indicators for this group. Three researchers carried out *ad libitum* sampling with naked eye observations and detailed descriptions of 20 Ostfriesisches Milchschaf ewes and their lambs, 5 days/week, 9 am to 4 pm, over a 4-month period at a dairy farm. We observed between June and September in Buenos Aires city (Argentina). The herd was outdoors during the day in a natural field. During the night, the sheep were confined indoors with *free access to water*. The animals were acclimated to humans. As in every ethogram design where data is registered by more than one observer, we settled several agreements: described the parts of the body involved in the behaviours, outlined the facilities and defined the terms in which every behaviour would be registered. Finally, we grouped the behaviours into functional categories. We accumulated 800 min of observation, describing 57 behaviours, grouped in 9 functional categories: Rest, Locomotion, Feeding, Maintenance, Excretion, Environment Interactions, Intraspecific Interactions, Interspecific Interactions, Other. We designed a spreadsheet for recording the frequency of behaviours. This spreadsheet is a double-entry table. Columns contain the functional categories and rows express the moment when the behaviour is observed, at pre-set times every minute of the 20-min sessions. We consider the number of recorded behaviours sufficient to begin with the next stage. We will use the resulting spreadsheet to determine the activity budget. This will allow us to assess the potential use of plasticity and variability of behavioural repertoire, use of space, social bonds and abnormal behaviours, as direct indicators of the well-being state of the herd. To carry out a study in a particular field, it is convenient to have an ethogram of that group to obtain relevant information. Since the ethogram is a descriptive study, it is not expected to be innovative, but it should be applicable to that group in those contexts. We didn´t find other studies with ethograms for dairy sheep in Argentina. Considering that welfare is the state of an individual in relation to its environment, we suggest that carrying out a behavioural inventory of this group in these particular conditions is relevant to the next step of AW assessment.

# Effects of multimodal pain management strategies on acute pain behavior and physiology in disbudded neonatal goat kids

*Whitney Knauer[1], Emily Barrell[1], Alonso Guedes[1] and Beth Ventura[2]*
[1]*University of Minnesota, College of Veterinary Medicine, 1365 Gortner Avenue, 55108, USA,*
[2]*University of Minnesota, Animal Science, 1364 Eckles Avenue, 55108, USA; bventura@umn.edu*

Disbudding is a common procedure in the dairy goat industry, typically carried out in the first week of life via a hot cautery iron without pain relief. This project aimed to identify a practical, efficacious pain management strategy for disbudding by monitoring changes in plasma biomarkers and behavior in goat kids disbudded with varying combinations of pain relief. 42 goat kids (1-13 days old) were obtained from a local farm and housed in group pens at the University of Minnesota large animal teaching barn. After a 3 d acclimation period, kids were balanced by breed and sex and allocated to one of the following 7 treatment groups (n=6/group) according to a randomized block design: negative control ('NC') of sham disbudding; 0.05 mg/kg IM xylazine ('X'); 4 mg/kg SQ buffered lidocaine ('L'); 1 mg/kg oral meloxicam ('M'); xylazine + buffered lidocaine ('XL'); xylazine + oral meloxicam ('XM'); and xylazine + lidocaine + meloxicam ('XLM'). All kids were shorn over their horn buds and jugular veins 1 day prior to disbudding. Treatments were administered 20 minutes before disbudding. Kids underwent thermal disbudding by the same trained individual, blinded to treatment; NC kids received the same handling except that the iron was cold. Jugular blood samples (3 ml) were obtained before (-20, -10, and -1 min) and after (1, 15, 30 min and 1, 2, 4, 6, 12, 24, 36, and 48 h) disbudding to measure plasma cortisol and prostaglandin E2 (PGE2). Pain sensitivity was measured via pressure algometry with a von Frey anesthesiometer at -20 min and at 4, 12, 24, and 48 h post-disbudding. Daily body weights were obtained prior to the morning milk feeding. Video cameras were mounted in each pen and ran continuously for behavior evaluation. A single trained individual blinded to treatment observed each kid during disbudding and recorded frequency of vocalizations and struggles. Video footage was also analyzed for periods of 10 continuous min at 1, 3, 6, 12, 36, and 48 h after disbudding to obtain frequency and duration of: self/allogrooming; body shaking; ear biting and flicking; head butting, scratching, and shaking; running, walking; standing; playing; lying; eating; drinking. Effect of treatment on weight gain after disbudding, mechanical nociceptive threshold (MNT), and behavior during disbudding was analysed using a linear mixed model with block, sex, breed and age accounted for as random effects, and repeated measures incorporated for MNT. Duration of disbudding did not vary among treatments. Post-disbudding weight gain and struggles during disbudding were not affected by treatment. However, treatment affected MNT (P=0.0264) such that M kids were more sensitive overall than NC kids (P=0.0061). Likewise, vocalization frequency was affected by treatment (P=0.0005), with vocalizations lower in XLM (P=0.0001), XM (P=0.0299), and XL (P=0.0184) kids compared to M kids. Vocalizations were also lower in XLM vs L kids (P=0.0188). This suggests that multimodal approaches may attenuate pain response during disbudding. Further analysis will elucidate treatment effects on blood chemistry and behavior in the 48 h following disbudding.

## Using faecal glucocorticoid metabolites as a method for assessing physiological stress in reindeer

*Grete Helen Meisfjord Jørgensen[1], Svein Morten Eilertsen[1], Inger Hansen[1], Snorre Hagen[1], Ida Bardalen Fløystad[1], Rupert Palme[2] and Şeyda Özkan Gülzari[1]*
*[1]Norwegian Institute of Bioeconomy Research NIBIO, Post Box 115, 1431 Ås, Norway, [2]University of Veterinary Medicine, Department of Biomedical Sciences, Unit of Physiology, Pathophysiology and Experimental Endocrinology, Veterinärplatz 1, 1210 Vienna, Austria; grete.jorgensen@nibio.no*

Non-invasive methods for assessing physiological stress have been developed for many animal species by measuring for example, faecal glucocorticoid metabolites (FGM). Due to expressed species differences in metabolism and timeframe for excretion of these stress hormones method must be validated for each species. The aim of this study was to investigate the delay time from induced stress (through an ACTH stimulation) until FGM appeared in elevated levels in reindeer faeces. The study was conducted in January 2018, in the fenced facilities of a private reindeer herder in Nordland County in Norway. The area was covered with snow and the ambient temperature ranged from -10 to -15 °C. The study was conducted in accordance with the regulation for use of animals in experiments, approved by the Ethics Commission on Animal Use by the Norwegian Food and Safety Authority (FOTS ID: 12274 03.04.2017). Eight male reindeer were injected intramuscularly in the neck with ACTH (0.25 mg/ml per animal: Synacthen$^{(R)}$). After injection, animals were moved to a fenced area where they were undisturbed, except for the collection of faeces from the ground or for feeding. The collection of faeces started two hours after the injection of ACTH and was repeated every hour until the 12$^{th}$ hour and every other hour until the 30$^{th}$ hour, in addition to a final sampling at the 44$^{th}$ hour. A total of 317 faecal samples was collected and extracted. The samples were analysed using enzyme immunoassays. The effect of hour since ACTH administration on FGM concentration was analyzed using a mixed model analysis of variance with Day (1-3), Hour (1-44) and Animal (1-8) as class variables. Animal was specified as a random effect in the model, taking repeated measures per animal in account. The mean levels of glucocorticoid metabolites found in the faecal samples was 788.18±132.13 ng/g (mean ± SEM) after two hours, 4,039.1±811.0 ng/g after seven hours and 1,639.4±346.9 ng/g after 24 hours from ACTH injection. There was an observed time lag of around six hours from ACTH administration until elevated FGM levels were observed. Peak levels were found 7 hours after ACTH administration. The FGM levels remained elevated until 16 hours after ACTH injection (1,648.8±301.8 ng/g), but at a decreasing level already after 12 hours (2,234.5±267.5 ng/g). Looking at the eight animals separately, we found a consistent increase in FCM after seven to eight hours after ACTH injection. Our study shows that FGM levels may be used as a valuable indicator for assessing physiological stress in a semi-domestic animal species such as reindeer. In conclusion, elevated levels of glucocorticoids were found in faeces of male reindeer after six to seven hours after induced physiological stress.

## Changes in saliva analytes reflect acute stress level in horses

*María Dolores Contreras-Aguilar[1], Séverine Henry[2], Caroline Coste[2], Fernando Tecles[1], Jose J. Cerón[1] and Martine Hausberger[3]*
*[1]Interdisciplinary Laboratory of Clinical Analysis (Interlab-UMU), University of Murcia, Animal medicine and surgery, Veterinary School, Campus of Excellence Mare Nostrum, Espinardo, 30100, Spain, [2]Université de Rennes, Université de Normandie, Laboratoire Ethologie animale et humaine, UMR CNR 6552, Station Biologique, Paimpont, 35380, France, [3]Université de Rennes, Université de Normandie, Laboratoire Ethologie animale et humaine, UMR CNRS 6552, Campus de Beaulieu, Bat.25, 263 avenue du général Leclerc, Rebbes Cedex, 35042, France; mariadolores.contreras@hotmail.com*

Behavioural components do not give access to the temporal dynamics of the stress response. Therefore, it could be useful to develop further physiological measures. Saliva is a non-invasive and easy way to obtain samples. Until now, most studies have focused on salivary cortisol, but there is a lack of knowledge about how other salivary analytes related to the autonomic nervous system (ANS) can behave in a stress condition. The aim of this study was to evaluate the possible changes in cortisol, salivary alpha-amylase (sAA), lipase, adenosine deaminase (ADA), total esterase (TEA) and butyrylcholinesterase (BChE), and their potential correlation with horses' behaviour after an experimental acute stress. For this purpose, saliva samples were collected in nine riding horses subjected to acute stress consisted in opening an umbrella in front of them released in an indoor riding arena. Horses' behaviour was recorded during 1 minute by scan sampling (30 scans/min). The experiment took place with a basal time (T1) in their box stalls, a second basal time (T2) in the arena before opening the umbrella, the stress time (T3) just after the umbrella is opened, and 30 min (T4) and 60 min (T5) after T3 in their box stalls. A Friedman test with a Dunn's multiple comparisons test was used to assess statistical differences between times, and a Spearman test was used to analyse correlations between saliva analytes and horses' behaviour. Cortisol (P=0.050), TEA (P=0.004), lipase (P<0.001) and BChE (P=0.006) increased at T3 from T1, with then a decrease at T4 in sAA (P=0.029), ADA (P=0.046) and BChE (P=0.004). Lipase increased at T5 compared to T1 (P=0.006). There were positive correlations between cortisol and BChE and the frequency of occurrence of alarm acoustic signals (snore/blow) (r=0.68, P=0.050; r=0.69, P=0.046, respectively), and of 'glances to umbrella with the left eye' (r=0.71, P=0.039; r=0.76, P=0.025; respectively). BChE was correlated with the index of emotionality calculated by combining the behavioural responses (r=0.66, P=0.050). Negative correlations were observed between 'sniffing at the ground' and sAA (r=−0.65, P=0.043), and the index of laterality (right eye (RE) – LE/ RE+LE, reflecting a preferential use of the right eye) and ADA and BChE (r=−0.68, P=0.041). In conclusion, BChE in saliva appears as more closely related to acute stress than cortisol, and a low sAA value could be used to assess quietness. So, this preliminary study opens the possibility of a wider use of saliva as a sample for evaluation of acute stress in horses by the measurement of analytes such as BChE and sAA, and to correlate them with other ANS assessments such as cardiac parameters to know better the temporal dynamics of the stress response.

### Test-retest reliability of the Animal Welfare Indicators protocol for horses

*Irena Czycholl, Philipp Klingbeil and Joachim Krieter*
*Kiel University, Institute of Animal Breeding and Husbandry, Olshausenstr. 40, 24118 Kiel, Germany; iczycholl@tierzucht.uni-kiel.de*

Objective tools for the assessment of animal welfare are needed. The present study analysed the test-retest reliability of the Animal Welfare Indicators (AWIN) welfare assessment protocol for horses as a further step towards an objective welfare assessment tool on-farm. Four farm visits on each of 14 farms in Northern Germany were conducted by the same trained observer. Farm visits 1 and 2 were three days apart, farm visits 1 and 3 six weeks apart and farm visits 1 and 4 twelve weeks apart. This experimental setup should enable a detection of a potential seasonal effect as well as temporal variations. Although by nature of the AWIN protocol, the assessments were carried out for individual horses, for further analysis, for each of the indicators, the percentages of affected horses per farm were compared, i.e. the results were compared at farm level between the different farm visits. For statistical analysis, a combination of different reliability and agreement parameters was calculated: Spearman's rank correlation coefficient (RS), intraclass correlation coefficient (ICC), limits of agreement (LoA) and smallest detectable change (SDC). The Qualitative Behaviour Assessment was analysed by means of principal component analysis. Most of the indicators demonstrated acceptable (RS≥0.4; ICC≥0.4; SDC:≤0.1; LoA $\epsilon$ (0.1; 0.1)) to good (RS:≥0.7; ICC:≥0.7; SDC:≤0.05; LoA: $\epsilon$ (0.05; 0.05)) reliability. Exceptions were the indicator swollen joints (median of statistical parameters: RS: 0.28, ICC: 0.13, SDC: 0.14, LoA: $\epsilon$ (-0.13; 0.16)) as well as the behavioural tests (e.g. median of statistical parameters of the Avoidance Distance Test: RS: 0.45, ICC: 0.18, SDC: 0.26, LoA: $\epsilon$(-0.20; 0.20)) which were of insufficient reliability. The Horse Grimace Scale, was of insufficient reliability for those indicators representing moderate presence of tensions in all of the facial regions (score 1) (e.g. median of statistical parameters of moderate presence of tensions in the ears (score 1): RS: 0.23, ICC: 0.02, SDC: 0.41, LoA: $\epsilon$ (-0.40; 0.46)) while reliability for those indicators representing severe presence of tensions (score 2) was acceptable (e.g. median of statistical parameters of obvious presence of tensions in the ears (score 2): RS: 0.48, ICC: 0.41, SDC: 0.10, LoA: $\epsilon$ (-0.10; 0.10)). The Body Condition Score was of acceptable reliability for the categorisation into 1, 2, and 5, but not for 3 and 4. Hence, the exact categorisation between 3 and 4 should be redefined. The Qualitative Behaviour Assessment was of sufficient reliability in the comparison of farm visit 1 and 2 (RS=0.87 for Principal Component (PC) 1 and RS=0.91 for Principal Component (PC) 2) however, this changed for the comparison of farm visit 1 to 3 (PC1: RS=0.09, PC2: RS=0.19) and 1 to 4 (PC1: RS=0.33, PC2: RS=0.19). Hence, for longer time frames, test-retest reliability was not given. In general, this study points out a good test-retest reliability for most of the indicators included in the AWIN welfare assessment protocol for horses, but also reveals some problematic indicators that need revision or replacement in the future. Therewith, this study contributes to the further improvement and refinement of animal welfare assessment tools for horses.

## A study into the impact of sweet itch on equine behavioural patterns

*Freddie Daw, Charlotte Burn, Ruby Chang and Christine Nicol*
*Royal Veterinary College, AWSE, Hawkshead Lane, AL9 7TA, United Kingdom; fdaw@rvc.ac.uk*

Sweet itch, a hypersensitivity to midge (*Culicoides*) bites that often results in pruritus, is a common condition affecting horses. The National Equine Health Survey (2018) recorded a prevalence of 7.3%, making it the 4th most common disease syndrome reported in the UK. Despite this, the impact it has on horses is poorly understood. Commonly, sweet itch causes horses to rub excessively across their entire body, causing irritation and lesions. The aim of this study was to examine how this disease may disrupt regular behaviour patterns to enable a fuller assessment of its effect on welfare state. The ongoing study takes focal behaviour observations of horses with sweet itch (SI) and non-affected control horses (NI). Each horse and its matched control were observed for consecutive 30 min periods on the same day. Observations are taken in summer and winter periods as midges, and so the symptoms of sweet itch, are supressed over winter. To date, 11 SI and 11 NI, each observed in both seasons, have been included. Further horses have been recruited for the 2019 season so at this stage we report descriptive results only. The direct observations to date show that over both seasons combined SI horses spent 1% (SD=3) of their time rubbing, whereas NI horses spent 0.17% (SD=0.65). Considering the summer season alone, SI horses spent 1.6% (SD=3.76) of their time rubbing and NI horses spent 0.26% (SD=0.84). This represents a disruption of normal behaviour patterns with an activity that may cause injury to the horse, with a noticeable difference between SI and NI horses' percentages. In the winter the percentages dropped to 0.07% (SD=0.19) for SI and 0.02% (SD=0.06) for NI horses. Other behaviours focused around the illness are tail swishing and head shaking, being directly related to the presence of insects. SI horses performed head shakes 0.18% (SD=0.44) of the time, while NI spent 0.07% (SD=0.17) head shaking, however SI horses spent 1.56% (SD=2.29) of their time tail swishing and non-affected horses spent 2.56% (SD=4.86). These behaviours can be used as an indicator of insect presence. As tail swishing drops from 2.32% (SD=5.21) across all groups in the summer to 0.43% (SD=1.1) in the winter it may be a useful behavioural tool for identifying when sweet itch risks are higher. Other key behaviours are grazing, sleeping and resting. Across both seasons, SI spent 59.3% (SD=34.3) of their time grazing and NI spent 71% (SD=29.6). Though further numbers are required before statistical analysis, this difference is of note, although the time spent resting and sleeping appears at this stage to be similar in NI and SI horses. The study is ongoing and will be supplemented with additional 24 h automated recordings from accelerometers with core behaviours identified from readings as well as behaviour observations from additional SI horses and their NI controls.

## Welfare of horses living in individual boxes: methods of assessment and influencing factors

*Alice Ruet[1], Julie Lemarchand[1], Céline Parias[1], Núria Mach[2], Marie-Pierre Moisan[3], Aline Foury[3], Christine Briant[1] and Léa Lansade[1]*
*[1]PRC, INRA, CNRS, IFCE, University of Tours, 37380 Nouzilly, France, [2]UMR 1313, INRA, AgroParisTech, University of Paris-Saclay, Vilvert, 78352 Jouy-en-Josas, France, [3]UMR 1286, INRA, University of Bordeaux, 71 Avenue Edouard Bourlaux, 33140 Villenave-d'Ornon, France; alice.ruet@inra.fr*

Welfare is the positive mental and physical state linked to the satisfaction of behavioural and physiological needs and individuals' expectations. Domestic horses' living conditions (*e.g.* individual boxes) might require strong adaptation abilities and induce welfare deterioration. Four behavioural indicators have been identified as constituting markers of compromised welfare: stereotypies, aggressiveness toward humans, apathetic attitudes and anxiety behaviours. These indicators are mainly assessed by three kind of protocols at the individual level: scans sampling, surveys and the AWIN Horse protocol. This study simultaneously assessed the aforementioned behavioural indicators and protocols on a total of 187 horses strictly living in individual boxes in the same stable and without access to paddocks, during two periods of observations. Per period, each horse was observed for 5 scans per day, during 25 days. The scans sessions (90 minutes) were equally distributed across 9:00 to 16:30. Two AWIN protocols were performed per horse (one per period) but only one survey could be completed. The temporal repeatability between the two periods was good for all indicators within the scans sampling (Spearman's rho; Aggressiveness: $r_s=0.51$, P<0.001, Apathy: $r_s=0.41$, P<0.001, Stereotypies: $r_s=0.34$, P<0.001, Anxiety: $r_s=0.17$, P<0.05) and measures assessing for aggressiveness (Fisher's exact test; $\chi^2=18.2$, P<0.001) and apathy (Fisher's exact test; $\chi^2=5.7$, P<0.05) within the AWIN protocol. Overall, the three protocols also provided convergent information, but depending on the indicator studied. For example, scans sampling and the AWIN protocol were not correlated regarding the assessment of apathy and anxiety, probably thus reflecting different emotional states related to the observation conditions. Scans sampling method showed the highest detection rates and inter-individual variability for all indicators. The co-occurrence between the indicators was also explored to provide new elements regarding the assessment of deteriorated welfare in horses. No positive significant correlation was found between them, indicating that welfare deterioration cannot be based on a single behavioural indicator. Finally, the effects of various factors (*e.g.* inherent parameters, housing, feeding, riding) on the expression of the four indicators was studied to suggest levers of action in the case where the welfare would be altered. Horses were more apathetic with age (Linear-Mixed effects Model: F=5.07, P=0.002), and a non-straw bedding increased aggressiveness toward humans (Generalized Linear Mixed Model: $\chi^2=5.61$, P=0.05) and locomotion-related stereotypies (Generalized Linear Mixed Model: $\chi^2=3.6$, P=0.05). Surprisingly, very few factors influenced the expression of the four behavioural indicators, indicating that a return to living conditions satisfying natural needs might be essential for horses whose welfare is deteriorated.

## Effects of cage size on the thermoregulatory behaviour of farmed American mink (*Neovison vison*)

*María Díez-León[1] and Georgia Mason[2]*
*[1]Royal Veterinary College, University of London, Hawkshead Ln, AL97TA Hatfield, United Kingdom, [2]University of Guelph, 50 Stone Rd E, N1G2W1 Guelph, Canada; mdiezleon@rvc.ac.uk*

In North America, heat stress in the summer can compromise the welfare of fur-farmed mink, particularly if growth of their winter coats is accelerated by melatonin implants. Mink cope with heat by becoming inactive, interacting with water, panting, and adopting spread out postures on the cage floor. However, the latter may be harder in small cages, especially for Pastel males (as large); and if pair-housed (the norm for growing juveniles). Larger floor areas could thus help mitigate heat stress by allowing more effective thermoregulation. We tested this hypothesis by comparing European (E) to Canadian (C) minimum floor areas (2,550 vs 2,225 cm²) during summer. We predicted that mink in C compared to E cages will have more body contact with cagemates when lying, and show more signs of heat stress, especially on hot days. We raised 64 unrelated melatonin-implanted Pastel male pairs from weaning in E (32) or C cages (32), alternating cage type across the facility. For 7 hot (>27 °C) and 7 cooler days (<27 °C), and when both mink were resting together on the floor cage, we recorded thermoregulatory behaviour (e.g. panting, proximity to the water line, adoption of spread out postures); the temperature of 'hot spots' (e.g. chin, belly, paws) via infrared (IR) thermography; and the extent to which cagemates were touching. On hot days compared to cooler days, mink spent more time by the water line (E: $F_{1,61}=39.5$, P<0.001; C: $F_{1,62.11}=26.4$, P<0.001); panted (there being no difference in panting between E and C cages); touched each other less (E: $F_{1,61}=69.8$, P<0.001; C: $F_{1,63.04}=9.1$, P<0.01); and adopted spread out postures more (E: $F_{1,61}=37.3$, P<0.001; C: $F_{1,63.2}=4.75$, P<0.05). Compared to E-housed mink, C-housed mink were more likely to have body contact with a cagemate (hot days: $F_{1,61}=126.4$, P<0.001; cooler days: $F_{1,60}=17.4$, P<0.001). This could explain why, on hot days, C-raised mink were less likely than E mink to adopt spread out postures ($F_{1,61}=30.3$, P<0.001), and more likely to be closer to the water line ($F_{1,61}=20$, P<0.001). For IR data, methodological problems reduced our sample size (n=7). Belly temperatures of C-raised mink were also 2.5 °C higher than E-raised mink when hot, although this result did not reach statistical significance. Results so far suggest that smaller cages prevent paired melatonin-implanted male mink from adopting spread out postures to dissipate heat, and may potentially elevate their temperatures in hot weather. If confirmed and found to be aversive (analyses of cage size preference are on-going), this would have important implications for cage size regulations.

## Efficiency of sheep cadaver dogs

*Inger Hansen and Erlend Winje*
*Norwegian Institute of Bioeconomy Research, P.O. Box 115, 1431 Ås, Norway; inger.hansen@nibio.no*

Around 100,000 ewes and lambs are lost every year on Norwegian rangeland pastures. Although inspection of the herds is good, it is very hard to find sheep cadavers. Dogs have been used for decades to follow the scent of animals and humans, dead or alive. However, only recently have dogs been trained to search for livestock cadavers. Norwegian Cadaver Dogs (NCD) has developed a course and an approval test for search dogs. To succeed, the dog has to find 3 out of 4 cadavers put out in a limited forest area. Per June 2016, 157 equipages (i.e. handler-dog teams) were approved. The aim of this study was to document the efficiency of dog equipages compared to searching for dead sheep without using dogs. Our hypothesis were that cadaver dog equipages (CDEs) were more efficient than humans alone to find cadavers and that old cadavers sited on open ground were easier for the dogs to find than fresh and hidden ones. Intensive searches were conducted in 10 'experimental fields' (0.5 km$^2$) with and without dogs. Within each field, 8 lamb cadavers weighing 2-4 kg were put out, of which 4 were 'fresh' and 4 'old' and half of them 'hidden' and the rest sited on open ground ('open'). Additionally, 6 'real fields' (4 km$^2$) were searched for cadavers, with and without dogs. A total of 16 approved CDEs and 16 persons without dogs participated and a total searching time of 8 hours per 1 km$^2$ area was given for both types of field. Km and min per cadaver were recorded in order to measure the search efficiency. All tracks and cadaver positions were GPS logged. Non-parametric statistics were used. In the experimental fields, the CDEs found 19 out of 80 lamb cadavers (23.8%), whereas persons without dogs found only 2 (2.5%, P<0.05). Five cadavers found by the dogs were fresh and open-sited, 4 were fresh and hidden, whereas 5 were old and open and 5 were old and hidden. Thus, the dogs found cadavers that were assumed the most difficult to detect (fresh and hidden) as often as the ones that were assumed easier (n.s.). Both cadavers found by people were old and open sited. There was no significant difference according to km search per field with or without dogs (9.5/10.5 km). Due to great variation in number of cadavers in the real fields and a limited number of fields, no significant differences with or without dogs were found regarding number of cadavers detected, whether they were fresh or old, sited open or hidden, km- and min search per cadaver. The experimental field results showed that it is more efficient to search for sheep cadavers with a dog than without and the dogs were superior regarding detection of hidden cadavers. However, the dogs did not find more than a quarter of the cadavers that were put out. The GPS logs showed that dogs had passed cadavers on several occasions, either without marking them or because the dog handler did not respond to the dog's signals. Results from this study may imply revision of NCD's course plans to ensure education of high quality CDEs. Based on experiences from this project, a guide for CDEs was outlined.

## Changes in behaviour of shelter dogs post adoption

*Vladimir Vecerek, Svatava Vitulova, Eva Voslarova and Iveta Bedanova*
*Faculty of Veterinary Hygiene and Ecology, University of Veterinary and Pharmaceutical Sciences Brno, Department of Animal Protection, Welfare and Behaviour, Palackeho tr. 1946/1, 61242 Brno, Czech Republic; vecerekv@vfu.cz*

The aim of this study was to analyse changes in dogs´s behaviour within six months after their adoption from a shelter. In order to collect information on behaviour of shelter dogs post adoption, questionnaires were provided to all shelters in the Czech Republic with a request to hand out a questionnaire and an explanatory letter to people adopting a dog from their shelter. The first questionnaire was completed in the first week post-adoption. The second questionnaire was sent to adopters 6 months after adoption. In both questionnaires, respondents were asked to indicate if they consider their dogs aggressive, fearful or sociable. Some chose more than one answer. The results were analysed using the statistical package Unistat 5.6. The actual and relative counts (frequencies) of dogs with behavioural traits (aggressiveness, fearfulness, sociability) observed first week and six months post adoption were calculated. Frequencies were compared on the basis of a chi-square analysis of 2×2 contingency tables with Yates correction. One hundred and ninety-two people who had acquired a dog within a 12-month period from 84 shelters in the Czech Republic responded to the survey. Significantly (P<0.001] more women responded to the survey. Respondents aged from 25 to 60 years were predominant (77.6%). Most respondents (88.5%) claimed to have previous experience with dog ownership. However, only 20.3% of respondents claimed a high level of experience with dogs including those exhibiting behavioural problems. The ratio of male and female dogs was equal. The median age of both male and female dogs was two years. Dogs of median size represented almost half of the dogs (47.9%) in the study. Most dogs (92.2%) were housed indoors. Personality traits in dogs as indicated by respondents the first week and six months after adoption were compared. Significant positive changes in the behaviour of the dogs were seen six months after leaving the shelter. The first week post-adoption, one third of dogs were reported to be aggressive, almost two-thirds were fearful, and only half were reported to be social. However, the number of fearful dogs significantly (P<0.001) decreased (61% vs 20%), whereas the number of sociable dogs significantly (P<0.001) increased (56% vs 93%), within six months of adoption. No impact (P>0.05) of adopters´ previous experience with dogs on changes to dogs´ personality traits was found. One issue that should be addressed by adopters is aggressiveness. No significant change was found in the prevalence of aggressiveness over the course of six months in our study. Dogs exhibiting aggression would therefore seem to be less likely to change their behaviour without appropriate intervention. Furthermore, the prevalence of aggressive dogs reported in our study seems to be higher in comparison with other shelter studies. Adopters might have considered even natural dog behaviours as an unacceptable aggression. However, adoption success depends on how the dog is perceived by its new owner and therefore, all issues compromising the dog-owner relationship need to be addressed.

## Factors affecting behaviour of dogs adopted from a shelter

*Eva Voslarova, Svatava Vitulova, Vladimir Vecerek and Iveta Bedanova*
*Faculty of Veterinary Hygiene and Ecology, University of Veterinary and Pharmaceutical Sciences Brno, Department of Animal Protection, Welfare and Behaviour, Palackeho tr. 1946/1, 61242 Brno, Czech Republic; voslarovae@vfu.cz*

The aim of this study was to assess behaviour of dogs adopted from shelters as perceived by their adopters and to investigate factors affecting dogs' behaviour. In order to collect information on the behaviour of adopted dogs, a questionnaire was handed out to people adopting a dog from Czech shelters. The questionnaire focused on the behaviour of dogs during the first week after adoption. Information regarding whether the respondents' dogs had exhibited any behaviours which they considered unacceptable, and if so what these were, was collected. Information about the characteristics of the adopters, the adopted dogs and adopters' level of prior experience with dogs was also collected. The results were analysed using the statistical package Unistat 5.6. Six independent variables were constructed for the characteristics of the dogs adopted by respondents to the survey: sex, age, size, experienced abuse, health problems, behavioural problems. Actual and relative frequencies of dogs in all categories according to the monitored independent variables were calculated and the effect of the dogs' sex, age, size, health status and experienced abuse on the prevalence of behavioural problems in dogs was tested by means of a chi-square analysis of k × m and 2×2 contingency tables. 192 people who had acquired a dog within a 12-month period from 84 shelters in the Czech Republic responded to the survey. Whereas health problems were indicated in only 14.1% of dogs, 71.9% of dogs exhibited behavioural problems. The most frequent behavioural problems in adopted dogs were aggression (24.0%), fearfulness (21.4%), destructiveness (16.7%), excessive vocalisation (15.1%) and separation anxiety (13.0%). 59 dogs exhibited more than one type of problem behaviour. No effect ($P>0.05$) of sex, age, size or health status was found. However, abuse was found to have a significant ($P<0.05$) impact on the occurrence of problem behaviours. Shelter dogs with a documented history of abuse exhibited problem behaviours after adoption more frequently than non-abused dogs. Despite the high occurrence of behavioural problems and limited previous experience with dogs, only a small number of the adopters in our study sought professional advice. No cases of returning adopted dogs back to the shelter were reported. In our study, 41.2% of dogs adopted from shelters by the respondents to our survey had a history of abuse (no details on abuse type were provided). The real ratio may be even higher as the history of some dogs was unknown. The results indicate that there is a high probability that, when adopting a shelter dog, it has been abused in the past and will manifest some kind of problem behaviour. Given the common presence of problem behaviour in shelter dogs and how often it was cited as the primary reason for relinquishment or returning dogs to animal shelter, providing help to adopters to remedy common shelter dog behaviour problems could significantly increase the rate of successfully adopted dogs.

## Associations between daytime perching behavior, body weight, age and keel bone deviation in laying hens

*Mirjana Đukić Stojčić and Lidija Perić*

*University of Novi Sad Faculty of Agriculture, Department of Animal Science, Trg Dositeja Obradovica 8, 21000 Novi Sad, Serbia; mirjana.djukicstojcic@stocarstvo.edu.rs*

Keel bone deviation (KBD) is assumed to result from extended perching behaviour during the dark period and resulting long-term pressure on the keel bone. This pilot study was conducted to investigate associations between perching behavior during the light period and body weight (BW) of laying hens on KBD in Big Dutchman enriched cages. A total of 120 Lohmann Brown laying hens were randomly chosen from the same commercial flock, and divided over four cages with 30 hens per cage. The keel bone of each hen was palpated, and the occurrence of KBD was noted based on palpation training at the Veterinary Public Health Institute, Bern CZ. All efforts were made to ensure that hens in the study were handled with minimal disturbance. In each cage, BW of each hen was measured. Based on BW, hens were divided into two subgroups. The first subgroup (a) consisted of hens whose BW was lower than 1,800 g, and they were marked with a yellow strip on both legs. The other subgroup (b) consisted of hens that were heavier than 1,800 g, and they were not marked. The measurement and observation of hens was done at 46 and 70 weeks of age. Perching behaviors were observed using scan sampling during the period from 10:00 to 13:00, on five consecutive days. Results revealed that hens at the end of the production cycle perched significantly more during the light period than hens in the middle of the production cycle, and had more KBD ($P < 0.05$). Lighter layers perched less (46 weeks of age 9.1% and 70 weeks of age 14.3%) compared to heavier laying hens (46 weeks of age 9.9% and 70 weeks of age 16.5%). Also, lighter hens showed a non-significant tendency to have less KBD compared to heavier hens (46 weeks of age 0.4 vs 0.5; 70 weeks of age 0.8 vs 0.9). The results suggest that. although heavier hens perched more during the daytime than lighter ones, especially when older, the differences were not large enough to detect an association with KBD incidence in a sample of 120 hens.

## Association of fearfulness at the end of lay with range visits during 18-22 weeks of age in commercial laying hens

*Manisha Kolakshyapati, Peta Taylor, Terence Sibanda and Isabelle Ruhnke*
*University of New England, 1/127 A, Kirkwood Street, Armidale, 2350, Australia;*
*mkolaksh@myune.edu.au*

Hen behavior on commercial farms is variable, such that some hens rarely access the range while others utilize the outdoors regularly. The relationship between fearfulness and early ranging behavior is relatively unknown. Therefore, the aim of this study was to investigate the relationships between range usage during the first four weeks of range exposure and fearfulness. A total of 624 hens across two flocks (Flock A; Flock B) on a commercial free-range farm were grouped based on their early range use between 18 and 21 weeks of age. The least range users (representing 20% of flock individuals) were grouped as 'Stayers' (average % of days that Stayers accessed the range: 7.04±0.52); and the most range users (representing 60% of flock individuals) were grouped as 'Rangers' (average % of days that Rangers accessed the range: 50.96±0.54). Individual range use was monitored daily until 72 weeks of age using a custom-built radio frequency identification (RFID) system. Subsequently, hens were randomly selected from each group (selected hens of group A, n=278; selected hens of group B, n=346) and subjected to two behavioral tests to assess fearfulness the novel arena test (NA) and novel object test (NO). Hens were placed into the centre of an isolated square arena (NA; 1.7 m$^2$) and left for 8 minutes. Immediately after, a novel object was introduced through a small door to minimise human contact and was left with the hen for 5 minutes (NO). Hen behaviour was continuously recorded using an overhead video camera and later analysed by one observer blinded to treatment. Activity was scored by splitting the arena into 16 equal parts outlined with a transparent grid and the number of lines crossed was calculated. Latency to step, number of lines crossed and escape attempts were assessed as indicators of fear for the NA. Approach/avoidance behavior towards the novel object, time spent within an area close to the novel object (<40 cm) and the number of escape attempts were assessed as indicators of neophobia and exploration during the NO. Statistical analysis was performed using SPSS statistics v.24. Stayers took longer to first step ($\chi^2_{(1,604)}$=8.1, P=0.004), crossed fewer lines (F$_{(1,614)}$=11.5, P=0.001) and were less likely trying to escape ($\chi^2_{(1,604)}$=13.3, P<0.001) than Rangers during the NA. There were no correlations between early ranging and any parameters investigated during NO test (P>0.05). These findings do suggest that general fearfulness may be associated with ranging behaviour. There was no evidence that range usage at 18-22 weeks of age was correlated to neophobia or explorative behaviour. Further investigation regarding the causative relationship on fearfulness and range usage is wanted.

## Feather corticosterone: a new tool to measure stress in laying hens?

*Jutta Berk[1], Ellen Kanitz[2], Winfried Otten[2], Julia Malchow[1], Joergen B. Kjaer[1] and Thomas Bartels[1]*

[1]*Institute of Animal Welfare and Animal Husbandry, Friedrich-Loeffler-Institut, Dörnbergstr. 25/27, 29223 Celle, Germany,* [2]*Institute of Behavioural Physiology, Leibniz Institute for Farm Animal Biology, Wilhelm-Stahl-Allee 2, 18196 Dummerstorf, Germany; jutta.berk@fli.de*

Numerous references describe stable deposition of corticosterone in feathers (Fcort) and a high correlation between corticosterone in feathers and peripheral blood. Fcort might be useful as an indicator in animal welfare research as it offers a non-invasive sampling procedure, and the option to record the stress accumulated over several weeks in one sample. In this context, it needs to be examined to what extent the amount of Fcort differs between poultry strains and housing systems. This work was supported by the Federal Ministry of Food and Agriculture [Grant number 2817901515]. Pilot experiments were carried out and, in trial 1, eggs from the layer strains Lohmann Selected Leghorn [LSL], Lohmann Brown [LB] and Lohmann Tradition [LT] were incubated under identical conditions. Pullets were kept in house A in a mixed flock [LSL: n=26; LB: n=21; LT: n=25] to avoid housing effects and need for many replicates in this limited pilot trial. Additionally, in trial 2, feathers from hens [n=24] of another LT flock (as part of another experiment), allocated in house B, were taken. Both houses had littered floor, perches over a slatted area, nests, artificial lighting with fluorescent tubes and free access to standard feed and water. One primary feather (P III) of the 2nd feather generation of each bird was taken. Feathers were cut in the area of the umbilicus superior, and stored dry and dark at room temperature until analysis. The sample material, which consisted of vane and rachis of the feather, was pulverised in a ball mill. Fcort was extracted by a methanol-based extraction technique, and Fcort amounts were determined by ELISA. Data was analysed with concentration of Fcort for individual birds as data points using a general linear model with trial 1: line (LB, LSL, LT) as fixed effect and only housing system A and trial 2: housing as fixed effect (housing A and B) and only LT birds. Trial 1 showed Fcort to be higher in LT compared to LB and LSL ($F_{2,71}=4.02$, P=0.0223, LT=17.9 pg/mg, LB=15.2 pg/mg and LSL=15.1 pg/mg). Possible correlations with other parameters were not considered due to the limited number of birds. Trial 2 showed housing system to differ for the strain LT tested in both barns ($F_{1,44}=21.35$, P<0.001, housing A=17.9 pg/mg and housing B=13.0 pg/mg). These pilot studies show that layer strains seem to differ in their level of Fcort when kept in the same environment and they also show that housing systems affect this level for LT. This could be a general house effect or can be attributed to a range of environmental and social factors that cannot be elucidated by the experimental design and scale used. Further research is needed to study the influence of biotic and abiotic factors (e.g. stocking density, handling, environmental enrichment, light, wavelength, etc.) on the deposition and/or degradation of Fcort before this method can be applied as a welfare indicator in poultry.

## A cross-sectional study into factors associated with feather damage in laying hens in Canada

*Nienke Van Staaveren[1], Caitlin Decina[2], Christine F. Baes[1], Tina Widowski[1], Olaf Berke[2] and Alexandra Harlander[1]*
[1]*University of Guelph, Department of Animal Biosciences, N1G 2W1 Guelph, Canada,* [2]*University of Guelph, Department of Population Medicine, N1G 2W1 Guelph, Canada; nvanstaa@uoguelph.ca*

New housing systems for laying hens are currently being introduced in Canada. The transition to these furnished cages and non-cage housing systems requires a change in management to address the potential impact of feather pecking leading to feather damage (FD). In an effort to develop a management plan for egg farmers to assess FD and identify practices to prevent/reduce FD, we surveyed Canadian egg farmers on their housing and management practices and estimated the FD prevalence in their flock. A total of 122 laying hen farms were invited to participate in the study (autumn 2017) and information on 65 laying hen flocks was collected across the country (response rate: 52.5%). Twenty-six flocks were housed in furnished cages, while 39 flocks were housing in non-cage housing systems. FD was estimated using a visual FD assessment tool instructing farmers to score 50 birds throughout the barn and record FD damage based on severity (0: intact feather cover, 1: FD smaller than 2-dollar coin, 2: FD larger than 2-dollar coin). Linear regression modelling in R was for univariable analysis (P<0.25) followed by multivariable analysis to identify factors contributing to the prevalence of FD in furnished cage and non-cage systems separately. The average prevalence of FD found among participating flocks in furnished cage systems was 21.9% (95%CI: 10.4-33.4%) and 25.9% (95%CI 15.6-36.2%) in non-cage systems. Factors associated with higher FD prevalence included increasing flock age, brown-feathered birds, midnight feeding, and lack of a scratch area in furnished cages (R$^2$: 77%), and increasing flock age, all wire/slatted flooring system, and limited manure removal in non-cage systems (R$^2$: 73%). Knowledge translation and transfer of findings to a Canadian feather pecking management plan is currently in progress. Foraging opportunity continues to be an important component of FD occurrence, with suggestion that factors of air quality, lighting cycle, and strain differences are also influential. Based on the findings presented, farmers may benefit from prioritizing litter availability for foraging, being diligent in the maintenance of good air quality (either through ventilation or manure removal), and abstaining from the use of abnormal lighting cycles such as midnight feeding, that may infringe upon birds' ability to rest and escape aggressors.

# Influence of ramp provision during rear and lay on hen mobility within a commercial aviary system

*Ariane Stratmann[1], Janice Siegford[2], Filipe Maximiano Sousa[3], Laura Candelotto[1] and Michael J. Toscano[1]*

*[1]University of Bern, Center for Proper Housing: Poultry and Rabbits, Burgerweg 22, 3052 Bern, Switzerland, [2]Michigan State University, Department of Animal Sciences, Anthony Hall, East Lansing, USA, [3]Veterinary Public Health Institute, University of Bern, Schwarzenburgstrasse 155, 3097 Liebefeld, Switzerland; ariane.stratmann@vetsuisse.unibe.ch*

Keel bone fractures (KBF) represent one of the greatest welfare problems in laying hens, and a primary cause in non-cage systems is believed to be collisions. Installation of ramps between tiers in multi-tier laying hen aviaries has been shown to reduce falls and collisions, though their optimization is not clear. The current study is part of a larger work comparing ramps introduced either during rearing or lay periods. Our experiment focused on comparing mobility of hens during the laying period across the four treatment combinations: ramps during rear and lay, ramps during rear only, ramps during lay only, or no ramps. Lohmann Selected Leghorn (LSL; n=2,400 chicks) were reared in four pens (n=600 chicks per pen) of a commercial aviary where two pens contained ramps from 10 days of age, providing access to all three tiers; the two remaining pens were without ramps except leading to the litter. At 16 weeks of age (WOA), chicks were transported to an on-site laying barn containing a Bolegg Terrace aviary system divided into 20 side-by-side pens each containing 225 hens. Of the 20 pens, eight were assigned hens based on rearing treatment, i.e. each pen contained all hens from a single rearing condition with half having been exposed to ramps during rearing. Half of the pens were then further assigned ramps or no ramps resulting in the four treatment combinations (n=2 pens per combination). Within each pen, 20 hens were selected as focal hens in a stratified manner and their location within the aviary recorded using a custom infrared tracking system at four time points (19, 21, 23 and 30 WOA). Infrared receivers were attached to the legs of focal hens and recorded zone-specific codes for each of the five zones (litter, lower tier, nest boxes, top tier, and upper perches) at a frequency of 1 Hz. Location data for individual hens were analyzed using (generalized) linear mixed effect models for duration of time spent in each zone, number of transitions in each zone, and the total and average distance traveled, using treatment, age, and their interactions as fixed effects. A treatment by age interaction was found for duration in the litter with hens having ramps in rear and lay spending more time in the litter with increasing age while in other treatment groups duration in the litter decreased over the same period. Surprisingly, the number of visits to the top tier decreased from 21 WOA for hens with or without ramps throughout the entire study period, whereas use of the top tier by other treatment combinations was relatively unaffected by age. The number of visits to the nestbox zone decreased with age independent of treatment. Total distance traveled was neither associated with WOA nor treatment. Our results suggest that rearing treatments do affect use of ramps during lay and further work is required to determine the full consequences on hen welfare and productivity.

## Piling behaviour of laying hens in Switzerland: origin and contributing factors

*Jakob Winter, Ariane Stratmann and Michael Jeffrey Toscano*
*University of Bern, Veterinary Public Health Institute, Division of Animal Welfare, Center for Proper Housing: Poultry and Rabbits (ZTHZ), Burgerweg 22, 3052, Switzerland; jakob.winter@vetsuisse.unibe.ch*

Piling behaviour (PB) in laying hens – a dense clustering of two or more motionless animals – may lead to smothering (i.e. death, most likely due to suffocation) and therefore is considered a welfare concern. Despite the hazard, little information exists about location, triggers, risk factors, frequencies, and other key details of piling events. To provide this information and determine an objective definition of piling, we conducted an exploratory study on 13 Swiss layer flocks (flock size: mean: approx. 5,000 animals, range: 1,100-15,750 animals) known to have repeated piling episodes. Data collection took place at 20 and 30 weeks of age, for one day each during the complete light period. For each flock, data collection included: video recordings of PB, environmental data (temperature, wind speed), two behaviour tests at the flock-level (novel object and stationary person) in areas known by the producer for PB (barn and winter garden), barn dimensions, and interviews with flock owners about barn management (frequency of barn visits, ranging management, providing enrichment). Based on the observed process and behaviour of animals involved in piling, we defined piling events as a cluster of mostly motionless laying hens standing in the closest possible proximity, with their heads mostly in the same orientation. The minimal number of hens required to fulfil the piling definition is two- which marked the beginning and end of a piling event. We used linear regression analysis to identify factors which related to the frequency, duration and the number of animals involved in a piling event. Further data was analysed descriptively. For the selected flocks, PB occurred repeatedly in the same corners of the barn and was caused by seven main triggers (i.e. sunlight, barn light spots, a person visiting the barn, distinct behaviours of a single hen e.g. pecking or resting at a wall which attracted other hens, fights between hens, increased local animal densities as result of sudden mass movements, and detection of a novel food item). Our results show that more piling events (PE) occurred in white colored flocks (mean: 38±17 PE/d) than in brown flocks (11±6 PE/d). More PE occurred in week 30 (29±21 PE/d) than in week 20 (20±15 PE/d). Most PE occurred at 5-10 h after dawn (AD) (13±12 PE). Only a low number of PE occurred in the winter garden (5±6 PE/d) compared to the barn (22±16 PE/d). Piles lasted longer in brown (24±26 min) than in white (14±13 min) and mixed (12±15 min) colored flocks. White colored flocks seem to have larger piles (36±29 hens) than brown (21±19 hens) and mixed (24±16 hens) colored flocks. Piles were larger at 0-5 h AD (30±28 hens) and at 5-10 h AD (35±33 hens) compared to 10-15 h AD (23±17 hens). We conclude that PB does not occur randomly but at certain locations in the barn with distinct flock age and time of day patterns and is likely caused by several triggers. Brown colored flocks seem to have a lower number of smaller, longer lasting PE, compared to white flocks. Future experiments should aim to confirm the identified triggers of PB and develop strategies to prevent PB and related smothering.

## Euthanasia – manual versus mechanical cervical dislocation for broilers

*Leonie Jacobs[1], Dianna V. Bourassa[2], Caitlin E. Harris[3,4] and R. Jeff Buhr[3]*
*[1]Virginia Tech, Animal and Poultry Sciences, 175 W Campus Dr, Blacksburg, VA 24061, USA, [2]Auburn University, Poultry Science, 260 Lem Morrison Dr, Auburn, AL 36849, USA, [3]USDA-ARS, US National Poultry Research Center, 934 College Station Rd, Athens, GA 30605, USA, [4]The University of Georgia, Poultry Science, 110 Cedar St, Athens, GA 30602, USA; jacobsl@vt.edu*

Manual cervical dislocation (CD) is commonly used to euthanize broiler chickens that are ill or lame, but can be challenging due to personnel fatigue, a lack of strength or training. Mechanical cervical dislocation is a potential alternative. The aim was to assess the onset of brain stem death based on the birds' loss of behavioral reflex responses and musculoskeletal movements for mechanical and manual cervical dislocation. Broilers from experimental flocks at the research farm were euthanized at 36 (n=60), 42 (n=80), and 43 days old (n=60), by CD or the Koechner Euthanizing Device (KED). We hypothesized that CD would result in quicker brain death, as both CD and KED aim to dislocate vertebrae from the skull, CD by stretching and rotating, and KED by stretching only. On day 1 random birds were euthanized prior to placement in a cone. On days 2 and 3, birds were euthanized after placement into cones and a modified KED treatment was included: the KED plus an extra head extension at ~90° (KED+). The onset of brain death was assessed by recording the cessation of musculoskeletal movements (sec; n=200) and two behavioral reflexes: the nictitating membrane (n=200), gasping reflex (n=196). The experimental setup did not allow blinding for treatments, which were equally divided over observers. External blood loss was recorded (y/n; n=200) and the gap size was estimated between the atlas cervical vertebra and the skull (skull-to-atlas gap; 1-cm increments; n=118). Data were analyzed with mixed models and Kruskal-Wallis Chi-Square tests. This experiment was approved by the Institutional Animal Care and Use Committee of Virginia Tech. After euthanasia, all broilers displayed clonic/tonic convulsions, indicating a state of unconsciousness immediately after application. CD resulted in shorter latencies to brain death compared to KED and KED+. Mean duration (±SEM) of the nictitating membrane reflex was 10±4 sec for CD, 62±4 for KED and 60±4 for KED+ (P<0.001). Mean duration for gasping were 20±3 sec for CD, 66±3 for KED, and 62±3 for KED+ (P<0.001). Cessation of musculoskeletal movements occurred after 89±7 sec for CD, 106±7 for KED and 109±7 for KED+ (P<0.001). No CD birds bled externally, yet external blood loss occurred in 80 and 88% of KED and KED+ birds (P<0.001). The skull-to-atlas gap was similar in CD and KED+ birds, with gaps of 1.8±0.1 and 1.7±0.1 cm respectively, which were both larger than with the KED (1.1±0.1 cm; P<0.001). The experiment showed that CD resulted in quicker brain death compared to KED or KED+. Although the KED+ resulted in a similar gap size as CD, reflexes and movements persisted. Therefore, CD may be the recommended method compared to KED or KED+. However, convulsions indicated that all broilers were unconscious and therefore unable to perceive pain immediately after application of each method. Additional research into these and other alternatives for euthanasia is needed.

## Use of routinely collected slaughterhouse data to assess welfare in broilers

*Xavier Averós[1], Enrique Cameno[1], Bernardino Balderas[2] and Inma Estevez[1,3]*
*[1]Neiker-Tecnalia, Arkaute Agrifood Campus, P.O. Box 46, 01080 Vitoria-Gasteiz, Spain, [2]Servicio Andaluz de Salud, San Juan de Dios 15, 18001 Granada, Spain, [3]IKERBASQUE, Basque Foundation for Science, María Díaz de Haro 3, 48013 Bilbao, Spain; xaveros@neiker.eus*

Routinely collected slaughterhouse data by EU mandatory legislation usually remain stored with no further processing. Such data however if analysed can create added value. Even if animals are transported during short (<8 h) journeys, transport conditions, especially during summer or harsh winters, affect broiler welfare and lead to increased deaths on arrival (DOA) and carcass rejections (REJ), closely linked to on-farm welfare conditions. Mixing loads at different farms may also be a risk factor for broiler welfare, although little attention has been put to this aspect. In order to prove the potential of the analysis of the collected information we processed the record sheets as collected in a Spanish slaughterhouse. We determined the effect of mixed loads, season, flock welfare status and bird cleanliness at arrival on DOA and REJ from 1-year records of a slaughterhouse in Southern Spain. Data from 1,927 journeys, transporting 2,002 flocks from 212 farms were collected. The veterinary services visually scored flock welfare status at arrival (WA) with a 2-level scale (Good (GW); Bad (BW)) and bird cleanliness (FC) with a 3-level score (Good (GC); Regular (RC); Bad (BC)). Mixed loads and transport season were identified. Number of loaded birds, DOA and REJ (n) were identified per flock, and risk factors and significant ($P<0.05$) 2-way interactions were assessed. Flock was the experimental unit in DOA and REJ models, and both variables were adjusted to a negative binomial distribution with the number of loaded and slaughtered birds being respective offsets. Models accounted for random journey variation and for unequal journey variance, the latter due to season in the DOA model and to bird cleanliness in the REJ model. Mixed loads (0.24±0.03%) tended to increase DOA with respect to unmixed loads (0.20±0.01%; P=0.061). A seasonal effect was detected on DOA (P<0.001), with autumn transports (0.26±0.02%) resulting in more DOA than spring transports (0.18±0.02%). FC affected DOA (P<0.001), with GC and RC (0.17±0.01% in both cases) resulting in less DOA that BC (0.35±0.05%). WA also affected DOA (P=0.015) although, surprisingly, visually assessed GW (0.25±0.02%) resulted in more DOA than BW (0.19±0.02%). REJ was higher in mixed transports (0.97±0.09%) as compared to unmixed transports (0.62±0.03%; P<0.001), with the detrimental effect of mixed loads being particularly evident during autumn (1.21±0.20%) and winter (1.15±0.16%) transports (mixed loads×season; P=0.003). These results prove the clear relationships that can be detected when more attention is paid to routinely collected slaughterhouse data, allowing for the detection of the interactive nature of some risk factors on welfare outcomes of transported broilers. Results indicate that, for farms sending birds to this slaughter plant, special attention should be paid to journeys transporting mixed loads, as well as to autumn and winter transports.

## Associations of lameness in broiler chickens to health and production measures

*Erik Georg Granquist[1], Guro Vasdal[2], Ingrid C. De Jong[3] and Randi Oppermann Moe[1]*
*[1]Norwegian University of Life Sciences, Faculty of Veterinary Medicine, Ullevålsveien 72, 0454 Oslo, Norway, [2]Norwegian Meat and Poultry Research Centre, Lørenveien 38, 0515 Oslo, Norway, [3]Wageningen University and Research, Wageningen Livestock Research, P.O. Box 338, 6700 AH Wageningen, the Netherlands; erikgeorg.granquist@nmbu.no*

Lameness in broiler chickens is a primary welfare concern as it is considered painful. Reduced growth and culling of lame birds also affects farm profitability. Footpad dermatitis may cause lameness and has been used as an indicator of welfare in chickens. Identifying risk factors associated with lameness, (such as pathological conditions associated with condemnation at post mortem inspection) may provide important tools for flock welfare assessment. The aim of this study was to explore lameness and the associations between lameness and health/ production measures of animal welfare in commercial broiler production, using the Welfare Quality® protocol for broilers. A total of 50 broiler flocks with similar management were included in the sample, and farm visits were conducted for lameness scoring at a mean age of 28.9 days. Animals were handled according to ethical standards and valid regulations, and farmer participation in the study was voluntary. Broilers (n=150) were examined from each flock (0.95% of the total population, n=7,500). The mean body weight of the broilers at visit was 1,544.64 grams (95% CI: 1,503.11, 1,586.18).The percentage of animals in the six different gait score (GS) categories were GS0: 2.53%, GS1: 44.19%, GS2: 33.84%, GS3: 16.32%, GS4: 2.36% and GS5: 0.53%. Production and other welfare data were collected for each flock after slaughter. Univariable and multivariable linear regressions were used to determine statistical associations between the flock mean (Log10) gait scores and production/health data. Higher gait scores were associated with increased hock burn score (P<0.02), increased footpad dermatitis score (P<0.01), reduced bird cleanliness score (P<0.01) and peat litter (P<0.01). Although not statistically significant, there was a tendency for increased flock gait score being associated with wet litter (P=0.07). In addition, condemnations at post mortem inspection were associated with increasing gait scores (P<0.05), indicating that at least a portion of the lameness cases display pathological changes on the carcasses. In conclusion, 19% of the birds showed moderate to severe lameness, which was associated with several production or health and welfare observations including feather cleanliness and condemnations as unfit for human consumption at slaughter. Although stocking density and growth rate are already known key factors for lameness, associations of lameness with hock burns, footpad dermatitis and cleanliness of the birds suggest that a suboptimal physical environment (e.g. litter- and air quality) may be detrimental to leg health.

## Testing by numbers: addressing the soundness of the transect method in commercial broilers

*Neila BenSassi[1], Xavier Averós[1] and Inma Estevez[2]*
*[1]Neiker-Tecnalia, Animal Production, Campus agroalimentario de Arkaute, 01080 Vitoria, Alava, Spain, [2]Ikerbasque, Basque Foundation for Science, María Díaz Haroko Kalea 3, 48013 Bilbao, Bizkaia, Spain; iestevez@neiker.eus*

The transect method for on-farm welfare assessment in meat poultry consists in walking the house within transects which are delimited paths between feeder and drinker lines to determine the frequencies of welfare indicators. This method was validated for turkeys by assessing all individuals in a flock at the end of rearing, an approach that is not feasible in broilers due to the large flock sizes. However, the soundness of the transect method could be investigated through capture-recapture techniques of a known subpopulation to estimate the probabilities of overlooking and/or repeating birds. This study was conducted in 11 flocks belonging to 3 farms. Eighty birds were captured and individually marked in groups of ten chickens in 8 house locations. Bird movement was tracked during two consecutive days by collecting the position of detected marked birds while walking along non-adjacent transects (4 samplings/house/day). Detection rate (DR) was calculated as: (N detected marked birds per house sampling)/(Total N of marked birds)×100; repetition rates/house (RRH): (N repeated marked birds per house sampling)/(N detected marked birds in the same house sampling)×100; and repetition rate within transect (RRT): (N repeated marked birds in a transect)/(N detected marked birds within the same transect)×100. The effects of flock density, transect number/house (6 vs 8) and sampling time (am vs pm) were determined using generalized linear mixed models. For repeated birds, the number of travelled transects (TT) was estimated and the effect of flock density, transect number and position where birds were first detected (wall vs central) analyzed. The population random distribution was tested by calculating a distribution index as: (N observed birds in transect z – N expected birds in transect z)$^2$/(N expected birds in transect z) (z: transect id). The effects of transect number, position and their interaction were analyzed. An overall DR of 64.76±0.87% (mean ± SE), a RRH of 23.85±0.77%, and a RRT of 1.66±0.58% were found. Higher RRH was observed in narrower houses (26.405±0.915 in 6-transect vs 20.531±1.048 in 8-transect houses; P<0.001). RRT was higher in am samplings (3.126±1.112 am vs 0.263±0.263 pm samplings, respectively; P=0.038). TT was higher in 8-transect houses (1.580±0.046 vs 1.881±0.068 in 6 and 8-transect houses, respectively; P=0.005) and at walls (1.438±0.036 vs 2.068±0.087, in central and wall transects, respectively; P=0.0001). Results indicate that DR was close to two-thirds of the subpopulation which could be considered a good estimation given its relation to the probability of the observer and the bird to coincide in the same location and time. Repetition rates were minimal for observations conducted in the pm, when movement was substantially reduced, and over transects separated by 3 transects in between. Bootstrapping results showed that sampling only 2 transects/house is sufficient for a representative welfare assessment. The transect method should then be conducted on 2 transects, being one wall and one central, and separated by at least 3 transects.

## Changes in broiler chicken behavior and core body temperature during heat stress

*Marisa Erasmus[1], Kailynn Vandewater[1], Matthew Aardsma[2] and Jay Johnson[2]*
*[1]Purdue University, Animal Sciences, 270 S Russell St., 47907, USA, [2]USDA-ARS Livestock Behavior Research Unit, 270 S Russell St., 47907, USA; merasmus@purdue.edu*

Heat stress is an important poultry health and welfare issue, resulting in major economic losses. Little information is available regarding the effects of heat stress on broiler chicken behavior and core body temperature (CBT). The objectives of this study were to examine changes in broiler chicken behavior and CBT as a result of heat stress. Mixed-sex broiler chickens were housed in groups of 15 in 2 littered floor pens. At 34 days, data loggers (iButton, Maxim Integrated Products, Inc.) were surgically implanted in 10 males and 10 females to continuously record CBT. Birds were maintained under thermoneutral conditions (22.5±0.1 °C) until 42 days, when cyclic heat stress (HS) was implemented. The HS cycle consisted of increasing the temperature from thermoneutral conditions to a mild HS period (26.3±0.1 °C), followed by a moderate (30.1±0.1 °C) and hot (34.5±0.2 °C) period. Each period was maintained for 1 h. Cyclic HS was repeated daily from days 42 to 45. Overhead video cameras recorded behavior on days 39 and 40 (pre-HS) and 42 and 43 (HS). Behavior (sitting, standing, drinking, eating, preening, dustbathing, walking and out of view) was analyzed using 10-min instantaneous sampling for the 17 h light period for focal birds with iButtons (n=6 males, 5 females) and birds without iButtons (n=5 males, n=4 females). The occurrence of panting was also recorded. All procedures were carried out in accordance with the guidelines of the Purdue University Institutional Animal Care and Use Committee. Data were analyzed using the GLIMMIX and MIXED procedures (SAS 9.4). The proportion of observations in which birds preened was higher pre-HS (0.07±0.009) than during HS (0.05±0.009; P=0.01) and was higher for females (0.07±0.009) than males (0.05±0.009, P=0.048). Males (0.76±0.02) sat more than females (0.70±0.02, P=0.03) and birds with iButtons sat less than birds without iButtons (0.76±0.02 vs 0.70±0.02, P=0.02). CBT differed among HS periods, males and females, behavioral categories and when birds were panting vs not. CBT increased with increasing ambient temperature (thermoneutral: 41.33±0.05°C, mild: 41.92±0.08°C, moderate: 42.46±0.08°C, hot: 43.22±0.08°C; P<0.001). Female CBT was higher than male CBT (42.53±0.06 vs 42.17±0.06, P<0.0001). CBT was higher when birds were panting (42.44±0.06°C) than when birds were not panting (42.26±0.06°C, P<0.0001). HS significantly affected broiler chicken behavior, with birds spending less time preening during HS. CBT differed depending on the behavior birds were performing. Further research is needed to identify early changes in behavior associated with increasing ambient temperature and CBT.

## Identifying welfare issues in turkey hen and tom flocks applying the transect walk method

Joanna Marchewka[1], Guro Vasdal[2] and Randi Oppermann Moe[3]

[1]Institute of Genetics and Animal Breeding Polish Academy of Sciences, Jastrzebiec, 05-552 Magdalenka, Poland, [2]Norwegian Meat and Poultry Research Centre, Lorenveien 38, 0515 Oslo, Norway, [3]Norwegian University of Life Sciences, Faculty of Veterinary Medicine, P.O. Box 8146 dep., 0033 Oslo, Norway; j.marchewka@ighz.pl

The varying lengths of the production cycle of turkeys requires producers to separate the two sexes, but still they remain housed in the open space of the same building, under equal management and production conditions. The two sexes differ with regard to behavioral and physiological characteristics, and are slaughtered at different ages, resulting in different growth patterns and final slaughter age. Most of the work on turkey welfare has focused on toms at the end of their production cycle, since this period is considered challenging, when the barn is at its maximum capacity with regards to number of birds, ventilation capacity, litter quality and animal care. Research on the welfare of toms and hens at the time before hens are slaughtered is currently lacking. The main aim of the current study was to measure on-farm health and welfare issues of commercially reared hen and tom turkeys at 11 weeks of age using the transect walk method. In turkey production systems, hens are typically reared for 12 weeks, while toms are reared for up to 20 weeks. The current study was conducted between November 2017 and March 2018 in 20 commercial turkey flocks on 16 different farms in Norway. Norwegian commercial turkey flocks are uniform with regard to stocking density and the absence of beak and toe trimming practices. On each farm, one barn, divided into tom and hen area, was evaluated using the transect walk method. An observer walked the transects in random order and recorded the total number of birds per transect that were: immobile, lame, with visible head-, tail- or wing- wounds, small, featherless, dirty, sick, terminal, or dead. In eight flocks, producers separated a small part of the rearing area in the barn in order to place any unfit birds which required treatment or separation from the whole flock, called the sick pen where birds were scored according to the indicators used in the transect walks method. To analyse the data we applied ANOVA and Spearman correlations in SAS software (v 9.3). The most commonly observed welfare challenge in both hens and toms were dirty birds, birds with featherless areas, tail wounds and wing wounds. Across sexes, poor litter quality resulted in more head wounds ($P<0.05$). Toms had significantly more tail wounds ($P<0.001$), there were more sick birds ($P<0.01$) and more terminal birds ($P<0.01$) compared to hens at the same age. No differences were found between prevalence of birds with reduced welfare in the production area and in the sick pens, neither in the production area of the farms with sick pens and without sick pens. Several of the welfare indicators were positively correlated between the sexes, including lameness, head wounds, wing wounds, tail wounds, dirty birds and dead birds, suggesting underlying environmental or management causes.

## The artificial blood feeding of *Aedes aegypti* mosquitos as an alternative to the use of live research animals

*Elisa Codecasa, Bertrand Guillet, Florian Frandjian, Patrick Pageat and Alessandro Cozzi*
*IRSEA, IRSEA Quartier Salignan, 84400, France; e.codecasa@group-irsea.com*

*Aedes aegypti* is one of the major vectors of dengue hemorrhagic fever causing millions of cases every year in the world. It can be reared in laboratory and the blood feeding is a fundamental part of routine protocols since it is required for egg production by adult females. For this, live hosts such as guinea pigs, mice, hamsters and chickens are amongst the species used for the blood meal. Despite his efficiency, this method is expensive, time consuming, and subject to government regulation. Furthermore, the increasing awareness about animal welfare and the "3 Rs" principles (reduction, refinement and replacement) in the scientific use of live animals also need to be considered when feeding mosquito species. In response to that, numerous artificial systems have been developed and described in literature, which are successful and show very similar results to the feeding method using live animals. Nevertheless, these devices are often complex to use and data on their long-term effects on the colonies are scarce. We present a simple and low-cost device and add information about its long-term efficiency on the maintain of a lab *A. aegypti* colony. The feeding rate, fecundity, hatchability and pre-adult development were analyzed in order to evaluate the performance of each stage derived from an artificial blood meal. The mosquitos belonged to a population artificially fed for more than a year (equivalent to 20 generations). Larval, pupal, and adult stages were maintained at 27±1 °C, 75±5% relative humidity and 12:12 h photoperiod and the adults were housed in acrylic cages to mate and fed with sugar solution until the blood meal. The generation used for the first study produced the stages (eggs and larvae) involved in the 3 others. The artificial feeder was assembled with common laboratory materials and consisted of 2 separate blocks: a heating element in contact with the blood reserves put on the mosquito cage's net. The blood came from a pig slaughterhouse next to the experimental facility. For the feeding rate test, 622 females were fed and killed to detect the presence of blood in their abdomens. In the fecundity test, 58 females were single-housed in order to analyze each egg laying. Finally, 900 and 638 eggs were maintained to study the hatchability and the preadult development. About results, the average number of blood fed females was 85.51% ±6.52%; 56 of 58 single-housed females laid eggs and the average number of eggs per female was 103.50±51.02; the hatchability (the percentage of larvae hatched per number of eggs) was 85.05%±7.33% and the average number of preadults was 73.57% ±2.02. Results were in agreement with those found in literature for both artificial and natural feeding settings. In conclusion, our findings give relevant information about replacement of vertebrate hosts for the feeding of laboratory-reared mosquitos, answering thus to the topic 'Animal lives worth living'.

## The EU Platform on Animal Welfare and its subgroups – strengthening ISAE's voice in Europe

*Moira Harris[1], Birte Nielsen[2], Mette Herskin[3], Emma Baxter[4], Janne Winther Christensen[3] and T. Bas Rodenburg[5]*
*[1]Harper Adams University, Newport, TF10 8NB, United Kingdom, [2]INRA, Jouy en Josas, 75231 Paris, France, [3]Aarhus University, Blichers Allé 20, 8830 Tjele, Denmark, [4]SRUC, West Mains Rd, EH93JG Edinburgh, United Kingdom, [5]University of Utrecht, Yalelaan 2, 3584 CM Utrecht, the Netherlands; mharris@harper-adams.ac.uk*

The EU Platform on Animal Welfare is an interactive network, established by the European Commission in 2017, aimed at promoting dialogue among competent authorities, businesses, civil society organisations and scientists on animal welfare issues of relevance for the European Union. ISAE became involved by responding to a general call for representation from professional organisations with expertise in animal welfare. ISAE's appointed representative (the ISAE President) attends regular meetings of the EU Platform alongside representatives from other stakeholder groups and independent experts (many of whom are also ISAE members). At the central level, we try to further ISAE's agenda on global development of applied animal behaviour and welfare science, together with partners such as OIE. There are two official Platform subgroups supported by the EC, on Animal Transport and Welfare of Pigs, as well as four voluntary initiatives. Thanks to the diversity of expertise among our members, the ISAE has representatives on both official subgroups and three of the voluntary ones – Responsible Ownership and Care of Equidae, Welfare of Fish and Welfare of Pullets. On each, ISAE's experts provide objective, science-based expertise. Each of the subgroups met initially to establish their priorities and determine the types of output they would produce followed by subsequent meetings and out-of-meeting activity to refine priorities and develop outputs. The Animal Transport subgroup focuses on extreme temperatures, cattle exports outside the EU and unweaned animals, aiming to improve knowledge transfer for those working with animals in transport, develop tools to improve implementation and enforcement of rules, complete a knowledge/technical gap analysis and investigate technical innovations. The focus of the Welfare of Pigs subgroup is tail docking and the assessment of risk factors associated with tail biting, aiming to produce a document on how authorities can instruct official vets to assess risk parameters. The Equine subgroup aims to develop a guide to good animal welfare practice for the keeping, care, training and use of horses as well as a similar guide for donkeys, asses and mules. The Welfare of Fish subgroup is focusing on water quality and handling (factors of importance across all production stages, transport and slaughter), and is developing general and species-specific guidelines on these. The Welfare of Pullets subgroup aims to provide guidelines on the welfare of pullets (young pre-laying hens) as no specific EU legislation currently exists. The ISAE's presence on the EU Platform and its subgroups strengthens the Society's voice in the European Union, continuing to ensure that it fulfils its aim to make expertise in applied animal behaviour and welfare available to policymakers and other stakeholders.

## Above the minimum: integrating positive welfare into animal welfare policy

*Nicki Cross*

*Ministry for Primary Industries, Animal Welfare, Pastoral House, 25 The Terrace, Wellington 6011, New Zealand; nicki.cross@mpi.govt.nz*

The Animal Welfare Act (1999) is the primary legislation relating to the care of animals in New Zealand and requires that owners and persons in charge of animals provide for their physical, health and behavioural needs. The Act allows for the development of codes of welfare by the National Animal Welfare Advisory Committee (NAWAC). The codes of welfare provide, in more detail than the Act, the requirements that people need to meet to care for their animals. Codes contain minimum standards that have a legal effect and must be complied with. Codes also contain guidance material and best practice recommendations that have no legal effect, but are included to raise the bar of animal welfare. In 2015, the Animal Welfare Act was amended to formally recognise that animals are sentient beings. Although NAWAC had always performed its duties with the understanding that animals are sentient, with the recent recognition of animal sentience within the Act, the concept also needed to be reflected within the codes of welfare to encourage those caring for animals to provide opportunities for them to have experiences and emotions that will provide them with a positive quality of life. The code of welfare for horses and donkeys was issued by the Minister in 2016 and was the first code of welfare issued since the Animal Welfare Act was amended. The code contains 15 minimum standards around equine management. In order to integrate science based indicators of positive welfare into the code, MPI commissioned research to identify practical indicators of positive welfare in horses. These indicators were added to the guidance material in the code of welfare to encourage people to provide positive experiences for their equids. Indicators described the behavioural signs of relaxation in horses such as the horse lying down when housed, or the neck hanging low, and horses exhibiting a drooping lower lip. Indicators of discomfort and anxiety, such as hind leg kicking, flared nostrils, tail swishing and the white of the eye being visible, were also added to the guidance material as behaviours that should be sought to be minimised when handling horses. NAWAC continues to integrate the concept of animal sentience in its work, including codes, standards and opinion pieces. The issue and implementation of a number of directly enforceable regulations around horse management in 2018 were intended to further increase the welfare of horses in New Zealand. The effect that the introduction of these regulations has had on horse welfare will be discussed and compliance data presented. The equine industries are also becoming increasingly aware of the public scrutiny directed towards ensuring the welfare of horses competing in events such as horse racing, and are taking proactive steps to improve the welfare of horses in equine sports. The concept of positive welfare is being integrated into both the New Zealand legislative system, and into non-legislative initiatives, to encourage those caring for animals to provide them with positive experiences and a quality of life above the minimum required by the Act.

## Building an informed audience: engaging youth in animal behavior and welfare education

*Melissa Elischer and Chelsea Hetherington*
*Michigan State University, Extension, Justin S. Morrill Hall of Agriculture 446 W. Circle Drive, East Lansing, MI 48824, USA; elischer@msu.edu*

4-H is the youth development program of land grant universities in the United States. The mission of 4-H is for youth to learn through hands-on, experiential activities, such as raising livestock to experience the practical aspects of animal husbandry. Youth have access to countless resources about nutrition, physiology, disease prevention, and basic care recommendations to guide their efforts; however, there has been a lack of science-based information regarding animal behavior and animal welfare in the 4-H program. To address this gap, 300 youth throughout the state of Michigan aged 8 to 19 years old participated in a program focused on introducing animal behavior and welfare concepts. Topics covered included the Five Freedoms, the 'Three Circles' view of animal welfare, animal perception and senses, low stress handling methods, animal ethics case studies, and interactive activities that encouraged youth to experience their environment from an animal's perspective. Species of focus included beef cattle, dairy cattle, sheep, goats, horses, rabbits, cavies, swine, and chickens. Initial program delivery occurred from March to June 2018. Following the conclusion of the program, 166 participants (55%) voluntarily completed a survey assessing interest, engagement, attitudes, and aspirations about science based on the program. Nineteen evaluations were removed from the data set due to incomplete information, leaving 147 completed responses ((49%); 89 female, 56 male, 2 refuse to answer). Descriptive statistics were performed to assess the effectiveness of the program. Following animal behavior and welfare program delivery, 91% of respondents indicated they agreed or strongly agreed with an interest and engagement in science related activities. Seventy percent of evaluation respondents agreed or strongly agreed they had positive attitudes about science, see science in their future, and recognize the relevance of science.

## Impact of frame reflection on veterinary student perspectives toward animal welfare and differing viewpoints

*Kathryn Proudfoot[1] and Beth Ventura[2]*

*[1]Ohio State University, Veterinary Preventive Medicine, 1920 Coffey Rd, Columbus, OH 43210, USA, [2]University of Minnesota, Animal Science, 1364 Eckles Avenue, St. Paul, MN 55108, USA; proudfoot.18@osu.edu*

To improve patient welfare, veterinarians must develop and maintain a good relationship with their clients. Communication skills like reflective listening and empathy can improve relationships, but are difficult to teach and are not always included in veterinary curricula. One approach may be to use 'frame reflection' to encourage students to understand different perspectives on contentious animal welfare (AW) issues. Frame reflection is a learning process in which students place themselves in the 'shoes' of another and move beyond their own framing of a situation. This study explored the effects of a frame reflection assignment on veterinary students' perspectives of AW and attitudes toward differing views. Second year veterinary students at Ohio State University (n=160) completed a series of exercises in which they: articulated their own AW values (Q1), interviewed someone with whom they held differing AW values, and reported their interviewee's views as if they were their own. Students then shared their perceptions toward their interviewee and discussed applications of the exercise in their future careers (Q2). Content analysis was applied to student responses in two stages: (1) both authors independently hand-coded, then discussed identified themes until a mutually agreed upon scheme was reached; and (2) NVivo was used to identify deeper patterns in responses. The student population was 79% female and 21% male, with small animal (56%), large animal (11%), mixed animal (16%), equine (4%) or individualized (13%) career areas of emphasis. Approximately 50% of students selected affective state as their primary AW value in Q1, with 40% choosing biological functioning and 10% emphasizing natural living. Most students interviewed a classmate or friend, followed by family members, significant others and veterinary mentors or colleagues. Interviews most commonly focused on elective procedures in pets (24%), veganism/vegetarianism (17%), using animals in research (9%), and euthanasia decisions (8%). Preliminary analysis of Q2 responses showed that students had both productive and critical views of their interviewees, though productive views were more common and were clustered under the following themes: acknowledgement of shared values between the student and interviewee; description of actions taken during the interview (e.g. reflective listening, learning, frame reflection); and improved knowledge of, belief in the legitimacy of, and empathy toward disparate AW views. Critical responses included: a perceived lack of shared values; judgment of interviewee's personality; and assumption of interviewee's ignorance as reason for disagreement. A majority of students reported that the exercise would benefit their careers by improving their readiness to engage and communication strategies with clients, with a subset articulating that these skills would directly benefit patient welfare. Preliminary findings suggest that this assignment may improve veterinary students' perception of those that disagree with them, which may aid in improved communication and relationship-building with future clients.

## Comparison of patterns of substrate occupancy by individuals versus flocks of 4 strains of laying hens in an aviary

*Janice Siegford and Ahmed Ali*
*Michigan State University, Dept. of Animal Science, 470 S Shaw Ln., Lansing, MI 48824, USA;*
*siegford@msu.edu*

Aviaries are intended to improve hen welfare by providing more space and resources such as litter. However, aviaries house hens in relatively large flocks, which may affect individual welfare. Design of aviaries is based on the assumption that different strains of hens and individuals within flocks respond equally to the given resources, with similar impacts on welfare. Research indicates distinct differences in resource use among different strains of laying hens in aviaries, with impacts on welfare and production. However, we do not know the degree of variability at the level of the individual hen. We examined substrate occupancy (i.e. floors, litter and perches) inside aviaries during peak lay (28 wk of age) at the level of individual and flock for 4 strains of laying hens (B1: Hy-Line Brown; B2: Bovans Brown; W1: DeKalb White; and W2: Hy-Line W36). Observations were conducted over 3 consecutive days in 16 aviary units (4 units/strain, 144 hens/unit, 7 focal hens/unit) twice at 3 times of day (morning, midday and evening), and hens' roosting locations were recorded at night. At each observation, focal hens were observed for 3 min and an instantaneous scan sample of the substrates occupied by all hens the flock was conducted. Duration of time spent on each substrate was calculated as % of total observation time per focal hen, and roosting location was calculated as % of total observations. Strain differences were assessed using GLMM and multiple comparisons in R 3.3.1 with $\alpha$ set at 0.05. During the day, brown focal hens spent more time on wire floors (B1: 50; B2: 46%) and nests (B1: 20; B2: 23%) than white hens (W1: 22, 11; W2: 22, 14% for floors and nests respectively; all $P \le 0.05$). White focal hens spent more time on litter areas (W1: 44; W2: 41%) and perches (W1: 18; W2: 18%) compared to brown hens (B1: 22, 6; B2: 25, 5%, all $P \le 0.05$). At night, brown focal hens were observed on lower tier wire floors more frequently than white hens, which were observed on the upper tier resources more frequently. Similar patterns of resource use were recorded for flock distribution during the day, with brown hens again occupying wire floors and nests in greater numbers (B1: 46, 16; B2: 44, 16%) than white hens (W1: 22, 10; W2: 37, 11%, all $P \le 0.05$). At the flock level, higher numbers of white hens were again observed occupying perches and litter areas (W1: 18, 44; W2: 18, 38%) compared to brown hens (B1: 9, 25; B2: 9, 28%, all $P \le 0.05$). During the night, more brown hens were on lower tier wire floors (B1: 34; B2: 34%) than white hens (W1: 2; W2: 3%, $P=0.01$), which occupied the upper tier at higher numbers. Observing focal hens during the night revealed variability among individual hens within the same strain that could not be distinguished by scanning flock distribution. For instance, 30 of 56 focal hens of white strains retained their roosting location on the top tier across all observation days, while the other birds were inconsistent in roosting location. Though general patterns of resource use were similar between individuals and flocks within a strain, observing individual hens revealed variability not found at the flock level, suggesting individual hen observation is still important within large flocks to ensure welfare of all hens.

## Using synthetic video for automated individual animal behavior analysis

*Joshua Peschel*

*Iowa State University, Agricultural and Biosystems Engineering, 2348 Elings Hall, Ames, IA 50011, USA; peschel@iastate.edu*

This presentation describes the use of synthetic video clips of livestock animals to train deep neural networks for individual animal behavior characterization in real videos. Current approaches to training artificial neural network classifiers are usually based on feature extractions from a training data set and can be limited by a paucity of representative real-world video data from which to train (e.g. lameness, or tail and ear biting). A mathematical scheme is developed that includes a formalized problem definition and framework for evaluating synthetic video parameters including perspective, illumination, and pose. A case study for swine and beef cattle was performed to validate the approach. The results demonstrated that it is possible to produce a deep neural network using the synthetic video that has satisfactory performance recognizing behaviors in real video; further, classification performance can be improved through supplementing with real video data. Both cases were shown to perform better than using only real video data as a training set. This work is of importance to production animal researchers and practitioners, as well as those in the fields of computer vision, machine learning, and artificial intelligence, because the use of synthetic video is a fast and inexpensive way to increase the accuracy of automated visual sensemaking which will improve overall livestock animal monitoring.

## Some cows don't like to get wet: Individual variability in voluntary soaker use in heat stressed dairy cows

*Lori N. Grinter, Magnus R. Campler and Joao H.C. Costa*
*University of Kentucky, Department of Animal and Food Sciences, 407 W.P. Garrigus Building, 40546, USA; magnus.campler@uky.edu*

Soaking is an efficient methods of cooling dairy cows, which is normally performed at the milking parlour waiting area or feed bunk. With the trend towards automatization in dairy farms, group soaking opportunities are scarce. Also, dairy cow response to an increase in heat load varies from one to another. Voluntary soaking stations for heat stress abatement provides freedom of choice and cow self-management, shifting focus towards individual cow needs instead of the group. Thus, the aim of this study was to assess the individual behavioural response of dairy cows to a voluntary soaking opportunity. Fifteen lactating Holstein cows were enrolled over 8 1-wk periods in a random block design consisting of 2 treatments: (1) free access to a voluntary soaker (NMS); or (2) two mandatory 5 s soakings/d administered when exiting the milk parlor in addition to free access to the voluntary soaker (MS). All cows were trained and acclimated to the soakers during a 4-week period prior to the study. Hourly and daily mean THI was calculated using data obtained from a portable weather station on site. The number of soaker visits, number of soaker displacements, and area of cow wet (rump, neck and head, back) and licking soaker was recorded manually from video observations. Cows were able to position themselves to get fully or partially soaked. No effect of treatment on soaker use was found ($F_{1,14}$=3.03; P=0.1) thus all individual variation data was analyzed for the entire study period. A large variation in voluntary soaker use was observed ranging from 0 to 227 soakings or 13±30 (mean±SD) soakings per day. Four cows had a maximum use of less than 10 soakings/d, while four other cows had a maximum of more than 100 soakings/d. Total voluntary soaker use was predominantly on the back area (54.6±9.4%, mean ± SD) followed by the side (27.5±3.2%), rump (12.2±2.5%), neck and head (4.7±1.1%), and finally licking soaker (1.0±0.45%). At the end of each pre-programmed 5 s soak, the majority of cows continued to voluntarily use the soaker by reactivating the soaker (64.7±9.1%). Thus, approximately 30% of cows left the soaker after the first soak cycle of 5 s was completed due to displacement (by personnel (bringing cows to the milking parlour): 4.6±0.6%, or other cow: 1.3±0.3%), leaving by choice (14.1±2.0%) or unidentifiable reason (15.3±2.8%). Cows with a greater frequency of soaker use were more likely to displace another cow from the soaker ($F_{1,14}$=9.28; P<0.01) or be displaced from the soaker ($F_{1,14}$=30.61; P<0.001). It is possible that individual cows experience heat stress differently and choose to seek alleviation accordingly. Voluntary soakers may therefore be useful as a heat abatement strategy as cows can be flexible in seeking heat stress alleviation. Therefore, a voluntary soaker may offer a good opportunity for particularly heat intolerant cows to alleviate heat stress. Further research however should investigate different voluntary cow soaker designs, individual motivation for voluntary soakings, and the combination with other cooling strategies.

## Visualizing behaviour of individual laying hens in groups

*Elske N. De Haas[1,2], T. Bas Rodenburg[1] and Frank A. Tuyttens[2,3]*

*[1]Utrecht University, Department of Animals in Science and Society, Faculty of Veterinary Medicine, Yalelaan 2, Utrecht, the Netherlands, [2]Institute for Agricultural and Fisheries Research (ILVO), Animal Sciences Unit, Melle, Belgium, Animal Sciences Unit, Scheldeweg 68, 9090 Melle, Belgium, [3]University of Ghent, Faculty of Veterinary Medicine, Salisburylaan 133, 9820 Merelbeke, Belgium; elske.dehaas@ilvo.vlaanderen.be*

Tracking individual hens in a flock is extremely challenging. With tracking devices mounted on the birds it is possible to asses where a hen is or has been. In this presentation I will discuss how we can visualize hen's behaviour patterns with the use of location data obtained with ultra-wide band tracking. With data from indoor and outdoor kept hens we are able to asses time spent in certain zones, location preferences, distance travelled, (stability of) daily patterns and locomotion speed. We have also attempted to asses proximity by looking at distance to other flock members, so far only conducted under small experimental groups. I compared heat maps and line patterns of video software (Ethovision) and tracking software (Tracklab) and found a high consistency after several corrections in the Tracklab software package. But the consistency between video image and UWB tracking indicates that the tracking does reflect the movement of an animal in a set indoor space. In an outdoor space where precipitation and vegetation can influence signal strength, location detection was found to be less accurate. However, hens were tracked over several days which allowed us to assess missing data points by estimation using the location before and after the missing sample points. With these longitudinal data, we were able look at daily patterns, observation rate in specific zones (which can be used for preference) and several activity measurements. Visualizing patterns of movement of individual hens with longitudinal data can make figures very blurry and indescribable. Downsizing the dataset to consistent data over birds over only a few specific days can help for visualization, but should not be used for data analysis or solely for data interpretation.

## Using experimental data to evaluate the effectiveness of tail biting outbreak intervention protocols

*Jen-Yun Chou[1,2,3], Keelin O'Driscoll[3], Rick B. D'Eath[2], Dale A. Sandercock[2] and Irene Camerlink[4]*
*[1]University of Edinburgh, Royal (Dick) School of Veterinary Studies, Easter Bush, EH25 9RG, United Kingdom, [2]SRUC, Animal & Veterinary Sciences Research Group, Roslin Institute Building, Easter Bush, EH25 9RG, United Kingdom, [3]Teagasc, Pig Development Department, Animal & Grassland Research Centre, Moorepark, Fermoy, P61 C996, Co. Cork, Ireland, [4]University of Veterinary Medicine, Institute of Animal Welfare Science, Veterinärplatz 1, 1210 Vienna, Austria; jenyun.chou@ed.ac.uk*

In the EU routine tail docking is banned as a way to control tail biting, and thus much recent research has focused on finding solutions to reduce, predict and prevent the occurrence of tail biting among undocked pigs. Preventative measures are essential in attempting to prevent tail biting although outbreaks still occur. It is therefore crucial to investigate how to manage tail biting outbreaks (TBOs) when they happen. Currently very little scientific literature provides data that specifically evaluates the effectiveness of different TBO intervention protocols. This is primarily because biting outbreaks are unpredictable, which creates difficulty in terms of data collection; most outbreak control advice for farmers is based on past experience and anecdotes. This study aims to evaluate the effectiveness of different intervention protocols to control TBOs. TBO data were collected from two experiments on undocked pigs in fully-slatted systems, in which 1,248 undocked pigs from 96 pens were followed from weaning to slaughter. All pigs had similar records of individual growth and tail lesion scores, so comparison is possible across experiments. In total, 40 TBOs were recorded with regard to the day of onset, length and severity of the TBO, pigs' involvement, and interventions used. An acute TBO was defined as when three or more pigs in a pen of 12-14 pigs had open tail wounds and fresh blood clearly visible (i.e. severe tail lesions). If 1 or 2 pigs suffered from severe tail lesions for 72 hours consecutively, or if three or more pigs suffered from moderate but prolonged tail lesions for 72 hours, it was defined as a chronic TBO. When an outbreak was identified, one of three intervention protocols (removing the biters, removing the victims, or providing extra enrichment: 3 hemp ropes) were randomly deployed. Biters were identified by continuous observation for 10 minutes beside the pen as soon as a TBO was identified. Pigs' daily histories of performing or receiving tail bites were also recorded and used to assist in identifying the biters. If the first intervention failed to control tail biting, a second intervention was randomly assigned and then the third until tail biting ceased. Data are analysed combining quantitative and qualitative methods such as survival analysis. To provide practical advice for farmers, cost-benefit analysis is also be included with the calculation of costs from the intervention/treatment methods and growth reduction. This study aims to provide practical outbreak management recommendations based on experimental data and to generate advisory information for further dissemination.

## Factors influencing farmer willingness to reduce aggression between pigs

*Rachel Peden[1], Faical Akaichi[1], Irene Camerlink[2], Laura Boyle[3] and Simon Turner[1]*
*[1]Scotland's Rural College, Roslin Institute Building, EH25 9RG, United Kingdom, [2]University of Veterinary Medicine, Institute of Animal Welfare Science, Veterinärplatz 1, 1210 Vienna, Austria, [3]Teagasc, Pig Development Department, Animal & Grassland Research and Innovation Centre, Moorepark, Fermoy, P61 C997, Ireland; rachel.peden@sruc.ac.uk*

Aggression between pigs remains an important animal welfare issue despite several decades of research. Aggression is primarily caused by the unstable social structure created by regular regrouping, as pigs fight in order to re-establish dominance relationships. Several strategies to reduce the occurrence or intensity of aggression at regrouping have been identified. However, these strategies are insufficiently adopted in commercial practice. Uptake of livestock welfare research relies on various stakeholders being willing to recommend or adopt changes to farm structure or management (e.g. veterinarians, researchers, farmers). We surveyed 122 UK and Irish pig farmers on their attitudes and practices regarding aggression between growing pigs. A structural equation model was used to investigate the factors that influence farmer willingness to implement aggression control strategies, in order to identify targets for initiating a change in practice. The majority of farmers mixed pigs at least once during each production cycle, and had tried at least one strategy to reduce aggression in the past. Strategies were considered to be moderately useful, but farmers expressed limited willingness to use them again in the future. Willingness to implement aggression control strategies was directly influenced by (1) their beliefs about the outcome (P<0.01), and (2) their perceived possibility to make a change (P<0.01). Furthermore, willingness was indirectly influenced by (3) their perceptions of the problem (P<0.001) and (4) relevant stakeholder groups (P<0.01). Veterinarians had the greatest influence on farmer behaviour when compared to all other stakeholder groups (P<0.05). We recommend that researchers employ a combination of approaches to encourage a change in practice. Aggression control strategies should be tested outside of the highly controlled research setting, in order to establish their outcomes and ease of practical implementation under commercial conditions. The economic consequences of different strategies should be calculated, in order to advise farmers on the most cost-effective solutions and their impacts on farm profitability. Finally, information on the consequences of aggression, how to recognize it as a problem and how to control the issue should be effectively transferred into industry by researchers. Knowledge transfer should be directed not only at farmers but at various stakeholder groups, with special emphasis on veterinarians. The Structural Equation Model approach could be useful to understand farmer decisions with regards to other specific and entrenched animal welfare issues.

## Reduce damaging behaviour in laying hens and pigs by developing sensor technologies to inform breeding programs

*T. Bas Rodenburg[1,2], Lisette Van Der Zande[1], Elske N. De Haas[2], Lubor Košťál[3], Katarina Pichová[3], Deborah Piette[4], Jens Tetens[5], Bram Visser[6], Britt De Klerk[7], Malou Van Der Sluis[1], Jorn Bennewitz[8], Janice Siegford[9], Tomas Norton[4], Oleksiy Guzhva[10] and Esther D. Ellen[1]*
*[1]Wageningen University, P.O. Box 338, 6700 AH Wageningen, the Netherlands, [2]Utrecht University, Department of Animals in Science and Society, Faculty of Veterinary Medicine, Yalelaan 2, 3584 CM Utrecht, the Netherlands, [3]Slovak Academy of Sciences, Institute of Animal Biochemistry and Genetics, Ivanka pri Dunaji, 840 05 Bratislava, Slovak Republic, [4]KU Leuven, M3-BIORES, Kasteelpark Arenberg 30, 3001 Leuven, Belgium, [5]Georg-August University, Functional Breeding Group, Department of Animal Sciences, Burckhardtweg 2, Göttingen, Germany, [6]Hendrix Genetics, Spoorstraat 69, 5831 CK Boxmeer, the Netherlands, [7]Cobb Europe, Kruisstraat 5, 6674 AA Herveld, the Netherlands, [8]University of Hohenheim, Institute of Animal Science, Garbenstr. 17, Hohenheim, Germany, [9]Michigan State University, Department of Animal Science, Anthony Hall, 474 S. Shaw Lane, East Lansing, MI 48824-1225, USA, [10]Swedish University of Agricultural Sciences, Dept. Biosystems and Technology, Box 103, 230 53 Alnarp, Sweden; lisette.vanderzande@wur.nl*

The COST Action GroupHouseNet aims to facilitate the prevention of damaging behaviour in group-housed pigs and laying hens. One area of focus is on how genetic and genomic tools can be used to breed for animals that are less likely to develop damaging behaviour. The behaviours we are focusing on are feather pecking in laying hens and tail biting in pigs. Both species are kept in groups, and identifying actual performers of this behaviour (peckers and biters), and tracking them at the individual level remains challenging, but is essential for breeding programs. It is possible to use traditional behavioural observation, but this is time-consuming and costly. Sensor technology is a rapidly developing field and may offer solutions for phenotyping animals at the individual level. We propose that sensor technology combined with genomic methods may be useful in solving the problems of damaging behaviour in group-housed pigs and laying hens. When evaluating the sensor technologies used until now, for laying hens RFID and accelerometer-based approaches seem most promising. In pigs, computer vision is already used to record technical performance, and there seems to be potential for expanding this approach to the recording of damaging behaviour. If sensor signatures and genomic fingerprints of individual animals can be combined, this would significantly improve our possibilities to reduce damaging behaviour through genetic selection.

## Using thermal choices as indicators for fish welfare

*Sonia Rey Planellas[1], Simon MacKenzie[1] and Felicity Huntingford[2]*
*[1]University of Stirling, Institute of Aquaculture, Stirling FK9 4LA, United Kingdom, [2]University of Glasgow, Institute of Biodiversity, Animal Health and Comparative Medicine, University Avenue, Glasgow G12 8QQ, United Kingdom; felicity.huntingford@glasgow.ac.uk*

Wild fish, being ectotherms, regulate their internal temperature by voluntarily moving within natural temperature gradients. In this way they make thermal choices (varying with developmental stage and in accordance with biological rhythms) that promote effective functioning of various biological systems. For example, selecting appropriate water temperatures enhances digestive efficiency in sharks and gonadal maturation in seabass and salmon. One manifestation of thermal choice of particular relevance to welfare and welfare indicators is the case of behavioural fever, in which pathogenic infection causes a short-term increase in preference for higher water temperatures. Behavioural fever occurs in invertebrates and in ectothermic vertebrates, including fish. In functional terms, enhanced internal temperatures induced by behavioural fever cause targeted up-regulation of immune function in infected fish, and strikingly improved survival rates. Ensuring that cultured fish have access to a temperature gradient will enhance fish welfare through improved health; this has been demonstrated in Tilapia ponds. Behavioural fever also offers the possibility of using thermal choices as an early-warning indicator for disease states. Another context in which the thermal choices are potentially informative about welfare status is the case of emotional fever, a temporary increase in body temperature shown in response to a variety of stressors. Physiologically-induced emotional fever is well documented in endotherms, where it serves the function of promoting effective responses to challenge. Behaviourally-induced emotional fever has recently been demonstrated in zebrafish, stressed fish in a temperature gradient showing a transient increase in preference for warmer water following exposure to an acute stressor. Although this finding remains controversial, it potentially offers a sensitive behavioural indicator of stress and impaired welfare. In this talk we will illustrate the various beneficial effects of thermal choice in fishes and discuss their potential use as welfare indicators.

## The role of the serotonergic system as a welfare indicator in salmonids

*Marco A. Vindas*
*Norwegian University of Life Sciences, Food Safety and Infection Biology, Ullevålsveien 72, 0454 Oslo, Norway; marco.vindas@nmbu.no*

Signaling systems activated under stress are highly conserved, suggesting adaptive effects of their function. Pathologies arising from continued activation of such systems may represent a mismatch between evolutionary programming and current environments. Here, we use Atlantic salmon (*Salmo salar*) in aquaculture displaying a depression-like state (DLS) as a model to explore this stance of evolutionary-based medicine, for which empirical evidence has been lacking. Affective states are defined as neural responses that occur when the brain detects challenging or rewarding stimuli and involves changes in arousal, cognitive function, autonomic and musculoskeletal responses. There is now ample evidence that fish are capable of displaying affective states (such as DLS and frustration). We sampled 24 DLS and 24 healthy (H) individuals in an aquaculture farm to characterize their monoaminergic and cortisol profile at basal and post-stress conditions. We found a significant difference (Two- way ANOVA followed by Tukey post-hoc test) between DLS and H fish at resting conditions, with DLS showing elevated brain serotonergic activation (DLS: 0.4±0.05, C: 0.2±0.02, P<0.01), increased cortisol production (DLS: 124±33, C: 10±1, P<0.01) and behavioral inhibition. Post-stress we found that while control fish increased their serotonergic activity to 0.3±0.03 and cortisol response to 79±9, we make the novel observation that the serotonergic system in DLS fish is unresponsive to additional stressors (0.4±0.04), yet a cortisol response is maintained (192±35) and is potentiated compared to H fish (P=0.009). Furthermore, in a laboratory experiment we found that sustained pharmacological 5-HT activation (fish were given the serotonin enhancer fluoxetine for 17 days) resulted in the DLS behavioral profile. That is, they increasingly spent more time (starting at day 9) at the top half of the tank (Repeated measures ANOVA: $F_{(9, 18)}$=10.3, P<0.001) and a reduction in feeding (starting at day 7, Repeated measures ANOVA: $F_{(9, 18)}$=35.5, P<0.001). The inability of the serotonergic system to respond to additional stress, while a cortisol response is present, probably leads to both imbalance in energy metabolism and attenuated neural plasticity. Hence, we propose that serotonin-mediated behavioral inhibition may have evolved in vertebrates by minimizing stress exposure in vulnerable individuals. Furthermore, stress and welfare may be directly inferred by quantifying serotonergic signaling and behavior and we propose that by measuring serotonin levels in the blood, we may be able to ascertain the welfare status of production fish and therefore devise practices which increase salmon welfare in aquaculture systems.

## Developing measures of positive welfare for fish

*Becca Franks and Isabel Fife-Cook*

*New York University, Department of Environmental Studies, New York City, NY 10003, USA; krf205@nyu.edu*

Historically, traditional animal welfare assessment paradigms have focused on maintaining physical health and mitigating negative impacts to wellbeing. In contrast, recent animal welfare science has recognized the importance of positive welfare and accordingly incorporated indicators of positive affect into welfare assessments for use by zoos/aquaria, laboratories, and animal agriculture. However, the development and monitoring of positive welfare requires an in-depth knowledge of species-specific behavior and biology, which necessitates the need for species-specific or, at a minimum, taxa-specific standards. Research on positive welfare in fish is lagging in this regard and therefore merits further consideration. Merging what is already known about positive welfare in other species with existing fish behavior research yields a plan of action for developing indicators of positive fish, which, with more research, will contribute to the development of optimal welfare standards and assessment strategies for fish. We begin by exploring the origins of positive welfare assessment, and then outline the physical, psychological and species-specific areas of inquiry that can be applied to fish. In addition to presenting current findings on fish motivation, emotion and indicators of positive affect such as fulfillment of motivational urges (establishing agency, engaging in exploration and learning) and play behavior, we also suggest ways forward in developing accurate and appropriate positive welfare indicators for fish.

## Using social behaviour as a welfare indicator in fish

*Tanja Kleinhappel, Tom Pike and Oliver Burman*
*University of Lincoln, School of Life Sciences, Joseph Banks Laboratories, Green Lane, Lincoln,*
*LN6 7DL, United Kingdom; oburman@lincoln.ac.uk*

Observing social behaviour can provide us with insight into our understanding of how animals respond to both negative and positive stressors, and thus be valuable as an indicator of welfare – especially in shoaling fish. Although measures of dyadic interactions such as aggression and affiliation are informative, there are a range of other behaviours and approaches that give opportunities for a more comprehensive representation of group structure and shoal dynamics. However, a particular challenge is that such social metrics can be time-consuming to record and, in some cases, difficult to interpret. Using zebrafish (*Danio rerio*) as a model, we will: (1) present an example of using social metrics to assess group response to environmental enrichment, in which we observed increased shoal expansion in response to enrichment provision (GLMM: $\chi^2(1)=4.04$, P=0.044); (2) describe the development of an automated approach to monitoring social behaviour in groups of fish; and (3) discuss whether sleep behaviour within fish shoals has potential as a measure of welfare. By highlighting these particular examples, we hope to stimulate discussion focusing on the identification of further novel indicators of fish welfare, future opportunities for fish welfare research and its application, whilst acknowledging potential challenges and ways to mitigate these.

## Engaging students in learning about production animal welfare assessment

*Marta E. Alonso[1], Alejandra Feld[2], Xavier Averós[3] and Melissa Elischer[4]*
*[1]University of Leon, Animal Production Department, Veterinary Faculty of Leon, Campus de Vegazana, 24071 León, Spain, [2]Buenos Aires University, Cátedra de Bienestar Animal, Veterinary Sciences Faculty, Av. Chorroarín 280, C1427CWO, Buenos Aires, Argentina, [3]Neiker-Tecnalia, Animal Production Department, Campus agroalimentario de Arcaute, 01080 Vitoria, Alava, Spain, [4]Michigan State University Extension, Anthony Hall 474 S. Shaw Ln., Rm. 1287H, East Lansing, MI 48824, USA; marta.alonso@unileon.es*

Consumers across the globe are requesting more information about the care and welfare of farm animals as fewer and fewer people have a direct connection to food animal production. This has necessitated more class time focused on farm animal welfare for animal science and veterinary students as only a limited number enter university with farm animal experience. Many students pursuing animal-related degrees are more focused on companion animals and it can be challenging to stimulate interest in agricultural species. In addition to having to build student interest, there are limited classroom exercises widely available for engaging students in the use and application of different welfare assessment protocols. Using methodologies and tools that are appealing to students to stimulate external motivation and interest in farm animal welfare is imperative to achieving learning outcomes. Here, we share teaching methodologies used to engage students in learning how to evaluate animal health and welfare in production animal species, with a focus on ethological parameters. Melissa Elischer uses a 'hands-on' approach with interactive demonstrations and activities that encourage students to experience the world from a farm animal´s perspective, with subsequent discussion about the students' reaction to the activities. Xavier Averós introduces the transect method for poultry welfare assessment on the farm, using an app for real-time data entry. The method gives a snapshot of the flock's current welfare status while minimizing time requirements and bird disturbance. It is a practical method that is relatively easy to learn and apply during routine flock inspections. Alejandra Feld explains methodology that allows students to develop their own checklist of welfare indicators and use it subsequently in real field experiences. Marta Alonso shares her interactive tool that combines written information and descriptive YouTube videos to prepare students to fill out checklists for assessing sheep and cattle welfare during practical exercises. Workshop participants will discuss these teaching methodologies and new ideas that emerge. The intended outcome will be increased motivation and awareness about the importance of practical, science based assessment of animal welfare among students and future professionals who need a working knowledge of farm animal welfare.

## What can we learn from studying the motivation underlying maternal behaviour of cattle?

*Maria Vilain Rørvang[1], Birte L. Nielsen[2], Mette S. Herskin[3] and Margit Bak Jensen[3]*
*[1]Swedish University of Agricultural Sciences, Biosystems and Technology, Sundsvägen 16, 23053 Alnarp, Sweden, [2]INRA, UMR Modélisation Systémique Appliquée aux Ruminants, AgroParisTech, Université Paris-Saclay, 75005 Paris, France, [3]Aarhus University, Animal Science, Blichers Allé 20, 8830 Tjele, Denmark; mariav.rorvang@slu.se*

Calving is a central event in dairy production, but may be overlooked from the perspective of animal welfare. In some countries, individual maternity pens are recommended based on scientific studies. Understanding the motivation underlying pre-partum behaviour may help manage cows to enable calving in a calm and secluded place, where they undisturbed can nurse and bond with their calf. Here we present an overview of a series of studies on pre-parturient behaviour of cattle including recent research on their use of individual maternity pens and secluded areas within group maternity pens. Based on a literature review we suggest that pre-parturient cows are motivated to locate an undisturbed birth site by means of a combination of distance and physical cover. The first study showed that although pre-parturient dairy cows kept in individual maternity pens had no specific preference for a particular level of physical cover, cows with prolonged labour chose a higher level of physical cover. In another study, inserting a self-closing gate at the entrance of an individual maternity pen, did not affect the proportion of cows calving there. However, social dominance increased the probability of calving in a gated pen, whereas presence of alien calves in the group area decreased this probability. Interestingly, calving in a pen protected the cow and calf from disturbances even when the pen was without a gate and thus open to pen-mates. In a later study with ungated calving pens, the cows preferred pens with wide sides to pens with narrow sides suggesting that increasing level of cover is attractive. In several of these studies, the choice of calving site was where another cow had previously calved. In these cases birth fluid was contained within the bedding from a previous calving and these are suggested to have an attracting effect on the parturient cow. Overall, this series of studies suggests that the physical environment, the social environment, as well as olfactory cues from birth fluid, interact and modulate maternal behaviour of cattle, and thus the selection of birth site. The finding that birth fluid attracts pre-parturient cows, potentially opens unexploited opportunities for using odour molecules from bovine birth fluid to stimulate selection of birth sites. A combination of opportunity to hide and attractive odours may stimulate cows to enter individual maternity pens. Based on the research outlined we encourage future studies to focus on pre-parturient cows' motivation for access to isolation provided by physical cover and distance from the herd, the ability to escape potential threats or disturbances, as well as attraction to birth fluids, or other olfactory cues.

## Assessing the welfare of dairy cows before and after giving birth

*Kathryn Proudfoot[1], Benjamin Lecorps[2], Daniel M. Weary[2] and Marina A.G. Von Keyserlingk[2]*
*[1]Ohio State University, Veterinary Preventive Medicine, 1920 Coffey Rd, Columbus, OH 43210, USA, [2]University of British Columbia, Animal Welfare Program, 2357 Main Mall, Vancouver, BC, V6T 1Z4, Canada; proudfoot.18@osu.edu*

Dairy cows face a number of challenges before and after giving birth that may compromise their welfare. The majority of research to date has focused on understanding the effects of housing and management on behavior and sickness around the time of calving. Our objective is to review two other approaches that may provide additional insights on how to minimize the negative effects of parturition: (1) developing a deeper understanding of the natural, adaptive behaviors of pre-parturient cows; and (2) assessing the affective state of dairy cows in the few days after calving. In the first part of our talk, we will describe research that has furthered our understanding of the natural behaviors of pre-parturient cows by determining their preference for different calving environments. For example, we have shown that when housed individually, indoor-housed cows have a preference to move to a secluded area to give birth, particularly when calving occurred during the day compared to night. This finding aligns well with other research on dairy cows kept on pasture or range land; cows, especially heifers, choose to leave the herd to find naturally secluded areas to give birth, such as areas with tree cover and tall grass. When indoor-housed cows are kept in pairs or groups, the decision to use a secluded area to give birth becomes more complicated. For example, we found that when two cows were provided access to one hiding space, the first cow to give birth avoided the hide during labor; the reason for this avoidance is unknown, but may be driven by competition over the resource. Other work using larger groups of cows has found similar results, suggesting that competition for a secluded birth space or social status may play a role in the cow's decision and ability to seek a desirable calving site. In the second part of our talk, we will describe a novel approach to measuring the affective state of cows after calving. A better understanding of the cows' experience after calving may provide insights into their welfare and risk of become sick. For example, in humans and laboratory rodents, some new mothers experience periods of intense negative affect ('post-partum depression'), that impact their well-being, increase their risk of disease, and compromise the care provided to their young. Depression is characterized by anhedonia, the loss of interest in normally rewarding experiences. By assessing the cow's motivation to use a mechanical brush, a normally rewarding experience, we provide preliminary evidence that dairy cows show behavior that may be indicative of anhedonia in the few days after calving. Moreover, these signs of anhedonia are exacerbated when dams are separated from their calves. Together, these novel research strategies have provided new insights into aspects of dairy cow welfare before and after giving birth.

## Alternative management strategies to help dairy cows achieve successful outcomes during the transition into lactation

*Peter D. Krawczel[1], Heather M. Dann[2], Erika M. Edwards[1], Katy L. Proudfoot[3] and Amanda R. Lee[1]*
[1]*The University of Tennessee, Department of Animal Science, 2506 River Dr, Knoxville, TN 37996, USA,* [2]*William H. Miner Agricultural Research Institute, 1034 Miner Farm Rd, Chazy, NY 12921, USA,* [3]*The Ohio State University, Department of Veterinary Preventive Medicine, 1920 Coffey Rd, Columbus, OH 43210, USA; krawczel@utk.edu*

The transition into lactation is a challenging, yet unavoidable, reality related to the welfare of dairy cows. The physiological changes associated with calving and lactogenesis, pain of calving, changes in diet, and changes in social group are all contributors to this. Recent research on alternative strategies, such as refining the calving environment and revaluating when cows and calves should be separated, provides insights on approaches that may alleviate some of the stressors. The aim of this presentation for the calving workshop is to review (1) dairy cows and heifers' preference for a calving environment and factors associated with their preference and (2) the potential positive effects of delaying the separation of cows and calves following parturition. For the calving environment, we will review the relevant literature on cattle calving within confinement facilities as well as present our own findings from cattle calving with access to housing and pasture options. Our findings suggest primiparous and multiparous cows selected the barn more frequently for calving, whereas nulliparous heifers utilized the natural forage more frequently. However, temperature also influenced preference, as cattle were more likely to utilize natural forage cover when THI was high (68<THI≤79). Our work also suggested that the provision of pasture may reduce the motivation to separate from the herd. Twenty-five percent of cattle 'separated' from the herd to calve. If a cow or heifer 'separated', a greater percentage of the herd was at least 2 sections away compared to those that did not, which is lower than the results of similar work in confinement. This review will provide an overview towards why this may occur and the overall potential importance of separating from the herd. Post-calving, the immediate separation of dams and calves remains a controversial issue. Survey data indicate that this is a management approach viewed negatively by the public as well as a substantial portion of farmers. Our efforts with partial commingling of the cows and calves, which was accomplished by grouping them together on pasture overnight, suggest that this can be successfully implemented without affecting the welfare of either. Cows did not decrease their productivity when provided nightly access to calves nor were major behavioral changes evident. No differences in calf growth rates or approachability were evident. These data and similar work evaluating the effects of commingling dams and calves, both positive and negative, will be discussed. Overall, accommodation of dairy cows and heifers preferences for calving environment may provide a means to partially mitigate the stressors associated with parturition and improve the likelihood of a successful transition into lactation. Similarly, reconsidering the immediate separation of dams and calves post-calving address public concerns of this management strategy without compromising animal welfare.

## Keeping cow and calf together: behaviour and welfare issues

*Marie J. Haskell*

SRUC, Animal and Veterinary Sciences, West Mains Road, Edinburgh EH9 3JG, United Kingdom; marie.haskell@sruc.ac.uk

In modern dairy farming, it is common practice to separate the calf and dam within 24 h of birth. Separation maximises the amount of milk available for sale, minimises exposure of calves to disease from adult animals and allows for easier handling of the adult cattle and calves. However, there is increasing unease in consumers across the world about this abrupt separation of mothers and neonatal offspring, and evidence to suggest that there are welfare benefits to keeping the calf with the cow. Keeping the cow and calf together is currently practised on many small-holder farms across the developing world. However, in countries where intensive dairying housing and management systems predominate, management and housing practices that allow cow and calf to stay together must be re-learnt or developed from scratch to allow dairy farming to take place at a commercially viable scale. In Europe there are a few dairy farmers keeping cows and calves together for periods of 3-5 months, after which they are weaned. Anecdotal evidence suggests that many more farmers would be interested in investigating this system, but there are a number of issues that need to be investigated or elaborated before these farmers could be encouraged to try the cow with calf system. There are a number of key points in the period between birth and weaning when behavioural research may be useful. In the first few days after birth, the cow and calf form a strong bond. The high levels of maternal protection that this bond creates may cause the cow to be aggressive to stockpersons, which is an issue for human safety. Cows must be habituated to human handling when the calf is present. Handling and milking facilities should be designed to allow cow and calf to be moved together. The layout of indoor housing must allow space for young calves to live safely with a group of adult cows. Consideration must be given to providing separate creep areas for the calves. Managing the weaning process is also a major issue. As the cow and calf have a strong bond, processes to manage weaning with as little distress as possible for cow and calf require development, as is also recommended for beef cattle. There is also a disease issue of keeping young calves in a group of adult cattle. Calves are directly exposed to the diseases of adult cattle, to which the adults are immune. However, all of the questions raised may be overcome by appropriate experimental research and field studies of successful cow-calf systems in current operation. This type of research will provide support for farmers interested in investing in this system and encourage further uptake of this system.

## State-of- the-art dam rearing of calves – a sector-wide assessment of scientific and practical knowledge on dam-rearing systems

*Cynthia Verwer and Lidwien Daniels*
*Louis Bolk Institute, Kosterijland 3-5, 3981 AJ Bunnik, the Netherlands; c.verwer@louisbolk.nl*

Among society and farmers the practice of mother-bonded or fostered calf rearing is attracting growing interest. Farmers seek to extend the period in which dams or foster cows can spend time with a calf, providing both an opportunity to perform their natural (maternal) behaviour. This type of dairy farming requires knowledge of the natural behaviour of the animals, and needs an assessment of the essential aspects of, among others, animal husbandry, housing, feeding and human-animal interaction in relation to this natural behaviour, whilst still producing milk. In the Netherlands around 45 dairy farmers keep cow and calf together ranging from six weeks up to six months of age. Their experience with these dam-rearing systems vary from a year to over 25 years in which each farmer has developed his own method. In the last ten years several surveys and on-farm research projects have been performed on these farms, providing for numerous experience based suggestions of how to implement dam-rearing systems and the necessary requirements for success on, among others, cow-calf bonding and debonding, herd dynamics, milk production, health and welfare of both cow and calf. These studies have been extended with a survey in 2017 in which globally available scientific knowledge and Dutch stakeholder opinions were inventoried. An overview of the outcomes of these studies will be given with emphasis on the benefits that dam-rearing dairy farmers experience on their farm and in their herd. Wherever possible these experiences will be supported by findings reported in scientific literature. Special emphasis will be put on the state-of-the-art on how to manage dam and calf together the first days after calving regarding housing, feeding, health and welfare, including knowledge gaps that farmers experience. These knowledge gaps will lead to suggestions for future research.

**Using positive reinforcement to train laboratory pigs: benefits for animal welfare and research models**

*Dorte Bratbo Sørensen[1], Lisa Jønholt[1,2], Martin Carlsen[2] and Cathrine Bundgaard[2]*
*[1]University of Copenhagen, Section for Experimental Animal Models, Frederiksberg C, 1870, Denmark, [2]Novo Nordisk A/S, Novo Nordisk Park, Maaloev, 2760, Denmark; brat@sund.ku.dk*

According to the EU directive, laboratory animals should be given a 'degree of choice and control over their environment' and the facility shall 'set up habituation and training programmes suitable for the animals, the procedures and length of the project'. We have worked intensively with training of laboratory pigs for a little over two years and considering how easy these animals learn, much is to be gained by extending the use of operant conditioning techniques in the form of positive reinforcement training (PRT) in pig facilities. The assessment of the effects of PRT is often difficult in the laboratory settings due to a number of factors including constraints related to the study design, trainer skills and lack of control groups. These constraints will be presented and discussed in the presentation. Our preliminary studies in a Danish lab pig facility have shown that pigs being trained to follow a target are much easier to move from one location to another and e.g. weighing is much faster. Data not yet statistically analyzed seem to indicate that the behaviour of PRT pigs is more bold than that of control pigs and observations in a modified human approach test and a 'novel floor'-test ('trust-test') indicate that behaviours such as tail wagging were different between PRT pigs and control pigs. Last, PRT allows the researchers to design studies that could not otherwise be done. Two examples will be presented. The presentation will discuss preliminary data, personal observations and anecdotes mainly in the form of video-material and discuss how we navigate through the sea of constraints and demands for 'scientific proof' to optimize animal welfare and research data in the lab pig facility.

## Creating an animal training handbook for livestock producers, veterinarians, and researchers

*Kristina Horback*
*University of California Davis, Department of Animal Science, Center for Animal Welfare, 1 Shields Ave, 95616, USA; kmhorback@ucdavis.edu*

The ability for an animal to learn the consequences of their behavior is fundamental to their ability to cope with environmental change. Associative learning practices can be used by animal caretakers to reduce unpredictability and provide opportunities for animals to exercise control over their environment. For example, livestock producers can use classical conditioning of an auditory cue to signal feeding time, or, use operant conditioning to train animals to voluntarily walk over a scale for weighing. In order to facilitate the use of associative learning practices in animal agriculture, this presentation proposes a discussion on the creation and distribution of a user-friendly handbook on practical animal training techniques. Elements of this handbook may include how to shape an animal's behavior toward a desired outcome using positive reinforcement, how to solve potential training issues, and, how to be mindful of unintentional signals (particularly related to aversive events). The speaker will provide examples from personal experience with operant conditioning of commercial sows to use electronic sow feeding stations, as well as experience with training sows to exhibit operant choice behavior in cognitive studies. The goal of this handbook would be to reduce the need for animal handling during routine husbandry practices (such as weighing and relocation) and to encourage the use of desensitization techniques to reduce the fear of unknown and unpredictable stimuli in the environment (such as equipment, people, and enclosures). In addition to benefiting animal welfare, this handbook could also enhance the quality of animal-based research by reducing variability of physiological and behavioral responses to environmental change using associative learning practices, and thus, potentially reduce the number of animals required for a given research project.

# Authors index

## A

| | | | |
|---|---|---|---|
| Aardsma, Matthew | 351 | Barth, Kerstin | 85, 216, 221 |
| Abreu Jorge, Yandy | 312 | Bartosova, Jitka | 234 |
| Abreu, Madalyn | 208 | Batchelli, Pau | 181 |
| Adcock, Sarah J.J. | 314, 317 | Bateson, Melissa | 272 |
| Adhikar, Sudip | 151 | Batista Costa, Leandro | 230 |
| Adigun, Taiwo | 188 | Batllori, Norbert | 122 |
| Agenäs, Sigrid | 320 | Bauman, Adrian | 306 |
| Ahmed, Abdelkareem | 202 | Baumgartner, Johannes | 88 |
| Akaichi, Faical | 206, 363 | Baxter, Emma | 354 |
| Akbari Moghaddam Kakhki, Reza | 105 | Bedanova, Iveta | 339, 340 |
| Akin, Ella | 228 | Bell, Carmen | 298 |
| Alberghina, Daniela | 271 | Bell, David J. | 282, 318 |
| Ali, Ahmed | 358 | Bellegarde, Lucille | 304 |
| Almeida Torres, Ana Beatriz | 322 | Belson, Susan | 139, 150 |
| Almond, Kayleigh | 101 | Benetton, Juliana B. | 104 |
| Alonso, Marta Elena | 194, 369 | Bennett, Darin C. | 96 |
| Altrichter, Bernadette | 193 | Bennewitz, Jorn | 364 |
| Alves, Luana | 264 | BenSassi, Neila | 168, 350 |
| Alzugaray, Irene | 279 | Bensoussan, Sandy | 227 |
| Amorim Franchi, Guilherme | 112 | Berenjian, Atefeh | 199 |
| Amparo, Martinez-Talavan | 190 | Berezowski, John A. | 251, 312 |
| Andersen, Inger Lise | 169, 170, 179, 211 | Berg, Charlotte | 79, 163, 285 |
| Andersson, Maria | 229 | Berke, Olaf | 344 |
| Andrejchová, Zuzana | 129 | Berk, Jutta | 201, 343 |
| Arfuso, Francesca | 271 | Bernardino, Thiago | 185, 264 |
| Arhant, Christine | 193 | Berry, Donagh | 90 |
| Arnott, Gareth | 253 | Bestman, Monique | 189 |
| Asher, Lucy | 154, 312 | Bibal, Mathilde | 304 |
| Asif, Daniel | 95 | Bienboire-Frosini, Cécile | 74, 226, 233, 260 |
| Asmar, Salma | 122 | Bigge, Ashley | 159 |
| Asproni, Pietro | 74, 260 | Bilcik, Boris | 88 |
| Averós, Xavier | 88, 348, 350, 369 | Biong, Stian | 212 |
| | | Bjørnson, Kim I. | 83 |
| | | Blackwell, Emily | 203 |

## B

| | | | |
|---|---|---|---|
| Bacon, Heather | 307 | Blanco Penedo, Isabel | 300 |
| Baes, Christine F. | 344 | Blatchford, Richard A. | 96 |
| Bagiova, Martina | 234 | Blavy, Pierre | 118 |
| Bain, Melissa | 81 | Blokhuis, Harry | 285 |
| Balderas, Bernardino | 348 | Blott, Sarah C. | 308 |
| Barabas, Amanda | 261 | Bøe, Knut Egil | 131, 293, 303 |
| Barrell, Emily | 331 | Bokkers, Eddie A.M. | 90, 111, 259, 265 |
| Barrett, David C. | 135 | Bolhuis, J. Elizabeth | 119, 128, 310 |
| Barry, John | 90 | Bolstad, Ingeborg | 212 |
| Bartels, Thomas | 343 | Bombail, Vincent | 232 |

| | | | |
|---|---|---|---|
| Kohari, Daisuke | 223 | Lensink, Joop | 301 |
| Kolakshyapati, Manisha | 108, 342 | Leszkowová, Iva | 276 |
| König Von Borstel, Uta | 93 | Li, Angeli | 137 |
| Korhonen, Hannu | 207 | Lichovnikova, Martina | 88 |
| Košťál, Lubor | 88, 364 | Li, Congcong | 220 |
| Krametter-Frötscher, Reinhild | 161 | Lidfors, Lena | 79, 241, 291 |
| Krause, Annika | 269 | Lindberg, Jan Erik | 100 |
| Krause, E. Tobias | 97 | Liu, Zhenzhen | 166, 177 |
| Krawczel, Peter D. | 257, 302, 372 | Llewellyn, Rick | 150 |
| Krieter, Joachim | 270, 334 | Loberg, Jenny | 95 |
| Kristiansen, Tore Sigmund | 71, 72 | Lomillos, Juan Manuel | 194 |
| Kulke, Katja | 175 | Lonis, Wendy | 142 |
| Kunze, Wolfgang | 99 | López-Pérez, Belén | 279 |
| Kwon, Isabelle | 137, 147 | Løvendahl, Peter | 302 |
| Kwon, Kyeong Seok | 319 | Luo, Lu | 128 |
| | | Lürzel, Stephanie | 215, 216 |

**L**

| | | | |
|---|---|---|---|
| Ladewig, Jan | 132 | **M** | |
| LaFollette, Megan R. | 214 | Mach, Núria | 336 |
| Lafont-Lecuelle, Céline | 74, 226, 260 | MacKay, Jill | 237 |
| Lagoda, Martyna | 326 | MacKenzie, Simon | 365 |
| Lambton, Sarah | 94, 167 | Macrae, Alastair | 76, 142, 318 |
| Lammers, Aart | 98 | MacRae, Amelia Mari | 313 |
| Langbein, Jan | 225, 269 | Madaro, Angelico | 71, 72 |
| Lange, Annika | 215, 216 | Mainau, Eva | 157 |
| Langford, Fritha | 237 | Makagon, Maja M. | 96 |
| Lansade, Léa | 336 | Makowska, Inez Joanna | 313 |
| Larsson, Malin | 242 | Malak-Rawlikowska, Agata | 181 |
| Lavery, Michelle | 68 | Malchow, Julia | 201, 343 |
| Lawrence, Alistair | 158, 232, 254 | Mancera, Karen F. | 109 |
| Lay Jr, Donald | 144 | Manriquez, Diego | 174, 268 |
| Leclercq, Julien | 226 | Manteca, Xavier | 157 |
| Lecorps, Benjamin | 141, 259, 371 | Marcet-Rius, Míriam | 226 |
| Lee, Amanda R. | 302, 372 | Marchant-Forde, Jeremy N. | 102 |
| Leeb, Christine | 182, 200 | Marchewka, Joanna | 326, 352 |
| Lee, Caroline | 136, 139, 150 | Marini, Danila | 150 |
| Lee, Dong Hyeon | 319 | Mariti, Chiara | 275 |
| Lee, Jun Yeob | 319 | Mason, Colin | 282, 318 |
| Lees, Angela | 136 | Mason, Georgia | 68, 337 |
| Lee, Sang-Ryong | 319 | Mathew, Kehinde | 188 |
| Legrand, Amélie | 304 | Matos, Raquel | 204 |
| Lehenbauer, Sandra | 193 | Matthews, Jennifer | 94 |
| Leliveld, Lisette M.C. | 133, 134 | Maulbetsch, Freija | 154 |
| Lemarchand, Julie | 336 | Maximiano Sousa, Filipe | 312, 345 |
| Leme, Paulo Roberto Pedroso | 75 | Mazzola, Silvia Michela | 238, 239 |
| Leming, Ragnar | 88 | McBride, Peter | 147 |
| Lemos Teixeira, Dayane | 322 | McClure, Jennifer | 90 |

Printed in the United States
by Baker & Taylor Publisher Services